Technology IN ACTION

Second Edition

Brad and Terry Thode

Glencoe
McGraw-Hill

New York, New York   Columbus, Ohio   Woodland Hills, California   Peoria, Illinois

NOTICE

Publisher does not make any representation, warranty, guarantee or endorsement of any of the products, their use or the methods or techniques described in this book. Publisher has not, and expressly disclaims any obligation to independently test any products or to verify facts, information, methods or techniques described in this book.

The reader is expressly advised to consider and use all safety precautions described in this book or that might also be indicated by undertaking the activities described herein. In addition, common sense should be exercised to help avoid all potential hazards.

Publisher assumes no responsibility for the activities of the reader or for the subject matter experts who prepared this book. Publisher makes no representation or warranties of any kind, including but not limited to, the warranties of fitness for particular purpose or merchantability, nor for any implied warranties related thereto, or otherwise. Publisher will not be liable for damages of any type, including any consequential, special or exemplary damages resulting, in whole or in part, from reader's use or reliance upon the information, instructions, warnings or other matter contained in this book.

SAFETY FIRST

Follow the safety rules listed on pages 42-43 and the specific rules provided by your teacher for tools and machines.

Publisher does not necessarily recommend or endorse any particular company or brand name product that may be discussed or pictured in this text. Brand name products are used because they are readily available, likely to be known to the reader, and their use may aid in the understanding of the text. Publisher recognizes other brand name or generic products may be substituted and work as well or better than those featured in the text.

Glencoe/McGraw-Hill

A Division of The McGraw-Hill Companies

Send all inquiries to:
Glencoe/McGraw-Hill
3008 W. Willow Knolls Drive
Peoria, IL 61614

ISBN 0-07-822489-6 (Student Edition)
ISBN 0-07-822490-X (Teacher's Resource Guide)
ISBN 0-07-822491-8 (Performance-Based CD-ROM)

Printed in the United States of America.

3 4 5 6 7 8 9 10 071/071 04 03 02 01

Acknowledgements

Doug Walrath, Technology Teacher
Wood River Middle School
Hailey, Idaho

Dr. Greg Vogt
Teaching From Space Program
NASA Johnson Research Center
Houston, Texas

John Cvetich, Principal
Wood River Middle School
Hailey, Idaho

Ray Grosvenor, Vice Principal
Wood River Middle School
Hailey, Idaho

John Dominick, Principal
Ernest Hemingway Elementary School
Ketchum, Idaho

Dr. Jim Lewis, Ed.D
Superintendent of Schools
Blaine County School District
Hailey, Idaho

CONTENTS IN BRIEF

CONTENTS

Chapter 1 Getting Started in Technology

Chapter 2 Using Technology

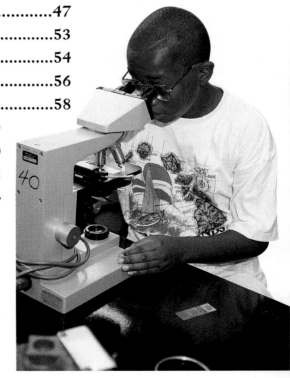

Chapter 3 How Technology Affects You

Chapter 4 Introducing Computers

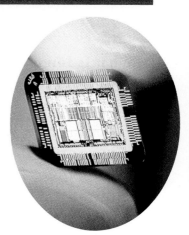

Chapter 5 Using Computers

Section

"Bunny" © 1998 Blue Sky Studios, Inc.

Chapter 6 Inventing Things

Section

Chapter 7 Making Things

Section

Chapter 8 How Things Work

Section

Chapter 9 Designing Things

Chapter 10 Exploring Automation

Chapter 11 How Business Works

Chapter 12 Building Things

Chapter 13 Using Energy

Section

Chapter 14 Moving Things

Section

Chapter 15 Finding & Using Information

Chapter 16 Producing TV/Radio Programs

Chapter 19 Exploring Chemical & Bio-Related Technology

Section

Chapter 20 Technology and Your Future

Section

CONNECTIONS

EXPLORING CAREERS

ACTION ACTIVITIES

How Do I Use Technology in Action?

THINGS TO EXPLORE

- Describe the organization of the text.
- Use the text effectively to know and apply technology in your world.

TechnoTerms
InfoLink
TechCheck
TechnoFact

Technology in Action offers you the opportunity to explore the fascinating world in which you live. The 20 chapters explore many different aspects of technology today and in the future. This introductory "chapter" shows how to use this textbook to get the most out of your adventure.

How Is the Book Organized?

Your exploration of technology will be much more fruitful if you pay attention to how *Technology in Action* is organized. First, each of the 20 chapters is divided into sections like this one that explore technology and provide hands-on activities. You'll read about a topic, answer **TechCheck** questions, and then do the Action Activity.

Headings. The written material is organized by levels. For example, the first heading in this introduction is "How Is the Book Organized?" This main heading starts a major topic. The topics are further divided into smaller parts by subheadings. Breaking the content down into smaller chunks makes it easier to make sense of things as you read.

As you begin a new section, you might want to start by reading the headings. This will give you an overall idea of what you will be learning and doing. It's like a road map.

Things to Explore. After the title, the sections start out with a list of Things to Explore, as you can see at the top of this page. These are the learning goals, or objectives, for this section. The objectives show what you will know and do by the time you finish the section.

TechnoTerms. Next to the list of Things to Explore are the TechnoTerms. These are important terms, so be sure you know their meaning. They are easy to spot within the sections because they are printed in **boldface** type. You can look up the TechnoTerms in the glossary at the end of the book. The glossary is an important tool for helping you understand and apply the content in this book.

INFOLINK

InfoLinks direct you to other parts of the book that have more information about the topic you are reading about. For example, to see the glossary, take a look at pages 476-492.

SECTION TechCHECK

1. Why is it important to pay attention to the book's organization?
2. What do the Things to Explore, or objectives, show you?
3. **Apply Your Knowledge.** Compare this book to one of the books you use in another subject area. How do you think technology might apply to the subjects you study in that class?

CONNECTIONS

Science, Mathematics, and Communication

Each chapter of the book includes a Science, Mathematics, or Communication feature that looks similar to this one.

Each feature relates the study of technology to one of these subjects and includes an activity.

ACTIVITY

Turn to page 26 to see a Science Connection feature. What is the topic?

ACTION ACTIVITIES

Practicing Design and Problem Solving

Real World Connection

In the activities throughout this book, you will build things and practice design and problem solving. You will also learn how technology applies to real world situations.

Notice that near the title of each activity is a reminder that tells you to fill out a TechNotes form. An example is shown on this page. Your teacher will give you the form to fill out.

Design Brief

Each activity begins with a design brief. This statement provides an overall goal, or objective, for the activity.

Materials/Equipment

This will be a list of equipment and materials you will need to complete the activity.

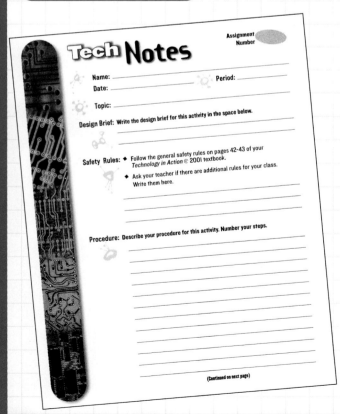

Tech Notes

Assignment Number

Name: _____ Period: _____
Date: _____
Topic: _____

Design Brief: Write the design brief for this activity in the space below.
_____

Safety Rules: ◆ Follow the general safety rules on pages 42-43 of your Technology in Action © 2001 textbook.
◆ Ask your teacher if there are additional rules for your class. Write them here.
_____

Procedure: Describe your procedure for this activity. Number your steps.
_____

(Continued on next page)

Procedure

1. The procedures tell you what to do to complete each activity. Sometimes they tell you exactly what to do. Others are general guidelines for creating a design.
2. After you complete the activity, your teacher may assign the evaluation questions.

Evaluation

1. How will you know what to do in each activity?
2. How will you know what materials and equipment to use for the activities?
3. How can you be sure that you will safely complete each activity?
4. **Going Beyond.** Check out the Action Activities throughout the book. What do you think will be the most exciting thing about working on the activities?

REVIEW & ACTIVITIES

CHAPTER SUMMARY

The summary states the key concepts covered in each section. This is an important tool for reviewing what you've read.

REVIEW QUESTIONS

These are questions you should be able to answer from the chapter. They will help make sure you've understood what you read.

CRITICAL THINKING

Knowing how to think critically is an important skill. These questions require you to think at a higher level. For example, you might be asked to apply what you've learned to a related situation or to something in your life.

CROSS-CURRICULAR EXTENSIONS

In these activities, you apply what you've learned to other subject areas. You will see how technology relates to mathematics, science, social studies, health education, and communication.

EXPLORING CAREERS

This part of the chapter review pages describes two careers that relate to the topics in the chapter, plus there's an activity for further exploration. You will explore how you might use technology in a future career.

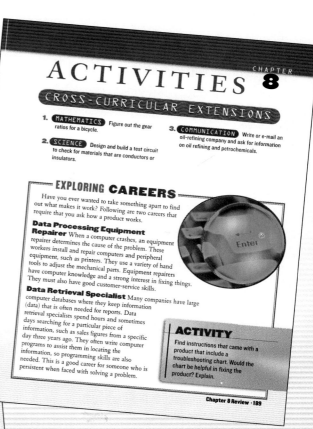

CHAPTER
8

REVIEW &

CHAPTER SUMMARY

SECTION 1
- A system is a combination of parts that work together as a whole.
- The systems model includes input, process, output, and feedback.
- The five basic systems used in technology are mechanical, electrical, fluid, thermal, and chemical.
- Trying to find the problem in a system is called troubleshooting.

SECTION 2
- Mechanical systems often include levers, gears, chains, cams, flywheels, springs, and other parts.

SECTION 3
- Electronics involves the movement of electrons through conductors, insulators, semiconductors, and superconductors.
- Electronic components such as diodes, capacitors, transistors, and resistors are connected in three basic circuits: series, parallel, and series-parallel.

SECTION 4
- Fluid systems are either hydraulic or pneumatic.

SECTION 5
- Chemical systems include batteries and petroleum products.
- Thermal systems control the temperature of things.

REVIEW QUESTIONS

1. What systems can you find in a flashlight? Do they work together or independently?
2. What is the difference between a single-acting cylinder and a double-acting cylinder?
3. What is a superconductor?
4. List three materials that are insulators and three that are conductors.
5. Why is it important that mechanical parts be made in standard sizes?
6. What parts of a car would be considered thermal systems?

CRITICAL THINKING

1. Name a situation in which pneumatic systems would be better to use than hydraulic systems. Explain why.
2. Research ways to use a computer to control a part of your robot.
3. Make a chart identifying different kinds of mechanical fasteners by name.
4. Design a hand-held foam cutter. Sketch your design on paper, and discuss it with your teacher. Make and test your hand-held cutter with the help of your teacher.
5. Design a flywheel-powered car.

ACTIVITIES

CHAPTER
8

CROSS-CURRICULAR EXTENSIONS

1. **MATHEMATICS** Figure out the gear ratios for a bicycle.
2. **SCIENCE** Design and build a test circuit to check for materials that are conductors or insulators.
3. **COMMUNICATION** Write or e-mail an oil-refining company and ask for information on oil refining and petrochemicals.

EXPLORING CAREERS

Have you ever wanted to take something apart to find out what makes it work? Following are two careers that require that you ask how a product works.

Data Processing Equipment Repairer When a computer crashes, an equipment repairer determines the cause of the problem. These workers install and repair computers and peripheral equipment, such as printers. They use a variety of hand tools to adjust the mechanical parts. Equipment repairers have computer knowledge and a strong interest in fixing things. They must also have good customer-service skills.

Data Retrieval Specialist Many companies have large computer databases where they keep information (data) that is often needed for reports. Data retrieval specialists spend hours and sometimes days searching for a particular piece of information, such as sales figures from a specific day three years ago. They often write computer programs to assist them in locating the information, so programming skills are also needed. This is a good career for someone who is persistent when faced with solving a problem.

ACTIVITY
Find instructions that came with a product that include a troubleshooting chart. Would the chart be helpful in fixing the product? Explain.

Getting Started in Technology

What Is Technology?

THINGS TO EXPLORE

- Define technology.
- Explain what a technologically literate person can do.

TechnoTerms
technologically
literate
technology

You live in a "high-tech" world. Fig. 1-1. Tech (pronounced tek) is short for *technology*. What exactly is technology? Is it robots, satellites, lasers, and computers? Does it include tools, such as saws and hammers? All of these are products of technology, but technology is a lot more!

Most definitions agree that **technology** is the use of knowledge, tools, and resources to help people solve problems. It depends on a combination of people like you, your ideas, and the tools you use. It involves both thinking and doing.

Technology is fast-paced, exciting, challenging, and fun! As you learn about technology you will be
- Using knowledge from science, mathematics, and other subjects to solve problems
- Designing, inventing, and making things based on your creative ideas
- Discovering how products of technology have helped make your life better

The Effects of Technology

The effects of technology are not always good for society or for the environment. Fig. 1-2. Some advancements in technology have caused problems such as acid rain. Other technologies are being developed to help solve those problems.

Fig. 1-1. Technology has enabled us to do so many things! Can you name some of the technologies shown here?

OPPOSITE Technology has had many impacts on our world. One of those impacts has been on the environment.

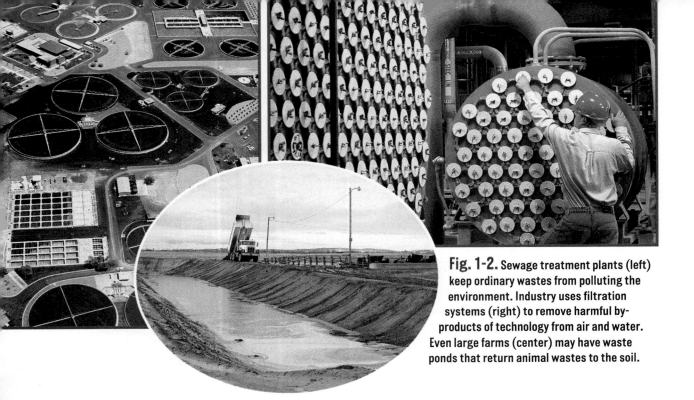

Fig. 1-2. Sewage treatment plants (left) keep ordinary wastes from polluting the environment. Industry uses filtration systems (right) to remove harmful by-products of technology from air and water. Even large farms (center) may have waste ponds that return animal wastes to the soil.

Because technology has an impact on people and the environment, you need to know how technology affects you. A person who understands the effects of technology is **technologically literate**. If you are technologically literate, you will be able to make decisions about your future and the future of technology based on facts. You will be able to

- See how technology has changed through time
- Understand the newest uses of technology
- Think through a problem and use the tools of technology to solve it
- Decide whether a specific technology is good or bad for people or for the environment

SECTION 1
TechCHECK

1. What is technology?
2. What is the definition of a technologically literate person?
3. List four things a technologically literate person is able to do.
4. **Apply Your Knowledge.** Write your own definition of technology.

Technology Is Changing Fast!

THINGS TO EXPLORE

- Tell why technology is growing so fast.
- Explain how the history of technology can be organized.
- Explain the exponential rate of change.

TechnoTerms
exponential rate of change
knowledge base
scientific method

Over 90 percent of all technologies we have today were invented in the last 25 or 30 years. Fig. 1-3. But do you know when technology really began?

How Technology Got Started

Many people think that technology began in the 1700s with the first factories. However, technology really began over one million years ago. Prehistoric people used simple tools, such as clubs and axes, to work on different materials.

One way to organize history is to divide the past into periods based on the materials people used. This method shows how people developed new technologies to meet their changing needs.

Fig. 1-3. Closed-circuit television enables this student to attend a class that is actually far away. What new uses for TV have you heard about recently?

- **The Stone Age** (2,000,000 B.C. to 3000 B.C.) During the Stone Age, prehistoric people used tools made mostly of stone, animal bones, and wood. The tools were important as weapons or for gathering food.

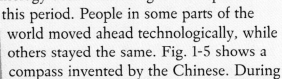

Fig. 1-4.

- **The Bronze Age** (3000 B.C. to 1200 B.C.) During the Bronze Age, people learned how to mix copper with tin to make a stronger metal called *bronze*. Bronze is an *alloy*. An alloy is a mixture of two or more pure metals. Bronze was used to make better tools and weapons. Fig. 1-4 shows the bronze head of a lance, a type of throwing weapon.
- **The Iron Age** (1200 B.C. to A.D. 500) Iron is harder and holds a cutting edge better than bronze. Iron ore was also easier for early people to find, and refining it was less costly. Iron inspired the design of more kinds of tools and is still used in industry today.
- **The Pre-Industrial Age** (A.D. 500 to 1750) Few changes in science and technology occurred during the first part of this period. People in some parts of the world moved ahead technologically, while others stayed the same. Fig. 1-5 shows a compass invented by the Chinese. During

Fig. 1-5.

SCIENCE CONNECTION

The Scientific Method

The process of scientific problem solving is called the *scientific method*. The scientific method was developed by the Italian scientist Galileo Galilei (1564-1642) and the English philosopher Francis Bacon (1561-1626).

The scientific method is applied generally to problems rather than followed strictly. The five steps of the scientific method are:

1. Recognize the problem.
2. Form a *hypothesis* (an educated guess) about the correct solution to the problem.
3. Make a prediction as to how your solution will work.
4. Experiment to test your solution.
5. Organize the hypothesis, prediction, and results of your experiment into a general rule.

this period, they also invented movable type for printing. These technologies were introduced in Europe at a later time.

During the second part of the Pre-Industrial Age, technology and science began to bring change. Several important scientific instruments, such as the microscope, were invented. Scientists started using the **scientific method** to find answers to their questions.

- **The Industrial Revolution** (1750 to 1900) During this period, many inventions brought changes that affected all of society. Using new inventions and machines, people set up factories that could produce goods cheaper and faster.
- **Recent Times** (1900-Today) Since the early 1900s, technology has grown rapidly. Fig. 1-6 shows an airplane built by the Wright Brothers. They achieved the world's first controlled flights.

Recent history can be divided into periods, or ages, based on developments in technology. Some of the recent periods include the Atomic Age, the Jet Age, the Space Age, and the Information Age. These ages overlap, and later discoveries are often based on earlier ones. Today, in the Information Age, skills such as finding and using information are important.

Fig. 1-6.

Scientists and technologists face the same traps in trying to solve problems. Everyone has a set of ideas about how the world works. These ideas often block new ideas or restrict our thinking to one path. Scientists must keep an open mind about their investigations because their results have to be based on scientific facts. Scientists can't let their results be changed by what they think the results *should* be. Another important reason for keeping an open mind is that many scientific discoveries happen by accident!

ACTIVITY

You know a boat is made to float. Form a hypothesis about why, and design an experiment to test your hypothesis.

Fig. 1-7. A few short years ago, putting an entire encyclopedia on computer disks wasn't practical. Floppy disks couldn't hold enough information. Today, all the information, plus colorful pictures, can be put on one CD. Find out how much information a CD can hold compared to a floppy diskette.

Technology's Rate of Change

Technology touches every part of our lives. It changes the way we do things, the way machines work, and the way we think. These changes are coming faster and faster. More people are adding new ideas and inventing more new tools to meet our changing needs. Fig. 1-7. But they don't throw away the old technologies; they build on them. When people combine ideas, we have even more new machines and tools. Some people say our **knowledge base** (all the facts known to people today) doubles every two to three years. This is what is called an **exponential rate of change**.

Because our knowledge base is so large and is growing so fast, it is important for you to learn where and how to find information.

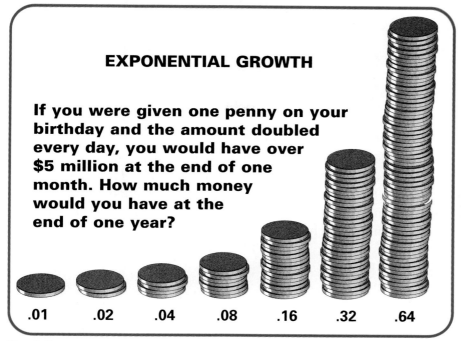

EXPONENTIAL GROWTH

If you were given one penny on your birthday and the amount doubled every day, you would have over $5 million at the end of one month. How much money would you have at the end of one year?

.01 .02 .04 .08 .16 .32 .64

Fig. 1-8. Technology is growing at an exponential rate. It is doubling every few years. Also, the time it takes to double is getting even shorter.

Exponential Growth If someone gave you one penny on your birthday and told you that the amount would double every day for a month, you might be disappointed at first. On the first day you would have only one cent. On the second day you would have double that amount—two cents. On the third day the two cents would be doubled to four cents. That still doesn't sound like much. By the end of 30 days, however, you would have over $5 million—$5,368,709.12 to be exact! Fig. 1-8.

SECTION 2
TechCHECK

1. List the technological ages of the recent time period.
2. Why is technology changing so fast?
3. What is meant by an exponential rate of change?
4. **Apply Your Knowledge.** Make a bulletin board display showing technologies you commonly use.

ACTION ACTIVITY

Making a 3D Model of Exponential Growth

Be sure to fill out your TechNotes and place them in your portfolio.

Real World Connection

Exponential growth is easy to understand if you make a model of it. Graphs are often used to model mathematical information (numbers). Fig. A. But in this activity, you will build a three-dimensional (3D) model showing the exponential growth of the world's knowledge base.

Design Brief

Working in groups of three or four students, design and make a 3D model that shows the exponential growth of the world's knowledge base. Use only recycled materials such as empty milk cartons, soda cans, or scrap paper. Fig. B.

Materials/Equipment
* recycled packaging or other materials
* scissors
* tape
* glue
* markers

SAFETY FIRST

If you use soda cans, watch out for sharp edges. Be sure to read and follow all the safety rules on pages 42-43, at the end of this chapter.

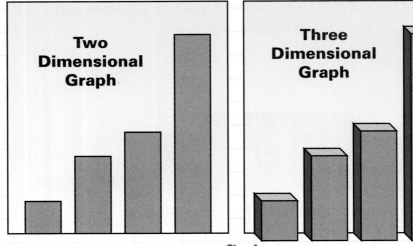

Two Dimensional Graph

Three Dimensional Graph

Fig. A

Procedure

1. List possible materials you can use for the 3D graph.
2. Make a sketch of how your graph will look.
3. Gather materials.
4. Build your model.
5. Show your model to other groups.

Evaluation

1. How many doublings could you create before the model became too big?
2. Did everyone in class have the problem of the model becoming too large at some point?

3. **Going Beyond.** Use a computer to create a three-dimensional chart showing exponential growth.

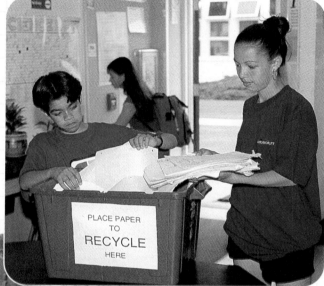

Fig. B

Solving Problems Step by Step

THINGS TO EXPLORE

- Explain the basic steps in problem solving.
- Tell how a design brief is different from a problem statement.
- Apply research skills in locating information.

TechnoTerms

design brief
parameters
problem-solving
strategy
problem statement

Have you ever had a problem that you wanted to solve but you didn't know where to start? Problems aren't always bad or complicated. A problem can be as simple as deciding whether to eat a chocolate chip cookie or an oatmeal cookie. But most of the time, the problems you will be solving in technology class are more complex than that. If you have steps to follow in trying to solve a problem, getting an answer is easier.

Problem-Solving Strategy

A step-by-step plan for solving problems is called a **problem-solving strategy**. Once you learn a basic problem-solving strategy, you can use it to solve all kinds of problems throughout your life. Fig. 1-9. Any problem in school, at home, or with your friends can be tackled.

Fig. 1-9. Problem-solving methods helped these students analyze the pollution in a pond (upper left), just as they helped people develop wind "farms" in California (lower left) that capture wind energy. How do you think problem solving resulted in the use of robots to weld cars (right)?

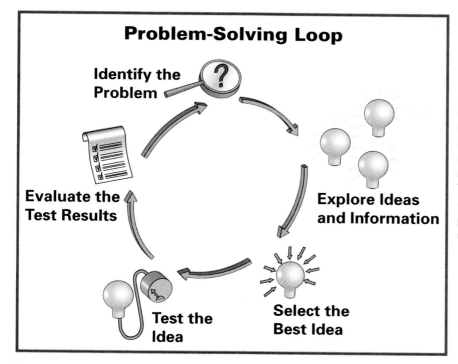

Problem-Solving Loop
Identify the Problem
Explore Ideas and Information
Select the Best Idea
Test the Idea
Evaluate the Test Results

Fig. 1-10. The problem-solving process is shown as a loop, or circle, because you may need to repeat the steps to arrive at the best solution. Think of a problem you solved recently. How many of the problem-solving steps did you use?

Identifying the Problem

Most problem-solving strategies have some things in common. Fig. 1-10. Usually, the first step in solving a problem is being able to state exactly what the problem is in your own words. This definition of the problem is sometimes called the **problem statement**. Here are some examples of problem statements.

- Design a solar-powered vehicle.
- Build a bridge to span a river.
- Produce a television show.
- Program a robot to move objects.

A more detailed problem statement is called a **design brief**. A design brief gives the **parameters** (specific details) of the problem. Parameters include such things as the amount of money you can spend and the time available. Fig. 1-11.

> Design Brief: Solar-Powered Vehicle
>
> Design a solar-powered vehicle that will carry a minimum of two passengers at least 100 miles without needing recharging. The design budget is $10,000. The vehicle must meet the federal safety standards for use on public highways.

Fig. 1-11. This design brief shows the parameters for a solar-powered vehicle.

TechnoFact

TSA The Technology Student Association (TSA) is for students who are enrolled in or who have completed a technology education course. Over 100,000 elementary, middle school, and high school students are members. Through its conferences, contests, and other activities, the TSA promotes technological literacy, leadership, and problem solving.

INFOLINK

Many companies have special research and development (R&D) departments that seek new product ideas or improve existing products. See Chapter 11 for more information.

Exploring Ideas and Information

Once the problem is identified, you need to explore ideas and gather information related to your problem. You need to use as many different resources as possible. Fig. 1-12. You also need to be sure your information is current.

You will need to evaluate the information and decide whether or not it is useful in solving your problem. Is the particular information "need to know" or just "nice to know"?

Experimentation is another way to do research. By finding out how things work you can then change or improve technology that already exists. Fig. 1-13.

INFOLINK

See "Where Do Ideas Come From?" in Chapter 6.

SOURCES FOR EXPLORATION

Internet · Magazines · LIBRARY · Texts · Books · CD-ROMS · Encyclopedias

Experts · PEOPLE · Friends · Teachers · Parents

Fig. 1-12. These are common resources. Which do you use regularly?

Selecting the Best Idea

Next you select what you think is the best idea to try first. Many times your first idea will not work, and you must choose a different idea that might work better.

Testing the Idea

The fun part is testing your idea or solution. Testing usually requires taking measurements or accurate notes about what's happening.

Evaluating the Results

The most important part of the entire problem-solving strategy is evaluating what happened during testing. Sometimes you want so much for your idea to work that you do not look at the results carefully. However, the chances of your first idea being totally successful are slim.

Remember that there may be many possible solutions to a problem. You may have to try many of them before you reach your final solution. If your first idea did not work, it doesn't mean you failed. It just means you learned what doesn't work. That is a big step toward solving the problem!

Fig. 1-13. This research scientist wears special gloves that are built into a sealed chamber. The chamber can be kept very clean, or a vacuum can be created in it, and work can still go on.

SECTION 3
TechCHECK

1. List the five basic steps in problem solving.
2. What is the difference between a problem statement and a design brief?
3. What different resources might you use to research a topic?
4. **Apply Your Knowledge.** Write a design brief for building a bridge across a river. Remember to add the specific parameters.

ACTION ACTIVITY

Doing Technology Research

Be sure to fill out your TechNotes and place them in your portfolio.

Real World Connection

Many of today's jobs require you to research information. In this activity you will research information on three topics from companies all over the country. To find the company names and phone numbers, you will use resources in the library or from your teacher. Fig. A.

Design Brief

Research three technology topics of your choice from the list provided. Find information on companies that make products or do research related to your topics. Request information either by phoning or writing to the companies.

Materials/Equipment

- telephone
- books
- videos
- magazines
- the Internet
- newspapers
- computer word processor or paper and pencil

Fig. A

Procedure

1. Choose any three of the following technology topics:
- composite materials
- computer controls
- conveyor systems
- fasteners
- fiber optics
- hydraulics
- lubricants
- plastics
- pneumatics
- robotics
- soldering
- superconductors
- welding
- any other technology-related topic your teacher approves

2. Research your topics using the resources in your library or those available from your teacher. Record information about the resources you used in a chart like the one shown in Fig. B.

3. Get your teacher's permission to call each company (use toll-free phone numbers if possible), or write a letter to request information. Ask them to send their catalog or product information to you at the school address. Don't forget to say "please" and "thank you." Good manners are appreciated everywhere.

Evaluation

1. Which of your three topics were easy to research?

2. How many companies sent you information on their products? Share the information you received with the members of your class.

3. If you owned a business, what would you be willing to do to help students learn more about technology?

4. **Going Beyond.** Talk to your media specialist at school or the local librarian to see what new research tools are available for you to use. Make a list to share with your classmates.

5. **Going Beyond.** With your teacher's permission, research a topic in three different resources on the Internet. Compare your information and evaluate whether the sources are current and accurate.

SAFETY FIRST
Check with your teacher before looking up information on the Internet.

TOPIC 1:	TOPIC 2:	TOPIC 3:
COMPANY NAME:	COMPANY NAME:	COMPANY NAME:
PHONE: (800)	PHONE: (800)	PHONE: (800)
RESOURCE:	RESOURCE:	RESOURCE:
PAGE:	PAGE:	PAGE:

Fig. B

Putting Your Abilities to Work

THINGS TO EXPLORE

- Explain why working in groups makes solving problems easier.
- Tell why brainstorming is important in problem solving.
- Tell why being creative is important to problem solving.
- Work as part of a group to solve a problem.

TechnoTerms

brainstorm
idea bank

The ability to work as part of a group is important to you in solving problems both in school and later in a job. Fig. 1-14. The old saying "two heads are better than one" is really true. Problems are often easier to solve with the help of other people. Those with different experiences and backgrounds can bring more information and a new way of looking at the problem. Each person's talents can be put to good use. Some of us are better at tossing out new ideas. Others are better at putting ideas into action. You will have a chance to do both as a member of a group.

Fig. 1-14. When people work together, more skills and ideas are contributed to solving a problem. Think of the last problem you and your family solved together. What did each person contribute to the solution?

Brainstorming

Many good ideas can be lost if someone puts an idea down before it has a chance to be discussed. When you **brainstorm** ideas, you list as many ideas as possible. During brainstorming the ideas are not judged as either good or bad. Even those that might sound silly are given a fair hearing.

During brainstorming, everyone in the group has a chance to contribute to the **idea bank** (all the ideas presented). From that idea bank your group selects the best idea.

Fig. 1-15. A film animation director inserts a computer-generated hummingbird into a piece of film with real people. Describe some recent cartoons you've seen that you think are very creative.

Being Creative

A person who is creative has the ability to come up with a new idea or way of doing things. Everyone can be part of the creative process.

To be a creative thinker, you need to use facts, feelings, experiences, and knowledge. You might have to break your usual rules for thinking or create some new ones. Fig. 1-15.

It is helpful to ask yourself "what-if" questions. For example, what if there were a pill that would make anyone smarter? Who should take it?

Remember that the more creative ideas you have to choose from, the better your chances of finding a good solution to the problem.

SECTION 4
TechCHECK

1. Name one advantage to working in groups to solve problems.
2. What is brainstorming? Why is it important in problem solving?
3. What can you do to be more creative?
4. **Apply Your Knowledge.** Brainstorm ideas for designing a classroom of the future. Write your ideas down to share with others.

ACTION ACTIVITY

Working in a Group to Solve a Problem

Be sure to fill out your TechNotes and place them in your portfolio.

Real World Connection

An important skill in the business world is being able to work as part of a team. You might be part of a design team or a construction team.

In this activity, you will work in groups using a step-by-step procedure to solve a problem.

Design Brief

Design and build a model of a skyscraper structure. Your goal is to build the tallest structure possible using only the materials listed below. The finished structure must support the weight of this textbook without falling down.

Materials/Equipment

- straws
- paper clips
- pliers
- scissors
- paper

SAFETY FIRST

Follow the safety rules listed on pages 42-43 and the specific rules provided by your teacher for tools and machines.

Procedure

1. Brainstorm at least four or five designs with your group. Fig. A.

2. Sketch all the designs for the structure. Evaluate the designs, and as a group select one to construct.

3. Use scissors to cut the straws to the lengths needed for your structure.

4. Bend the paper clips as needed to attach them according to your design.

5. Measure the height of your structure from its base to the very top. Write its height on your sketch.

6. Test your structure by balancing this textbook on top.

7. Problem-solving strategy reminder: Did your group try different ideas if the first one didn't work?

Evaluation

1. What were the things that held you back when solving this problem?

2. How could you change your structure to make it stronger?

3. Did your group members work well together to complete the project? Explain.

4. **Going Beyond.** Working in a group or team, design and make a model of a school locker that would be useful to a wheelchair-bound student. Your design should provide storage for coat, books, pencils, notebooks, and so on.

5. **Going Beyond.** Working in small groups, come up with a list of ten products you think will be useful in the future.

Fig. A

Sheets of paper can be attached to the frame with paper clips to form the "skin" of your building.

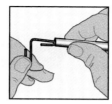

Bend paper clips at right angles to form fasteners that can be inserted into the ends of drinking straws.

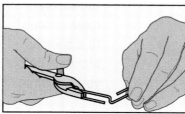

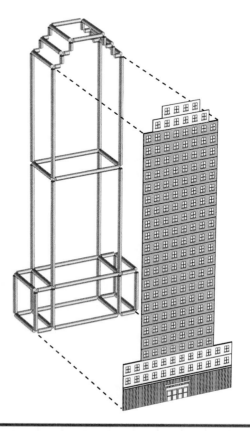

SAFETY

Throughout this book, you will be doing many activities that require you to use safety equipment and conduct yourself safely. These general safety rules should be followed at all times:

1 Follow the teacher's instructions. Do not fool around in the technology lab area.

THINK SAFETY FIRST

2 Pay close attention to what you are doing at all times. Do not let others distract you while you are using a machine.

3 Always wear eye protection. Special eye protection may be needed for some activities such as using a laser, welding, or using chemicals. Ask your teacher for help.

4 Never use any tool or machine without a demonstration by your teacher. Know the safety rules related to a specific machine before you turn it on.

5 Be careful not to wear loose clothing, jewelry, or other items that could get caught in a machine.

6 Always use the guards on each machine. Keep hands and fingers away from all moving parts.

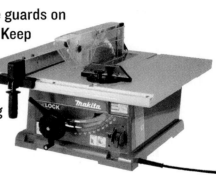

RULES

7 Keep the work area clean.

8 Report all injuries to your teacher at once.

INJURY REPORT

9 Put warning signs on things that are hot and could cause burns. Warn others of the hazards of laser light.

CAUTION
HOT
WIRE

10 Do not use electric tools near flammable liquids or gases. Store oily rags in a proper container. Know where the nearest fire extinguisher is and how to use it.

COLORS FOR SAFETY

COLOR	MEANING	COLOR	MEANING
RED	Danger or emergency	WHITE	Storage
ORANGE	Be on guard	GREEN	First aid
YELLOW	Watch out	BLUE	Information or caution

1 REVIEW &

CHAPTER SUMMARY

SECTION 1

- Technology is the use of knowledge, tools, and resources to help people solve problems.
- A person who understands the effects of technology is technologically literate.

SECTION 2

- Technology is rapidly changing the way our world works.
- The history of technology can be divided into ages according to the materials people used.
- Technology is growing at an exponential rate—doubling every few years.

SECTION 3

- It is easier to solve problems if you use a problem-solving strategy.
- The basic steps in most problem-solving strategies include: identify the problem, explore ideas and information, select the best idea, test the idea, and evaluate the test results.
- A design brief provides more specific information than a problem statement.
- Research and experimentation are important ways to find information.

SECTION 4

- It is often easier to solve problems when you work in groups.
- Creative thinking is important to develop new technologies.

REVIEW QUESTIONS

1. What does it mean if a person is technologically literate?
2. How are the time periods in technology's history organized?
3. Why is a problem-solving strategy important to use?
4. If you need to find information on a topic, what should you do?
5. What are the advantages of working in a group?

CRITICAL THINKING

1. During what part of technology history do you think the most important technological developments happened? Explain.
2. Create your own problem-solving strategy.
3. Write a design brief for a classroom of the future.
4. List as many other uses as you can for popsicle sticks besides making popsicles.
5. Write a design brief for a package that will protect a fragile (easily breakable) object.

ACTIVITIES **1**

CROSS-CURRICULAR EXTENSIONS

1. **SCIENCE** Design and test an earthquake alarm. The materials may include a simple battery-operated electric buzzer.

2. **COMMUNICATION** Asking "what-if" questions helps open your mind. Think about this one. "What would happen if people had twelve fingers instead of ten?" Write down your ideas and share them with your classmates in a brainstorming session.

3. **MATHEMATICS** Find a new way to show exponential growth. Make a graph or a model to show your results.

EXPLORING **CAREERS**

Almost all of our usual activities are being changed by technology. It's only natural that people's jobs and the duties they perform are also changing. Here are two careers that have been greatly affected by advances in technology:

Cyberlibrarian Online libraries make some books, periodicals, and reference materials easily available. They can be read by anyone, anywhere, anytime. Libraries are hiring more cyberlibrarians, those who have computer experience and Internet skills. These workers can conduct online searches and teach visitors how to navigate the Internet.

Automotive Technician When you start the engine, step on the gas, set the cruise control, change the radio station, and hit the brakes you are using computers. Today, all of these automobile functions, along with dozens more that take place under the hood, are controlled by computer chips.

Knowledge and training in automotive technology, combined with skills in computers and electronics, are necessary for this job.

ACTIVITY

As a class, think of a problem in your school, such as limited storage space in the library. Divide the class into groups of three. Use technology to find a solution to the problem.

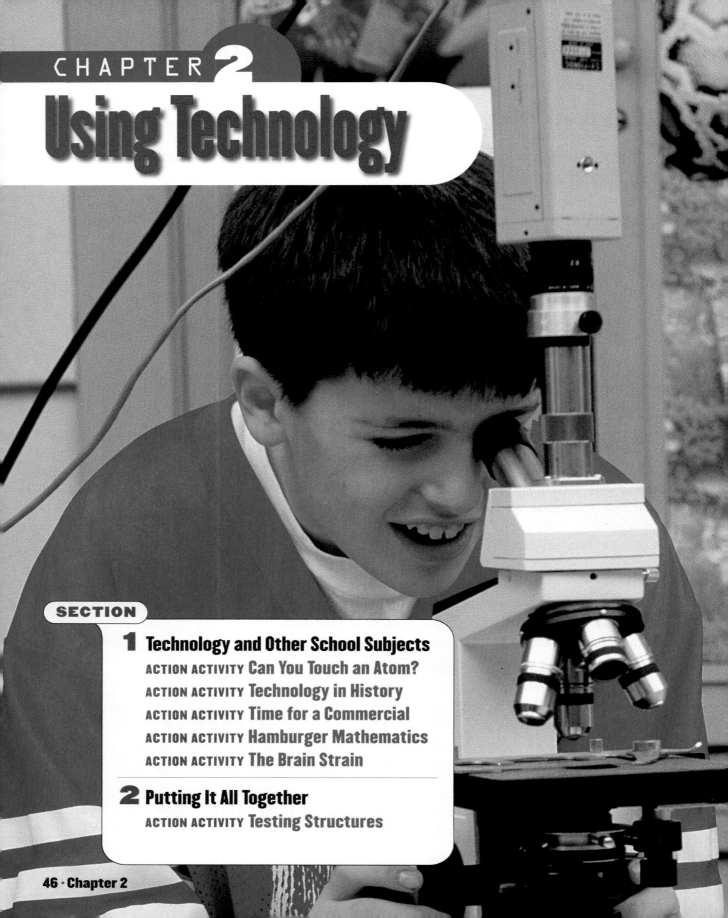

Technology and Other School Subjects

THINGS TO EXPLORE

- Describe how technology affects you in school.
- Explore technology's connections with other school subjects such as science, social studies, communication, mathematics, physical education, and health.

TechnoTerms
atom
theory
timeline

Technology touches almost everything. In your everyday life, you may take technology for granted. But common items such as your toothbrush and your shoes are very different from those of the past because of developments in plastics technology. Toothbrushes were once made of hog bristles instead of plastic. Today's running shoes, made of plastics, are very different in weight from the kangaroo leather running shoes of earlier days. Fig. 2-1.

Think About This

Did you ever stop to think that even candy bars change with technology? Hershey Foods Corporation has developed and patented a heat-resistant milk chocolate bar. The special chocolate bar is supposed to hold its shape at temperatures of up to 140°F.

OPPOSITE Technology has created high-powered microscopes that enable us to see things invisible to the naked eye.

Fig. 2-1. The design of running shoes has changed over time. Why do you think the weight of running shoes is important?

Fig. 2-2. Many schools benefit from technology. List all the technologies you use regularly at your school.

Technology is part of school, too. Think of the technologies that play a part in your classroom today. Your desks and chairs were designed to fit you instead of a third grader. Many classrooms provide access to VCRs, televisions, laser disc players, compact discs (CDs), and computers. Fig. 2-2.

We depend on technology to make our lives better. Because technology is so important to our world, it's important to understand what it is and how it works. School subjects can help you do that. That's one of the ways school subjects help prepare you to be an intelligent, useful member of our society.

Let's see how technology is related to such subjects as science, mathematics, social studies, language arts, and health or physical education.

SCIENCE CONNECTION

The Untouchables

MAGNETIC POLES

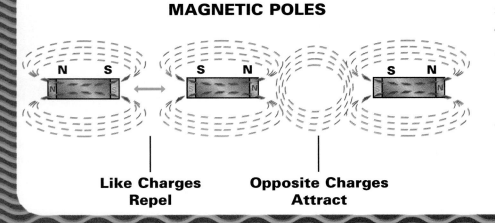

Like Charges Repel

Opposite Charges Attract

You know from studying science that electrons, one of the three basic parts of an atom, spin around the atom's center, or nucleus.

Technology and Science

Although technology and science are closely tied together, they are different. Science usually gives you the **theories** (ideas) about something. Technology lets you use your knowledge and resources to solve problems. Fig. 2-3.

Much of what scientists do is based on the scientific method. They start with a theory and then try to prove its truth. For example, for centuries people had theories about what matter was made of. Using tools of technology, scientists discovered that matter is made of **atoms**.

INFOLINK

See Chapter 1 for more information about the scientific method.

Fig. 2-3. First, scientists developed theories about traveling in outer space. Then technology gave us rockets so we could solve the problems of space travel. Watch the movie *October Sky*. How do the characters solve problems related to rocketry?

You also know that electrons have a negative electrical charge. But did you know that these negative charges keep one atom from touching another? This is because like charges repel, or push away, from each other, just as the like poles of two magnets do. So even though you might think that the atoms in something as hard as a brick are packed close together, they don't really touch at all!

ACTIVITY

Make a model of an atom that shows the electrons in orbit around the nucleus (neutrons and protons).

Technology and Mathematics

Mathematics and technology work well together. Technology has produced calculators and computers that can make many computations quickly and accurately. Fig. 2-4. However, you still have to know what operations—addition, subtraction, multiplication, and division—to use in solving a problem. You also must know how to enter the information correctly into a calculator or computer.

Technology and Social Studies

Most people think that technology is related only to science and mathematics. While this connection is easy to see, technology is just as much a part of social studies and other subjects.

People use history as a way of charting the present or planning the future. A study of history shows that technology definitely has changed with the times. Fig. 2-5. At each point in time, different technologies were important for what they could do to help us. Some technologies no longer exist because there's no use for them today. Other technologies have changed to better meet our needs. A graph called a **timeline** can show how the speed of technology change has increased in the past century.

Fig. 2-4. Most schools allow calculators in classes. Take a position either for or against this statement: Because of automatic calculators, many students are not developing adequate mathematics skills. Write a paragraph defending your position.

Technology and Communication

Being able to communicate with others is a skill that is important in all your school subjects. You may think of it only in terms of language arts or reading. However, in all your courses you need to be able to let your teachers and classmates know what your ideas are. Technology gives you many different ways to communicate using sound, the written word, and visual images. Fig. 2-6.

Fig. 2-5. Have you ever seen a phonograph like the one shown here? Manufactured around 1900, this model reproduced sound by means of a needle that followed a groove in a rotating disk, called a "record." Does the phonograph seem completely different from the CD player? Research how CDs are made and compare the two.

Fig. 2-6. Many people think that, if they know how to use a computer, they won't need to learn reading, spelling, and grammar. Not so. They must be able to correctly tell the computer what to do.

Technology and Health/Physical Education

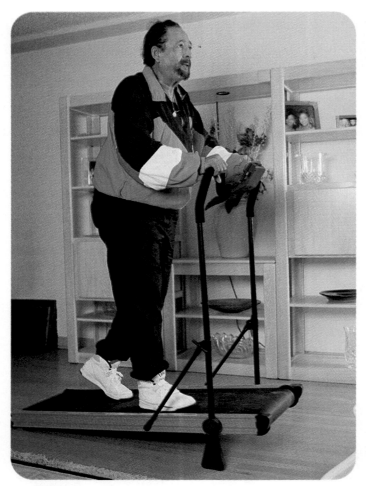

Fig. 2-7. Treadmills are an important innovation for people who must exercise every day. People can walk for as long as necessary without experiencing bad weather. Name some other recent health-related technologies.

Many recent developments in technology have taken place in the field of health. Technology has come up with new products and methods that help us stay healthy and fit. Fig. 2-7.

Technology will play an even more important part in your future. Thanks to advancements in medical technology, people are living longer and more active lives. As a result, it is common for joints to just wear out. Low friction, plastic replacement joints make it possible for many people to continue an active life. Do you know anyone who has had a hip, knee, or elbow joint replaced?

Medical technology may one day make it possible to grow new body parts. Experiments are being done to see if special cells can grow new tissue. This could help many people that have been injured in accidents or are suffering from disease.

SECTION 1 ✓ TechCHECK

1. What technologies do you use in your classroom today?
2. How are technology and science connected?
3. Name one technology you use both at home and at school.
4. **Apply Your Knowledge.** In small groups, brainstorm ways technology has changed your school and the way you learn. Share your ideas in a newsletter, a video production, a radio broadcast, or a chart.

ACTION ACTIVITY

Science—Can You Touch an Atom?

Real World Connection

Everything in our world is made of atoms and combinations of atoms called molecules. Scientists use high-powered electron microscopes to see as much as they can about the atom. However, many people have the wrong idea of what atoms are really like. Atoms are mostly empty space.

In this activity you will make sketches that show the size relationships among atoms and their parts.

> **Be sure to fill out your TechNotes and place them in your portfolio.**

Design Brief

In this activity, you will visualize the real size of atoms and their basic parts (electrons, protons, and neutrons). Fig. A.

Procedure

1. First, let's think about how small an atom really is. Divide a piece of paper into three equal spaces. Number the spaces 1, 2, and 3. In each space make a sketch showing sizes to scale.

2. In space 1, illustrate the fact that if a baseball were enlarged to the size of the Earth, its atoms would be the size of marbles.

3. In space 2, illustrate the fact that if a single atom were enlarged to the size of a fourteen-story building (140 feet tall), the nucleus would be the size of a grain of salt.

4. In space 3, make a sketch showing the atoms of your finger as your finger "touches" the top of your desk.

Evaluation

1. From what you have learned, can you really touch an atom? Explain your answer.

2. What is in the space between the nucleus and the electrons in an atom?

3. **Going Beyond.** Research the ways atoms move in solids, liquids, and gases. Make a graph or chart on the computer to share with the class.

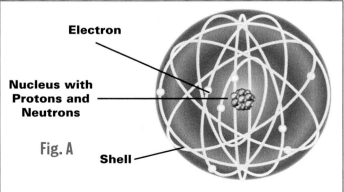

Electron

Nucleus with Protons and Neutrons

Fig. A

Shell

ACTION ACTIVITY

Social Studies—Technology in History

Be sure to fill out your TechNotes and place them in your portfolio.

Real World Connection

Advancements in technology have helped give us the quality of life we enjoy today. In this activity, you will organize and chart some of the events in history that have led to the present high level of technology.

Design Brief

Technology is sometimes divided into the following four groups: communication, construction, manufacturing, and transportation. In this activity, you will sort events in history into one of the four groups and make a timeline.

Materials/Equipment

- adding machine paper
- markers or pens
- meter stick

Procedure

1. Work in groups of four. Measure and cut 1 meter of adding machine paper.
2. Use a meter stick to draw one line on the paper for each of the four groups. Label each line.
3. Use a scale of 10 cm = 1000 years. Mark your timeline starting at 3500 B.C.
4. See the list on the next page. Chart each of the events on the appropriate line. Use circled numbers to represent each event.

Evaluation

1. How does your timeline show the rapid growth of technology? Explain.
2. During what period did most of the technological developments occur?
3. When did NASA launch the first Space Shuttle?
4. **Going Beyond.** Do some research on one development in the timeline list. Share what you find in a short report.
5. **Going Beyond.** List your favorite technology inventions. Research their development and place them on the completed timeline.

①	3500 B.C.	Writing first used by Sumerians
②	3000 B.C.	Egyptians created first book
③	1500 B.C.	Pulleys and simple machines used
④	A.D. 1045	Movable type used in printing
⑤	A.D. 1450	Printing press invented
⑥	A.D. 1712	Piston steam engine developed
⑦	A.D. 1835	Morse code/telegraph invented
⑧	A.D. 1876	Telephone invented
⑨	A.D. 1892	Reinforced concrete created
⑩	A.D. 1906	Radio developed
⑪	A.D. 1926	Television invented
⑫	A.D. 1926	Liquid-fueled rocket developed
⑬	A.D. 1933	FM broadcasting system introduced
⑭	A.D. 1946	ENIAC computer developed
⑮	A.D. 1947	Transistors invented
⑯	A.D. 1957	Sputnik (first artificial satellite) put into space
⑰	A.D. 1960	First laser operated Fig. A.
⑱	A.D. 1961	First man flew in space
⑲	A.D. 1966	First soft landing made on the moon by Luna 9
⑳	A.D. 1969	Neil Armstrong became first man on the moon
㉑	A.D. 1977	The Apple II started the personal computer industry
㉒	A.D. 1977	Fiber-optic cable first used in commercial communication
㉓	A.D. 1977	MRI (magnetic resonance imaging) first used by doctors
㉔	A.D. 1978	The 5 1/4-inch disk became the standard format for storage of computer data
㉕	A.D. 1981	Reusable spacecraft, U.S. Space Shuttle *Columbia*, made first flight
㉖	A.D. 1982	Synthetic insulin, the first drug manufactured using recombinant DNA, was sold
㉗	A.D. 1985	British Antarctic survey team discovered a hole in the ozone layer
㉘	A.D. 1986	Karl Muller and Georg Bednorz discovered a ceramic material able to superconduct at 35° Kelvin, a new record for high-temperature transmisson
㉙	A.D. 1988	The U.S. Patent Office approved a patent for a genetically altered mouse
㉚	A.D. 1988	A voice-operated typewriter recognized dictated words
㉛	A.D. 1988	The world's first public maglev system went into operation in West Berlin
㉜	A.D. 1990	A new line of biodegradable plastics was developed
㉝	A.D. 2000	Scientists deciphered the genetic code of the fruit fly

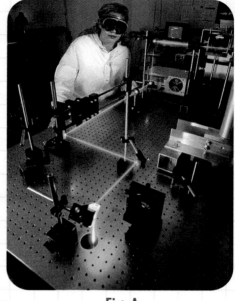

Fig. A

ACTION ACTIVITY

Communication—Time for a Commercial

Be sure to fill out your TechNotes and place them in your portfolio.

Real World Connection

If you are like many people, you see or hear dozens of commercial messages each day on television or radio. They often involve many forms of communication. However, most people don't think about how commercials are made.

In this activity, you will use communication skills and technology to produce your own commercial.

Design Brief

Write and produce a video (TV—something seen) or audio (radio—something heard) commercial for an imaginary product. Fig. A.

Materials/Equipment

- blank audio or video tapes
- props
- video camera
- VCR
- television
- audio tape recorder
- sound effects equipment
- computer (optional)

SAFETY FIRST
Follow the safety rules listed on pages 42-43 and the specific rules provided by your teacher for tools and machines.

Fig. A

Procedure

1. Work in groups of four or five students. Elect someone in your group to be the director. The director will organize the production of your commercial.

2. Brainstorm an imaginary product. Make a sample of your product to be used as a prop if you are making a video commercial. You might make a label that could be glued to a box, for example.

3. Write the script for your commercial. You may include everyone in your group, but you will need one person to operate the camera or the audio recorder. Your group might need to use sound effects (SFX) or video effects (EFX). All actions, sounds, or dialogue (talking) must be a part of your script. Use the format shown in Fig. B.

4. Gather any other props you need, and plan to bring any special clothes or costumes for your rehearsal and taping. The director of your group should schedule the use of the video or audio equipment with the teacher.

5. Rehearse and revise your production so that it lasts exactly 60 seconds. Record the final version.

6. Play your finished commercial for the class.

Evaluation

1. Survey the class to see how effective your commercial was.

2. How would you change your commercial to make it better?

3. Why is it necessary to write a script to use during production?

4. **Going Beyond.** Create a longer production such as a news program. Videotape your show so you can share it with others.

5. **Going Beyond.** Write a script and produce a public service announcement.

SOUND	VIDEO	DIALOGUE	
LOUD ROCK MUSIC	FADE IN: TWO GIRLS TALKING	MARY:	ARE YOU GOING TO THE DANCE SATURDAY NIGHT?
		KEISHA:	NO, HECTOR NEVER ASKED ME.
PHONE RINGS	ZOOM IN: KEISHA ON PHONE	KEISHA:	OH, HI, HECTOR. WE WERE JUST TALKING ABOUT YOU.

Fig. B

ACTION ACTIVITY

Mathematics—Hamburger Mathematics

Be sure to fill out your TechNotes and place them in your portfolio.

Real World Connection

Every fast-food restaurant must try to keep costs down to be competitive. Fig. A. The fast-food industry has used technology to help produce meals as efficiently as possible. That makes the job of a business manager very challenging!

In this activity you will act as the business manager of a restaurant.

Design Brief

Your fast-food restaurant plans to sell 2 billion hamburgers during the next year. As business manager, you will solve several problems involving mathematics.

Following are amounts needed to make one hamburger:

- beef, 113.5 g
- ketchup, 2.1 mL
- mustard, 1.5 mL
- salt, .19 g
- mayonnaise, 2.76 mL

Fig. A

Materials/Equipment

- paper
- pencil
- calculator or computer and spreadsheet software

Procedure

Answer the following questions using a calculator or computer spreadsheet.

1. If the average cow yields 175 kg of ground beef, how many cows will be needed for you to reach your 2-billion-hamburger goal?

2. If a tank holds 10 m³ (cubic meters), how many truckloads of ketchup, mustard, and mayonnaise will be needed?

3. How many tons of salt should be ordered? (*Hint:* You will need to find out how many grams are in a pound and how many pounds are in a short ton.)

Evaluation

1. If your restaurant sells hamburgers for $1.49 and they cost $1.19 to make, what will your annual profit be?

2. Would your place of business survive in your own hometown? Why or why not?

3. How many grams are in 1 pound?

4. **Going Beyond.** What would your annual profit be if you raised the price of a hamburger to $1.75?

5. **Going Beyond.** Ask someone at a local fast-food restaurant how many hamburgers are sold on an average day. Make a comparison chart to show your findings.

ACTION ACTIVITY

Health Education—The Brain Strain

Real World Connection

It is important to exercise both your muscles and your brain. Why not do both at the same time? In this activity, you will design an exercise and study cell where you can do homework while exercising your muscles.

Design Brief

Work in groups of two or three to design and sketch a combination exercise machine and study area. The exercise equipment might be similar to a stationary bicycle or a stair-stepping machine. The study area should include a place to write, a light, and a place for a computer. The design must let you exercise and study at the same time!

Materials/Equipment

- paper
- pencil
- equipment catalogs
- computer with graphics software (optional)
- exercise machine (optional)

> **Be sure to fill out your TechNotes and place them in your portfolio.**

> **SAFETY FIRST**
> Follow the safety rules listed on pages 42-43 and the specific rules provided by your teacher for tools and machines.

Procedure

1. Brainstorm possible solutions to the design problem with your group members.
2. Evaluate each idea and come to a *consensus* (agreement) on a practical design.
3. Use old catalogs or ads to find pictures of the equipment you would like to put into your exercise-study cell.
4. Design the exercise-study cell to be safe, quiet, and easy to use.
5. Make a sketch or a computer drawing of your product. Think of a name for it and estimate a retail (store) price.
6. Present your product idea to the class.

Evaluation

1. Does the exercise-study cell adjust to fit people of different sizes? Explain.
2. Could the exercise-study cell be used by a physically challenged person? Explain.
3. Does the product include pinch points where you or small children might become caught or injured?
4. Have the following items been considered in the design: fire resistance? low cost? environmental impact?
5. Would you buy such a product? Explain.
6. **Going Beyond.** Make a commercial about your product for radio or television.
7. **Going Beyond.** Visit your local fitness center or gym. Learn about the different pieces of equipment and the purpose of each.

Putting It All Together

THINGS TO EXPLORE

- Explain how solving real-world problems requires a combination of technology, mathematics, science, and communication.
- Put technology, mathematics, science, and communication to work together to solve a problem.
- Design and test an earthquake-resistant structure.

TechnoTerm
integrated

School subjects are often separated to make it easier for you to concentrate on certain things at one time. But in the real world, you don't use mathematics only from 1:00 p.m. to 2:00 p.m. and science from 2:05 p.m. to 3:05 p.m. Can you imagine an automotive engineer waiting until 1:00 p.m. to solve a math problem related to a car's design? Fig. 2-8. He or she must put many skills to work at the same time to get things done. The subjects you learn in school need to be **integrated** (used together) for you to solve problems.

INFOLINK

See Chapter 1 about how technology is created.

Technology Integrates Subjects

Your technology classes often provide good examples of how math, science, and language arts can be integrated. For example, you might be talented in mathematics or science. What

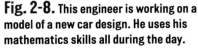

Fig. 2-8. This engineer is working on a model of a new car design. He uses his mathematics skills all during the day.

Fig. 2-9. Have you ever had trouble understanding what someone was trying to explain? Perhaps that person needed to work on his or her communication skills.

you know is valuable to your team, but you must know how to explain your idea. That means using communication skills, too. Fig. 2-9. In the next activity and in other activities in this book, you will need to integrate your knowledge and skills from technology, mathematics, science, and other subjects in order to solve problems.

SECTION 2
TechCHECK

1. What does integrated mean?
2. How does integrating technology, science, mathematics, and other subjects make it easier to solve problems?
3. While at school, why do you usually have a separate scheduled time for mathematics or science?
4. **Apply Your Knowlege.** Discuss with your teacher what science, mathematics, or communication concepts are useful in video production.

ACTION ACTIVITY

Testing Structures

Be sure to fill out your TechNotes and place them in your portfolio.

Real World Connection

You often hear about earthquakes in the news. They may occur close to home or in other parts of the world.

The Earth's crust is made of very large sections called plates. The plates move around, some sliding over or under the edges of other plates. The plates themselves have cracks called faults. As the plates move against one another, pressure builds up along these faults. When the stress gets too great, movement in the faults occurs. This movement is what we call an earthquake.

The buildings we live in can be designed to help stand up to the shaking forces of an earthquake. Reinforced foundations and braces in a structure can add strength and flexibility. The structure then moves with the earthquake instead of shaking apart.

Design Brief

Use the skills you have developed in many subject areas to design, build, and test a structure that will withstand the forces of a simulated earthquake. Ask your teacher for help.

Materials/Equipment

- earthquake simulator
- spaghetti (uncooked)
- gumdrops
- masking tape

SAFETY FIRST

Vibration can cause the earthquake simulator to fall off the edge of the table. Attach it using a C-clamp, Velcro hook and loop fasteners, or double-faced tape. Follow the safety rules on pages 42-43 and the specific rules provided by your teacher for tools and machines.

Procedure

1. Design and build both a short and a tall structure using spaghetti and gumdrops as shown in Fig. A.

2. Tape the structures to the base of the earthquake tester.

Test 1

1. Adjust the movement of the simulator to a slow speed. (The vibration speed is called the *frequency*.) Use a watch to set the speed at one or two cycles per second.

2. Watch the movements of the structures. Record the amount and direction of movement of each structure in your TechNotes.

Test 2

1. Slowly increase the speed of the earthquake simulator. Watch the effect on your structures.

2. Continue to increase the speed of the tester.

3. Note the amount and direction of movement in each structure. Record your observations in your TechNotes.

Test 3

1. Change the design of your structures by adding braces.

2. Repeat the tests while watching carefully. Record your results.

Evaluation

1. How did you use science, mathematics, and communication skills during this activity?

2. What does *frequency* refer to in an earthquake?

3. Why is it important that structures be designed to withstand earthquakes?

4. What do you think would happen in an earthquake if a short building were built too close to a tall building?

5. What is an earthquake fault?

6. **Going Beyond.** Videotape the tests. Play them back at slow speed to analyze structural failure, if any.

7. **Going Beyond.** Design, build, and test a structure that is as tall as you are.

8. **Going Beyond.** Research the history of earthquakes in your area.

Fig. A

2 REVIEW &

CHAPTER SUMMARY

SECTION 1

- One important way technology reaches you is through your school subjects.
- Technology plays a part in subjects such as science, mathematics, social studies, language arts, and health or physical education.
- Science usually gives you the theories (ideas), while technology lets you use your knowledge and resources to solve problems.
- Although technological devices like calculators make mathematics easier, a user must still know what to do with the information.
- At each point in history, different technologies were important for what they could do to help us.
- Technology has come up with new products that help us keep fit and healthy.

SECTION 2

- Technology education helps give you the "bigger picture" of how topics like mathematics, science, and language arts can be used together to find solutions.

REVIEW QUESTIONS

1. Name one way technology helps you in the classroom.
2. What tools of technology do you use in your mathematics classes?
3. How are science and technology different?
4. What can you learn from creating a timeline?
5. How does technology help us stay healthy?
6. Why do you need to integrate technology, science, mathematics, communication, and other subjects to solve problems?

CRITICAL THINKING

1. How does technology affect you most at home?
2. Name and explain one way in which technology and science work together to improve the environment.
3. How would your life be different if you could not use computers or television sets?
4. What medical developments have made your life better?
5. Why do you think calculators were invented?

ACTIVITIES 2

CROSS-CURRICULAR EXTENSIONS

1. **HEALTH EDUCATION** Research the role of technology in sports.

2. **SCIENCE** Explain what causes a curve ball in baseball or why tennis balls are fuzzy.

3. **MATHEMATICS** Research ten mathematics formulas that could be used in technology. Explain how they could be used.

4. **SOCIAL STUDIES** List ten inventions that led to the development of another invention. Choose one and make a presentation about it to the class.

EXPLORING CAREERS

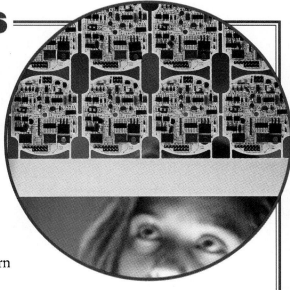

At one time people believed that computers would be of use only to scientists and engineers. They are now used by almost everyone. They have changed the way we receive and share information. Here are two careers that involve computers.

Computer Programmer Programmers write detailed instructions, called programs, that tell the computer what steps to take in order to perform a function. Programmers can find jobs in almost all areas of the work world. Mathematics and problem-solving skills are important. So is a willingness to learn new technology and keep your skills up to date.

Electrical/Electronics Technician Technicians are responsible for making sure that the designs an engineer develops actually work. If not, they study the problem and come up with a solution. These workers must be good at troubleshooting and have a strong background in computers and electronics. Helping a product move from the design stage to final production is one of the rewards of working in this career.

ACTIVITY

Select a career from the 1800s. Write down the changes that people from that period would find if they went to work in the same career today. What tools or methods are used now?

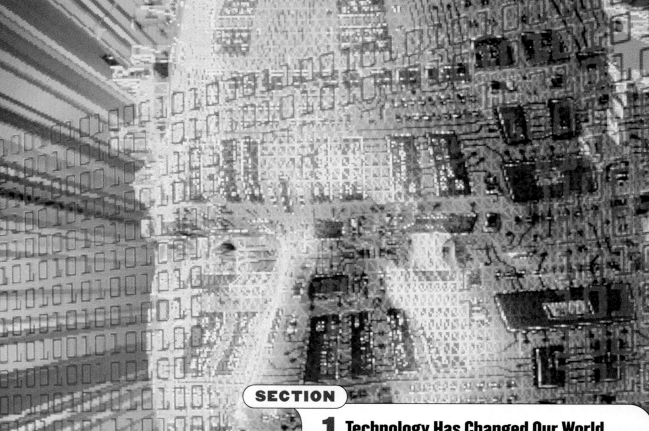

CHAPTER 3

How Technology Affects You

Technology Has Changed Our World

THINGS TO EXPLORE

- Explain how the rapid change in technology has changed society.
- Tell how changes in technology make other new things possible.

TechnoTerms
impact
palmtop

Chapter 1 explained the rapid changes that take place with technology. This growth has had many **impacts**, or effects. It has caused our society to change rapidly, too. For example, ask your parents or grandparents what they did for fun when they were your age. They probably played board games or listened to the radio. You might play video games or listen to CDs.

Benefits of Technology

Technology has improved some things and made other things possible. For example, before low-cost calculators were available, people used slide rules and expensive adding machines to do complex mathematics problems. Now, almost everyone owns a low-cost calculator. Before 1961, no one had ever seen the Earth from a distance. Now, thanks to new space technology, we know what the Earth, as well as the other planets, looks like. Fig. 3-1.

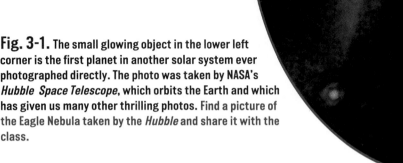

Fig. 3-1. The small glowing object in the lower left corner is the first planet in another solar system ever photographed directly. The photo was taken by NASA's *Hubble Space Telescope*, which orbits the Earth and which has given us many other thrilling photos. Find a picture of the Eagle Nebula taken by the *Hubble* and share it with the class.

OPPOSITE This image was created on a computer, which allows an artist to create many special effects.

Technology has also had impacts on your school life. In your classroom, you might be sitting in a chair designed by computer to fit your size. Maybe you use a computer or a word-processing typewriter to help you with your schoolwork. Teachers now teach with videos instead of movies in the classroom. Even the equipment you use in gym classes or football has been improved by technology.

Everyday Technology

Technology has even changed the way we buy things. Today it is common to get cash from an automatic teller machine, or ATM. You can search for the lowest prices and buy a car using a home computer. Many people find it easier to pay bills, send gifts, and even order pizza using a computer.

Computer technology is very important. Some people take their computers with them everywhere they go. Small computers called *laptops* are common. Even smaller pocket-sized computers called **palmtops** are popular. Many people use them to keep track of appointments and expenses. This type of technology will continue to get smaller and more powerful.

SECTION 1
TechCHECK

1. Name one way technology has changed society.
2. How has technology changed the way you learn at school?
3. What technology made seeing the Earth from a distance possible?
4. **Apply Your Knowledge.** Make a chart using pictures of the technologies you think affect you and your classmates most.

Evaluating Technology's Impacts

THINGS TO EXPLORE

- Evaluate technologies and their effects on you and the environment.
- Investigate the construction of an integrated circuit (IC).

TechnoTerms
biodegradable
evaluate
flowchart

Sometimes people invent a technology they think is needed for a certain area but forget to **evaluate** (judge) its impact on other areas. The impact on the environment is one example. Fig. 3-2. For this reason, some technological changes may be viewed as good and some as bad. However, we can't always predict how a technology will affect us or our world. We can only be sure that almost everything changes eventually.

Questions to Ask

As a member of society, your job is to constantly evaluate technology. How can it be used to obtain the most benefits for people and the environment while causing the least harm?

Fig. 3-2. When producing food in "tin" cans was first invented, everyone thought it was wonderful. Foods could be stored for long periods without spoiling. No one imagined then how much of a waste problem the empties would be.

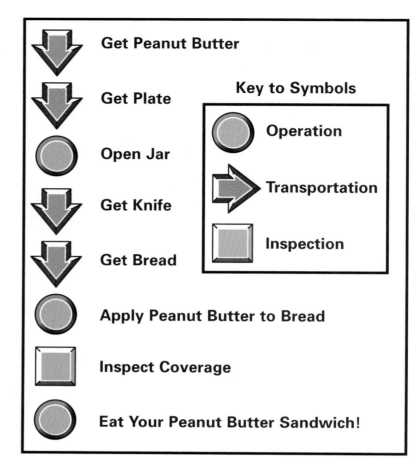

The process of making a sandwich has been analyzed in this flowchart. Do any of the steps cause harm? Are any of the steps wasteful?

COMMUNICATION CONNECTION

Analyzing a Process Using a Flowchart

The same questions about the impact of a product can be asked about the steps in the process used to produce it. Do any of the steps cause harm? Are any of the steps wasteful?

The steps in a process can become very complicated. There is an easy way to keep track of them. **Flowcharts** are used to illustrate the steps using symbols. Each symbol represents a special step in the process. For example, graphic symbols can represent various steps in the following processes:

• **Transportation:** Moving an object from one place to another

• **Operation:** Usually, using machines to change the shape of a product

Here are some questions that are useful to ask about a technology's impact.

- Does the technology require more or less energy or more or fewer natural resources than an existing technology?
- Will it damage the environment? For example, is it **biodegradable** (able to break down to natural materials)?
- Is it easier to use, or does it save time?
- Does it require special training to use?
- Does it put people out of work, create new jobs, or both?
- Is there a real need for this technology?
- Is this an appropriate use of technology?

SECTION 2
TechCHECK

1. What kinds of technology do you use most? List five. What impacts do you think they have?
2. What questions do you need to ask in evaluating the effects of a technology?
3. Why is it important to evaluate a technology?
4. **Apply Your Knowledge.** Evaluate a product using the suggested questions at the top of this page.

- **Delay:** Waiting for the next step

- **Inspection:** Checking the quality of a product

- **Storage:** Putting the product in stock

A flowchart can also show where a process might be stopped because an important step can't be completed. Flowcharts help in the planning process when deadlines must be met.

Even a simple process can be shown in a flowchart. See the flowchart on the opposite page.

ACTIVITY

Show the steps in the process of brushing your teeth. Can the steps you include prevent the wasting of water?

ACTION ACTIVITY

Analyzing Integrated Circuits

Be sure to fill out your TechNotes and place them in your portfolio.

Real World Connection

You may have seen the many electronic parts, called *components*, that make up a computer or a television. Today we can put millions of them on a single "chip" of silicon smaller than the nail on your little finger. These chips are called *integrated circuits* (ICs). Fig. A.

ICs are so small that people have a hard time assembling them into useful products. For this reason, two layers of plastic with electrical pin connectors are used to hold them. This special "sandwich" of plastic is called a dual in-line package (DIP).

Integrated circuits are found in computers, CD players, video games, wristwatches, cars, refrigerators, and even toasters. Without them, our lives would be very different.

Design Brief

Break apart a recycled integrated circuit. Use a stereo microscope to look at the chip and the wires connected to it.

Materials/Equipment

- ICs provided by your instructor; otherwise, take one from a donated electronic product
- chisel and hammer
- vise
- stereo microscope

SAFETY FIRST
- Follow the general safety rules listed on pages 42-43 and specific rules provided by your teacher for tools and machines. Be sure to wear safety glasses when breaking apart the ICs.
- Do **NOT** take apart TVs, computer monitors, or anything containing a picture tube. Dangerous voltages remain in these devices and can give you a shock even if the devices are not plugged in. Ask your teacher for help.

Fig. A

Procedure

1. Obtain an integrated circuit.
2. Use a chisel and hammer to break off the top layer of the DIP. Fig. B.
3. Study the actual IC and the connecting wires under the microscope.

Evaluation

1. What does "IC" stand for?
2. List five products that contain ICs.
3. What does "DIP" stand for?
4. **Going Beyond.** Research how ICs are made. Make a chart showing how they are made.
5. **Going Beyond.** Measure an actual IC and make a model by increasing its size at least ten times.
6. **Going Beyond.** Research the history of the first integrated circuits.

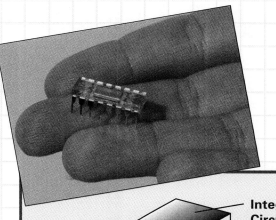

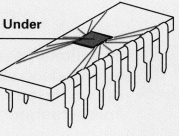

Integrated Circuit

Place in Vise and Break in Half with a Chisel

Fig. B

Look at the Semiconductor Under a Microscope

THINGS TO EXPLORE

- Explain what materials testing is and why it is important.
- Identify standard tests for materials.
- Test materials and compare data.

Have you ever seen a commercial on television stating that one product is better than another? Fig. 3-3. The reputation of products can determine whether a manufacturer is a success or failure. A company does a lot of testing of its products before it sells them.

Standards

Materials that are used to make products must meet specific **standards** (set of values or conditions) for properties such as strength, hardness, resistance to rust, and flammability (ability to catch fire). The materials must first undergo **materials testing** to see if they meet the standards. Then the products themselves are tested. Fig. 3-4.

Four out of five tasters prefer Fizz over Popsie.

In national taste tests conducted by an independent research organization, tasters preferred Fizz over Popsie four out of five times.

Fizz is America's favorite beverage!

Fig. 3-3. Comparisons are often made in print ads as well as TV commercials. What evidence does this ad give for its claims?

Fig. 3-4. Cosmetic items must be tested carefully. No one wants to use a product that might cause a harmful reaction. Look at the label for a cosmetic item at home tonight. What does it say, if anything, about the testing done?

Product Testing

Have you ever purchased a product that you thought was poorly made? Test results can help you decide if you want to buy a product. **Independent product testing** is testing done by government agencies or companies not involved in making the products. These test results help **consumers** (people who buy products or services) make wise choices when shopping.

SECTION 3
TechCHECK

1. What is materials testing?
2. Name three properties a material could be tested for.
3. Why do companies need to test products?
4. **Apply Your Knowledge.** Design a way to test paper towels for absorbency. Make a comparison graph on the computer.

ACTION ACTIVITY

Testing Pens for Quality

Real World Connection

Comparing products for quality and value is a never-ending job for consumers. No one likes to be cheated. That is why companies that make inferior products often go out of business.

In this activity, you will test different pens for quality.

Design Brief

Build a device to do a comparison test of writing pens. Evaluate the results of your test.

Materials/Equipment

* wood
* dowel rod
* paper roll
* pens
* bandsaw
* drill press
* nail, wood screws
* washers

> **SAFETY FIRST**
> Follow the general safety rules listed on pages 42-43. Ask your teacher how to build your test device safely. Follow the specific rules that your teacher provides on the safe use of tools and machines.

Procedure

Part 1 · Building the Test Device

1. Form groups of four or five students. Each group will build a testing device.
2. Cut the wood for the base of the test device. Ask your teacher for help.
3. Drill holes for the dowel rods, crank nail, and wood screws.
4. Assemble the parts as shown in Fig. A.

Part 2 · Comparing the Products

1. Place the test pens in the holes as shown in Fig. A.
2. Start turning the crank to begin the test. Make sure each pen is writing on the paper.
3. Crank to the end of the roll. Place a spacer washer between the wood base and the paper roll. Move the crank to the other dowel rod.
4. Crank the paper in the other direction making another set of test lines.
5. Repeat steps 3 and 4, inserting a spacer washer each time to move the paper roll enough to make a new line.
6. Watch to see which pen stops writing first.

Evaluation

1. What is a consumer?
2. Why is it important for businesses to make quality products?
3. What could happen to a company that makes inferior products?
4. **Going Beyond.** Use a 12-volt DC gear-head motor to make your test device work automatically.
5. **Going Beyond.** Design, build, and use a device to test the effectiveness of various powdered cleansers.
6. **Going Beyond.** Research the testing done at Underwriters Laboratories, Inc. Make a chart or a computer presentation showing your research.

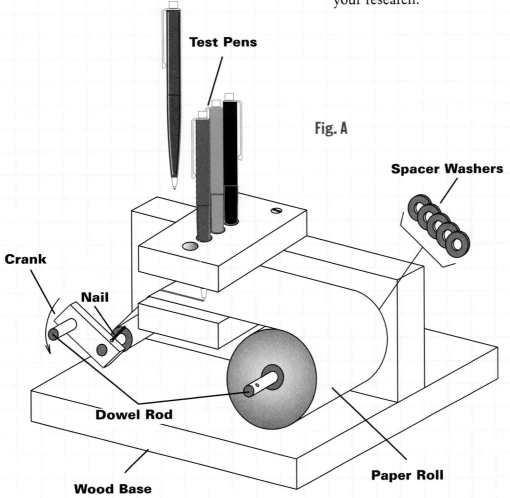

Test Pens

Fig. A

Spacer Washers

Crank

Nail

Dowel Rod

Paper Roll

Wood Base

The Ecology of a Product

THINGS TO EXPLORE

- Explain what ecology is.
- Tell why recycling is important.
- Compare the time it takes different materials to decompose.

TechnoTerms

ecology
landfill
recycling

We live in a "throwaway" world. When we're finished with a product, we often just throw it away instead of fixing it or **recycling** (reusing) it. Did you know that every man, woman, and child in the United States creates an average of 5 pounds of garbage a day? That's 230 million tons of garbage every year!

Ecology is the study of how things interact with the environment. Part of designing a product is planning ahead for what will happen to it after it is no longer useful. That includes every product from newspapers to jet airplanes. Fig. 3-5.

Fig. 3-5. Old cars and trucks are a big waste problem because they take up so much space. How does your community dispose of old vehicles?

TechnoFact

FROM OLD JUNKER TO SHINY NEW MODEL Nationwide, 20,000 cars are junked each day. Researchers are taking apart cars to experiment with reusing the materials. They are making glove compartments from waste products and using fabric scraps to make sound insulation. In the future, more plastics will probably be used in cars, because plastics can easily be reused.

Fig. 3-6. Materials like aluminum, which can be recycled, must be separated from other waste. **Research how recycling plants sort waste. Report your findings to the class.**

Why Recycle?

Do you ever stop to think what happens to the things you throw away? Many of them end up in **landfills** (garbage dumps) where they are buried or burned. Others are sometimes dumped into the oceans. Disposing of materials in these ways can eventually cause air or water pollution.

Whenever possible, the best alternative is to use materials that are biodegradable. Biodegradable materials break down naturally and return to the earth. Those materials that take a long time to decompose (break down), such as aluminum, plastic, and glass, should be recycled. Fig. 3-6. Some products, such as the batteries in flashlights, transistor radios, and other electronic devices, should always be recycled because they contain hazardous materials.

TechnoFact

MONEY THROWN AWAY The cost of throwing away our trash is going up fast because of new rules to protect the environment and because of lack of landfill space. It can cost $50 per ton to throw away trash. If you multiply the 230,000,000 tons of garbage we throw away each year by $50, we spend $11,500,000,000 just on garbage!

SECTION 4
TechCHECK

1. What is ecology?
2. Why is recycling necessary?
3. Name some items at school you can recycle.
4. **Apply Your Knowledge.** Contact the agency that operates your local landfill. Ask about their five-year and ten-year plans for managing the landfill.

ACTION ACTIVITY

Analyzing the Rate of Decomposition

Be sure to fill out your TechNotes and place them in your portfolio.

Real World Connection

When materials are put into landfills, it is hoped that they will decompose and return to the earth. In this activity, you will compare the rate at which different materials decompose when exposed to the weather.

Design Brief

Compare the rate of decomposition of packaging materials that are exposed to the weather.

Materials/Equipment

- packaging materials, such as cardboard, plastic wrap, grocery bags, and paper wrappings (Certain materials are supposed to be biodegradable. Include some of those.) Fig. A.
- sample board (1/4" x 36" x 36" plywood)
- tacks, glue
- measuring tools
- hand and power tools
- video camera (optional)

SAFETY FIRST
Follow the safety rules listed on pages 42-43 and the specific rules provided by your teacher for tools and machines.

Fig. A

Procedure

1. This activity will involve the entire class. Each student should bring in a sample of a packaging material that would normally be thrown away.

2. Cut your sample into two pieces. One half of your sample will be kept inside. The other half will be exposed to the weather conditions found in your area.

3. As a class, create two test boards on which to tack or glue the packaging materials. Make each board exactly the same way. The board to be kept inside is called the *control*. The second board, to be placed outside, is called the *variable*.

4. Each student should mount a test sample on each board using tacks or glue. Label all the samples so they can be compared later.

5. Store the control in a safe place inside. Put the variable outside in a place where it will be exposed to the weather and sun.

6. *Optional*: Videotape the start of the experiment and add to the tape each month during the test. In this way, a video record of the decomposition process can be shown at the end of the test. Another option is to take photographs.

7. Compare the variable to the control after each month. Make a log of the changes in each material.

Evaluation

1. Which materials break down the fastest in your area?

2. Which packaging material takes the longest to break down?

3. Make a list of five products that are commonly purchased in a grocery store. How could packages of each product be changed to help protect the environment?

4. **Going Beyond.** Write an advertisement that encourages people to recycle. This type of short "commercial" is called a *public service announcement* (PSA). Ask your teacher if your PSA could be read on a local radio station. Fig. B shows symbols that are found on some packaging. Find out what they mean. How might you use them in your PSA?

5. **Going Beyond.** Research who makes some of the packaging materials or products you tested. Write to them telling them the results of the test and inviting a response.

RECYCLABLE

Fig. B

RECYCLED

3 REVIEW &

CHAPTER SUMMARY

SECTION 1

• The rapid growth of technology has caused our society to change. Changes in technology make a lot of different things possible.

SECTION 2

• Part of your job as a member of society is to evaluate technology's effects on you and your environment.

SECTION 3

• Materials that are used to make products must pass specific standards. Materials are tested to see if they meet these standards.

• Product testing helps consumers make wise choices.

SECTION 4

• Ecology is the study of how things interact with the environment.

• Part of designing a product is planning ahead for what will happen to it after it is used.

• Recycling is one way to reduce garbage and save resources.

REVIEW QUESTIONS

1. Select a product you bought yourself. Would your parents have been able to buy the product when they were your age? Would your grandparents?

2. What is the impact of the product from question 1 on the environment?

3. Why do materials need to be tested?

4. What does the term *decompose* mean?

CRITICAL THINKING

1. Name two technologies that you have seen change in the last five years.

2. Explain why change is sometimes good and sometimes bad. Give examples.

3. What technology would you miss the most if it had never been developed? Explain why.

4. Design your own test for a product's quality or durability.

5. With your teacher's permission, gather several garbage cans from rooms in the school. Wear gloves and separate the garbage into recyclable and non-recyclable materials. Make a graph of your data to share.

ACTIVITIES **3**

1. **SCIENCE** Exchange a test sample board of materials with another school located in an area with a climate different from yours. After one semester, see if the climate makes a difference in how materials degrade.

2. **MATHEMATICS** Calculate the volume of one soda can or milk carton. Find out how many cans or cartons are used in your school each day. Calculate the total volume of space they would take up in a landfill. How would the volume differ if the containers were crushed?

3. **COMMUNICATION** Research new forms of biodegradable plastics. Make a radio commercial or video newscast to share the information.

EXPLORING CAREERS

Information and technology are big business and impact our lives every day. Advances in technology have opened doors for new companies that offer us more career choices than ever before.

Market Research Analyst Excellent mathematics skills are important for those interested in a job in market research. These workers collect information on products. Then the companies for which they work use this information to decide on what competing products and services to offer. This is a great job for someone who enjoys asking questions and digging for answers.

Webmaster Do you enjoy spending time in cyberspace? Companies employ webmasters to develop and maintain company websites. They handle the technical aspects of the site, including what information is found there and how fast you can access the information. Technical skills in computer programming and graphic design are needed.

ACTIVITY

Make a list of all the electronic and/or computerized products that you use from the time you get up until you go to bed. How many products are on your list?

Introducing Computers

Computer Basics

THINGS TO EXPLORE

- Describe ways computers are used and give examples.
- Identify basic computer components and peripherals.
- Explain the difference between hardware and software.
- Define ROM and RAM.

TechnoTerms

binary counting
  system
central processing
  unit (CPU)
hardware
microchip
peripherals
software

Computers are important tools for solving problems. A few major uses of computers include writing, drawing, finding and organizing information, calculating numbers, and controlling other devices. Fig. 4-1. In this chapter and throughout your technology study, computers will be put to use in a variety of ways. Just like any other tool, they are there to help you.

Computers in Your Life

Did you know, for example, that many events in your life are controlled by computers you may not even see? Fig. 4-2. For example, city traffic lights and telephone networks are computer controlled. Weather forecasters use computers to analyze information received from satellites above the Earth. In some hospitals, computers are used to draw precise diagrams for doctors to use in performing surgery. Even modern cars contain computer devices that control the engine.

Fig. 4-1. Designers of fabrics and even wallpaper may use computers to make their drawings. Find out what computer components artists need in order to make freehand drawings.

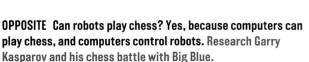

OPPOSITE Can robots play chess? Yes, because computers can play chess, and computers control robots. Research Garry Kasparov and his chess battle with Big Blue.

Fig. 4-2. At a shipping center, this computer with a touch screen helps control distribution of products.

A Brief History of Computers

The computers of the 1950s were huge machines that took up whole rooms. Fig. 4-3. These early computers were used only by the government and big corporations. It wasn't until 1977 that the first home computers became available. Since then, the power of the computer has increased many times, while its size has decreased. Today a computer that is more powerful than the computers of 40 years ago will fit inside a wristwatch!

INFOLINK

See Chapter 3, "How Technology Affects You," for an activity investigating integrated circuits.

The technology that led to smaller and more powerful computers was the **microchip**, or integrated circuit (IC). A microchip consists of a piece of silicon with hundreds of tiny electronic parts linked together to form circuits. Computers use ICs to store information.

MATHEMATICS CONNECTION

The Binary Counting System

The number system you generally use in everyday life is called the decimal system. It has ten digits from 0 to 9.

You can use these ten digits to write numbers larger than 9 by using a tens' column, a hundreds' column, a thousands' column, and so on. In other words, each place to the left of the ones' place increases in value by a power of 10. For example, 256 means 2 hundreds (10^2) + 5 tens (10^1) + 6 ones.

In the *binary system*, each place to the left of the ones' place increases in value by a power of 2. In order to write numbers larger than 1 in binary, you use a twos' column, a fours' column, an eights' column, a sixteens' column, and so on. In the chart, the decimal number 5 is 101 in base 2. That means 1 group of 2^2 + 0 groups of 2^1 + 1 in the ones' column.

Fig. 4-3. All of the power generated by this early room-sized computer is now compressed into a tiny microchip smaller than your fingernail.

Computer Code

The smallest unit of information used by computers is called a *bit*. The word *bit* stands for *binary* dig*it*. Eight bits together are called a *byte* (pronounced bite). Computers often process and store information one byte at a time.

Decimal			Binary			
Base 10			Base 2			
100s	10s	1s	8s	4s	2s	1s
		0				0
		1				1
		2			1	0
		3			1	1
		4		1	0	0
		5		1	0	1
		6		1	1	0
		7		1	1	1
		8	1	0	0	0
		9	1	0	0	1
	1	0	1	0	1	0

ACTIVITY

Write the following numbers in binary:
12
23

Why is this important? Computers use a code to represent letters, numbers, and punctuation. The code is written using the **binary counting system.** In binary, the two digits, 0 and 1, stand for electronic signals that are either "off" or "on."

When you hit a key on a computer keyboard, you send one byte of information in binary code. For example, if you typed TECH, the binary code would look like this:

The computer code used in the United States is called ASCII (pronounced ASK-ee). ASCII stands for American Standard Code for Information Interchange.

T	01010100
E	01000101
C	01000011
H	01001000

Bits and bytes

Memory and Storage

The data held inside a computer is stored in its memory. Computer memory is one of two kinds—RAM or ROM. Fig. 4-4. *RAM* stands for *random access memory.* RAM is temporary memory used for computer applications and your documents. *ROM* stands for *read only memory.* ROM is permanent memory built into the computer's electronic circuits. ROM contains operating instructions that can be used (read), but not changed by the user.

Information can be stored magnetically on a computer's hard drive. It can also be stored magnetically on computer disks called *floppy* disks or optically on compact discs called *CD-ROMs* (Compact Disc—Read Only Memory).

To keep from losing the work you do on a computer, save it to a hard disk or floppy disk frequently. You can also save information on recordable compact discs called CD-Rs (recordable only once) or CD-RWs (rewritable).

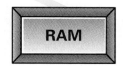

Fig. 4-4. Operating instructions for the computer are contained in ROM. They cannot be changed. The documents you create are contained in RAM. They can be changed as often as you wish.

Fig. 4-5. The main circuit board in a computer is called the mother board.

Hardware

The parts of a computer system are called **hardware**. One piece of hardware, the **central processing unit (CPU)** is the brain of the computer. It is a special kind of integrated circuit that, along with others, is mounted on a printed circuit (PC) board. Some computers have many printed circuit boards. The main circuit board that contains the CPU and connectors to other parts of the computer is called the *mother board*. Fig. 4-5.

Other basic hardware might include a monitor, a mouse, a floppy disk drive, a hard drive, and a keyboard. These are considered basic because they are needed to operate the computer. Other useful devices that can be connected to a computer system are called **peripherals**. Fig. 4-6.

Software

Computer hardware doesn't work by itself. You need to use some kind of **software**. Software is the coded instructions that tell a computer what to do. Software can be stored on floppy disks, compact disks, or a hard drive.

Software used for a specific purpose is called an *application*, or program. Applications are available for such things as word processing, databases, and spreadsheets. When software is used by the computer, documents are stored in RAM memory.

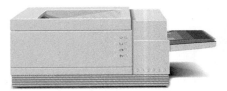
Printer. A machine that outputs text (words) and graphics (pictures) on paper.

MIDI (musical instrument digital interface). A device that lets you put music into a computer from an electronic musical keyboard.

Modem. A device that lets your computer communicate with another computer over a telephone line.

Scanner. A machine that copies text and graphics from paper into the computer.

Joystick. A device that changes hand movements into actions on a computer screen.

Fig. 4-6. Peripherals increase a computer's usefulness. Have you ever used one of these devices? Share your experience with the class.

SECTION 1
TechCHECK

1. What jobs can you do with a computer?
2. What are the parts of a computer system called?
3. Which form of memory, RAM or ROM, is permanently built into the computer?
4. **Apply Your Knowledge.** Look at the computer systems available for your use at school. Identify and list the basic parts, peripherals, and software you will be able to use with them.

Using a Word Processor

THINGS TO EXPLORE

- Define word processing and explain why it is useful.
- Explain what desktop publishing is.
- Create a word processing document.

TechnoTerms
desktop publishing
(DTP)
fonts
word processing

Writing text (words) with a computer is called **word processing**. Most businesses now use computers instead of typewriters. A word processor is faster, easier to use, and more efficient than a typewriter. Fig. 4-7.

Imagine you have just typed a 60-page report, but a paragraph on page 3 is missing and there is no room on the page to insert it. If you were using a typewriter, you would have to retype the report completely. On a word processor, you would simply fix the error on the computer screen and reprint the report. Think of the time and effort you just saved!

Creating a Document

Word-processing software lets you choose different **fonts**, or kinds of type characters (letters). Fig. 4-8. It lets you change the size of the characters and the style of type used. It even lets you change between single and double spacing easily. Most word-processing software will place page numbers, tabs, and paragraphs where you want them.

Fig. 4-7. Word processors have replaced typewriters. Locate an old typewriter and try it out. Describe your experience to the class.

Type Styles

This is plain text
This is bold text
This is italic text
This is outline text

Type Sizes

This is 10 point
This is 14 point
This is 24 point

Type Fonts

This is the Helvetica font
This is the Courier font
This is the Park Avenue font

Fig. 4-8. Several different type styles, sizes, and fonts are shown here. Make a printout showing an example of all the fonts available with the word-processing software you use at school.

Software is available that can point out spelling errors. Some word-processing software packages also include an electronic *thesaurus* that gives you synonyms (words that have the same meaning) to help you in your writing.

Desktop Publishing

After you create a document with word-processing software, you can organize it using desktop publishing software. **Desktop publishing** (DTP) software lets you put text and graphics together to make a report, newsletter, or newspaper. Desktop publishing has changed the way printed materials, from business cards to books, are published. Now anyone who has learned to use a computer can do things that once only trained professionals such as typesetters and printers could do.

SECTION 2
TechCHECK

1. What is word processing?
2. Why is word processing helpful to people?
3. What is desktop publishing?
4. **Apply Your Knowledge.** Using desktop publishing software, create a school bulletin that contains text and graphics.

ACTION ACTIVITY

Word Processing a Document

Be sure to fill out your TechNotes and place them in your portfolio.

Real World Connection

Businesses use word processing and desktop publishing every day to advertise or communicate. Word-processing software can help you become a better writer by making it easy to rearrange words or sentences. It can even check your spelling and grammar and let you add graphics.

Design Brief

Choose any one of the following projects or think of your own idea. Then use a computer and word-processing software to make and print your document. Fig. A. Choose one:

- Write a letter to a company requesting information about its products.
- Write a résumé. A *résumé* is a list of important facts about yourself. A résumé is given to employers when you apply for a job.
- Write your own definitions to the key terms in this chapter.
- Write a program for an event at your school, such as a band concert or football or basketball game.
- Write a book report needed for another class.
- Write an article for the school newspaper or a letter to parents about a school event.
- Write a video or audio script.

Materials/Equipment

- computer with word-processing software
- printer

SAFETY FIRST
Follow the safety rules listed on pages 42-43 and the specific rules provided by your teacher for tools and machines.

(Continued on next page)

**Brady High School
Senior Class Play**

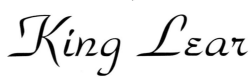

Shakespeare's

King Lear

**March 3 - March 10
8:00 p.m.**

**Murchison Auditorium
Brady High School
135 West Main Street**

**Main floor $10
Balcony $7.50**

Reserved seating only

Fig. A

Procedure

1. Boot a computer with word-processing software.
2. Type your document. Use one of the ideas suggested or get your teacher's approval for your own idea.
3. Learn how to rearrange words and sentences using the word-processing software.
4. Experiment with changing type sizes, styles, and fonts.
5. Check the spelling and grammar of your document using special software.
6. Have a friend *proofread* (check over) your work.
7. Save your document to a floppy disk or hard drive.
8. Print your document.

Evaluation

1. What was the easiest part of doing word processing?
2. Time yourself for one minute as you type. How many words per minute did you type?
3. **Going Beyond.** Ask your teacher, or use the *documentation* (instructions) for the word-processing software, to find out how to do the following: set tabs, change line spacing, make columns, and set or change margins.

> ### SAFETY FIRST
> • Avoid eye strain. Look away from the screen often. Refocus your eyes on distant objects.
> • Avoid strain on your neck, back, arms, or wrists. Stretch frequently. Get up and walk around at least once an hour.

SECTION 3 Using a Database

TechnoTerms
- data
- database
- field
- file
- record

THINGS TO EXPLORE
- Explain what a database is and how it is used.
- Identify a field, file, and record.
- Create a database.

Data is another word for "information." A **database** is a computer application that lets you find and organize information quickly and easily. This is why databases are sometimes called "electronic filing cabinets." For example, your school has a file for every student, including you!

Databases Save Time

Suppose the school needed to send a letter to the parents of seventh-grade girls taking technology classes. Without a computer, the school secretary would first have to find the files for all the seventh graders. Then he or she would have to sort out all the girls and go through each girl's file to see if she is in a technology class. This process could take a long time! However, a computer database application could do it in a manner of seconds.

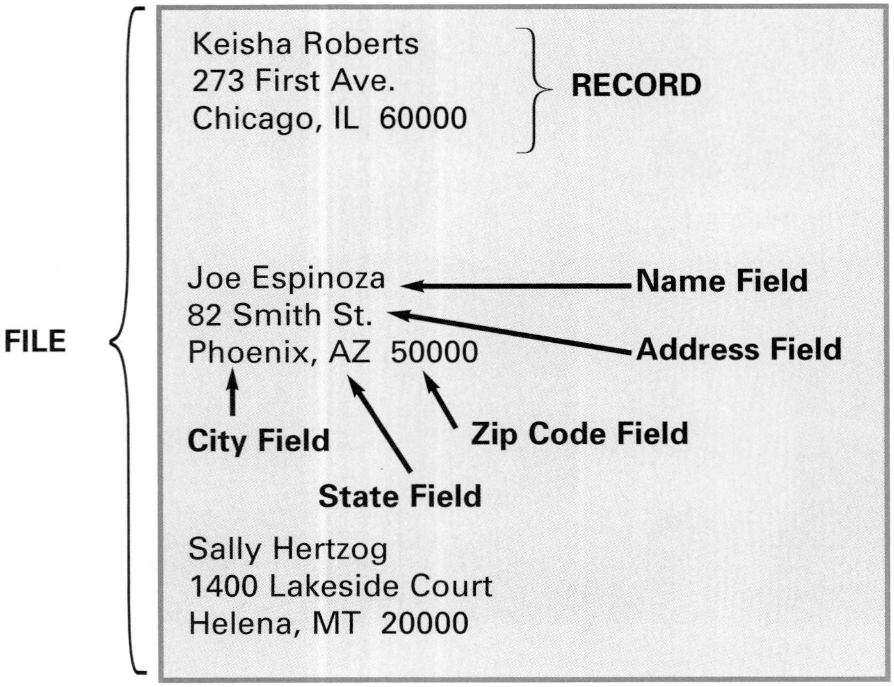
Keisha Roberts
273 First Ave.
Chicago, IL 60000
RECORD
Joe Espinoza
82 Smith St.
Phoenix, AZ 50000
Name Field
Address Field
City Field
Zip Code Field
State Field
Sally Hertzog
1400 Lakeside Court
Helena, MT 20000
FILE

Fig. 4-9. Many fields make up one record. Many records make up a file.

Setting Up a Database

To start a database, you need to make fields, records, and files. These are the major parts of most databases. Fig. 4-9.

- **Field:** One part of the data that helps describe the record. For example, a person's name would be one of many fields that make up a record.
- **Record:** All of the fields put together for one entry. For example, a person's record might include fields such as a first and last name.
- **File:** A group of records, such as all the records of the students in the seventh grade.

In the school records example, the fields would contain first name, last name, parents' names, address, class schedule, and so on for an individual student. A record would be the information in all the fields for one seventh-grader. A file would contain the records for all seventh graders.

Uses for Databases

In everyday life, databases are often used by companies to keep track of information on possible customers. Credit information, hobbies, sports interests, and travel preferences are some kinds of data stored about you and your family on computers. This kind of database sometimes leads to "junk mail" being sent to your home. Some people consider this use of computers as an invasion of their privacy. What do you think?

TechnoFact

IDENTITY THIEVES
You're 22, working full-time, and buying a new car. You apply for a loan and it's denied! Digging deeper, you find out that someone in another state has assumed your identity and run up big debts in your name. It may take years to clear your credit record. This can and does happen. Low-tech thieves steal driver's licenses or snatch bank statements from the trash. High-tech hackers find their way into databases, then copy and sell private information on hundreds of people!

SECTION 3
TechCHECK

1. What is a database?
2. What is a database field? Give examples.
3. How do people use databases?
4. **Apply Your Knowledge.** Make a database that includes everyone in your class.

ACTION ACTIVITY

Creating a Database

Be sure to fill out your **TechNotes** and place them in your portfolio.

Real World Connection

You have learned that the amount of information in the world, called our knowledge base, is doubling every few years. That's a lot of data!

In this activity you will work in groups to design a database. You will then use the database to organize and sort information.

Design Brief

Design and create a database that can be used to organize and find information. Fig. A. Choose any of the following topics, or think of your own:

- Events in the development of technology
- Space exploration
- Major inventions
- Instructional videos
- Technology-related magazine articles

Materials/Equipment

- computer with database software
- printer

SAFETY FIRST

- Follow the safety rules listed on pages 42-43 and the specific rules provided by your teacher for tools and machines.
- Avoid eye strain. Look away from the screen often. Refocus your eyes on distant objects.
- Avoid strain on your neck, back, arms, or wrists. Stretch frequently. Get up and walk around at least once an hour.

Procedure

1. Work in groups of three or four. Choose one of the ideas in the design brief, or get your teacher's permission to use your own idea.

2. Determine the fields that would make up a database record for your topic.

3. Show your database design to your teacher.

4. Research your topic, and enter the data into a database application on a computer.

5. Save your data for future use.

6. Try to use your database to find and sort information on a topic. Challenge other students in other groups to use your database to do research.

Evaluation

1. List 10 purposes for which stores or companies might use a database.

2. Do you think that a database with information about you should be given to companies, or should it be kept confidential (secret)? Explain.

3. **Going Beyond.** Use the documentation (instructions) for the database software to do the following:
   - Sort alphabetically
   - Sort numerically
   - Create a report
   - Edit (change) data already entered

Fig. A

Using a Spreadsheet

THINGS TO EXPLORE

- Define what a spreadsheet is and how it is used.
- Identify rows, columns, and cells.
- Make and use a spreadsheet.

TechnoTerms

cell
spreadsheet

The computer's ability to calculate is often called number "crunching" because it happens so fast! A **spreadsheet** application helps keep track of the calculations. For example, spreadsheets are often used to make budgets. Suppose you wanted to know how long it would take you to save enough to buy a stereo system. You might create a spreadsheet showing your monthly income like the one in Fig. 4-10.

Basic Parts of a Spreadsheet

All spreadsheets are made up of columns and rows. Usually rows are horizontal and numbered, and columns are vertical and given letter names. In the spreadsheet shown in Fig. 4-10, the income and expenses are in columns. The data for each month are in rows.

Each space in a column or row is called a **cell**. A cell can be identified by a combination of its column letter and row number. The cell in the upper left corner, for example, would be A1.

Cell A1   Column A

Fig. 4-10. Expenditures must be subtracted from income to find out how much is left. How much did this person spend all together?

Month	Income	Expenditures	Balance
September	$22.50	$10.35	$12.15
October	$25.00	$ 8.50	$16.50
November	$42.25	$ 7.25	$35.00
December	$65.00	$23.00	$42.00
		Cash Available	$105.65

Row 1

Spreadsheet Formulas

The advantage of using a spreadsheet is that it lets you answer "what-if" questions. For example, you might ask, "What if I earned twice as much money in January? How would that change my total income?" To find the answer, you would use a formula—in this case, addition—and the spreadsheet would show you the answer.

A formula might be as simple as adding numbers in a row or column. On the other hand, it might be more complicated. It could be an engineering calculation.

When balancing your checkbook, the spreadsheet application can automatically calculate how much money is left (balance) after each expense. You can even add your own formulas to a spreadsheet application. Fig. 4-10.

Super Computers

Today's super computers like the Cray T3E can make a trillion calculations per second. For example, if everyone in the United States (about 250 million people) wrote ten checks, it would take a super computer only a fraction of a second to balance all the checkbooks!

▲ When you first prepare a spreadsheet, try planning it out on paper before setting it up on the computer. How might this approach avoid problems?

SECTION 4
TechCHECK

1. What is a spreadsheet?
2. What is a cell?
3. How are formulas used in spreadsheets?
4. **Apply Your Knowledge.** Find out what spreadsheet applications are useful in schools. Check with your school secretary, the lunch program managers, and so on.

ACTION ACTIVITY

Working with a Spreadsheet

Be sure to fill out your TechNotes and place them in your portfolio.

Real World Connection

Whenever a lot of numbers need to be combined in any way, a computer spreadsheet can help. Spreadsheets are often used by businesses to keep track of money. Fig. A. In this activity, you will design and use a spreadsheet.

Design Brief

Make a spreadsheet that will do any of the following, or think of your own idea:

- Budget money to be spent on presents for family members and friends
- Keep track of money earned from odd jobs and allowances; keep track of expenses
- Keep track of products made by a certain company

Materials/Equipment

- computer and spreadsheet software
- printer

SAFETY FIRST
- **Follow the safety rules listed on pages 42-43 and the specific rules provided by your teacher for tools and machines.**
- **Avoid eye strain. Look away from the screen often. Refocus your eyes on distant objects.**
- **Avoid strain on your neck, back, arms, or wrists. Stretch frequently. Get up and walk around at least once an hour.**

Fig. A

Procedure

1. Work with a partner. Pick one of the topics listed, or think of your own. If you are using your own idea, get help from your teacher.

2. Design your spreadsheet on paper first.

3. Boot a computer that has spreadsheet software installed.

4. Enter the column and row headings that you need.

5. Enter the numbers in the correct cells.

6. With the help of your teacher, enter the formulas that you need in order to add, subtract, multiply, or divide.

7. Change the number in one of the cells to see how it affects the other cells.

8. Save your document.

9. Make a hard copy (printout) of your spreadsheet. Fig. B.

Evaluation

1. Think of five different uses for a spreadsheet in the real world.

2. What advantage is there in using a spreadsheet instead of a calculator?

3. **Going Beyond.** Design and make a spreadsheet that will do one of the following jobs:
- Calculate gas mileage from automobile trips
- Keep track of expenses for your household
- Organize information related to sports statistics in any sport you choose

Money Spent on Gifts

Person	Birthday	Anniversary	Total Per Person
Mom	$15.78	$21.96	$37.74
Dad	$18.24	$26.02	$44.26
Joe	$12.56		$12.56
Denise	$13.26		$13.26
Grandma	$22.37		$22.37
		Total Spent	$130.19

Fig. B. Here's an example of a completed spreadsheet. This spreadsheet helps keep track of the money spent on gifts.

CHAPTER 4

REVIEW &

CHAPTER SUMMARY

SECTION 1

• Major uses of computers include writing, drawing, organizing information, calculating numbers, finding information, and controlling other devices.

• The technology that led to smaller and more powerful computers was the microchip, or integrated circuit (IC).

• The parts of a computer system make up its hardware; the applications are called software. Peripherals are extra hardware components.

• ROM and RAM are different types of computer memory.

SECTION 2

• Writing text with a computer is called word processing.

• Desktop publishing (DTP) lets you put text and graphics together.

SECTION 3

• A database is a computer application that lets you organize and find information quickly and easily.

• The parts of a database include fields, records, and files.

SECTION 4

• A spreadsheet application keeps track of numbers and calculations.

• Spreadsheets contain rows and columns that intersect at cells.

REVIEW QUESTIONS

1. Why is the microchip such an important invention?

2. What is the difference between hardware and software?

3. If you are doing a project that involves adding many numbers, would you use a database or spreadsheet application?

4. What application can help you write a letter?

5. What is a database record?

CRITICAL THINKING

1. Use a word-processing application to write a long paper or report evaluating the effects of technology on you and your friends. Include a cover and title page made using a graphics application.

2. Create an acronym (a word made from parts of other words) that will remind you of the basic steps to follow in using a computer.

3. Explain how you think computers are used to forecast weather or control traffic.

4. Set up a spreadsheet to keep track of how you spend your allowance.

5. Suppose your school athletic program needed to prove that, on average, a certain number of students attended most events. How could they do this using a computer application?

ACTIVITIES 4

CROSS-CURRICULAR EXTENSIONS

1. **MATHEMATICS** Use the documentation for spreadsheet software to find out about other mathematics calculations that can be made besides addition, subtraction, multiplication, and division.

2. **SCIENCE** Make a display of electronic computer components and label each part using word-processing software.

3. **COMMUNICATION** Make a database of the favorite movies, sports teams, school subjects, and so on of your friends. Sort and place the results in a display.

EXPLORING **CAREERS**

As computers evolve, people who produce, operate, and repair them must become more knowledgeable. The following careers are important in the computer industry.

Technical Writer Technical writers create the user manuals that come with new computers or peripherals. Other technical writers create the words that appear on your computer screen. Technical writers need excellent writing skills. If they don't understand the technology, they must ask questions until they figure it out.

Computer Support Specialist
Computer support specialists take care of calls from customers who are having trouble with their computers or software. These specialists may also visit a site to help install computers or troubleshoot problems. Customers must be satisfied and treated well, so being able to work with people is important for this job. To be successful, specialists must keep up with current technology.

ACTIVITY

Write exact instructions for a person to use when placing a long distance phone call. Then ask a classmate to follow your instructions. Were your instructions clear?

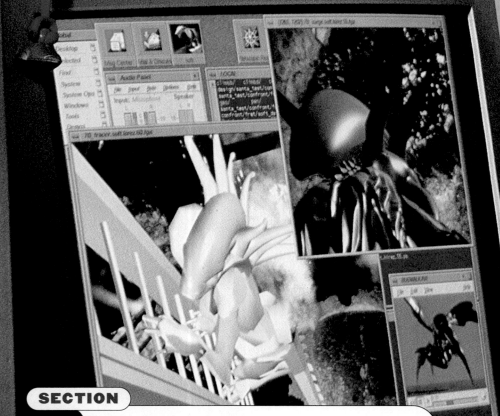

CHAPTER

Using Computers

Beyond the Basics

THINGS TO EXPLORE

- Identify real-world uses for the computer beyond the basic word-processing, database, and spreadsheet applications.
- Explain how the computer has changed the way we do photography.
- Describe a use for computer simulation.

TechnoTerms
digitizing
simulate
video conferencing

The computer is a very important tool in today's high-tech world. Beyond the basic applications of word processing, spreadsheets, and databases, you can use computers to do all kinds of things. For example, computers can be used to control input from temperature sensors in a research lab, train pilots, program robots, or access information from anywhere around the world.

Putting the Computer to Work

Computers have changed the way we do photography, the way we simulate, or model, real-world situations, the ways we communicate, and many other things.

In Photography You know that computers are good at number crunching (calculating numbers). It turns out photos can be changed into number values easily with a process called digitizing. **Digitizing** means to change into *digits*, or numbers. With a digital camera, you can capture still photos or video on a disk, memory card, or tape instead of film. Fig. 5-1. Because the photo or video is in digital form, you can use a computer to change it, print

INFOLINK

See Chapter 4, "Introducing Computers," and Chapter 15, "Finding and Using Information," for more about these uses for computers.

Fig. 5-1. Photos taken with a digital camera can be input into a computer. Compare the design of this camera with cameras used in the 1800s. How do they differ?

OPPOSITE Computers are creating excitement at the movies! By using computers, filmmakers can create almost anything they can imagine.

it, or send it electronically to another location. Another advantage is that you don't have to wait until an entire roll of film has been developed to see one particular picture. You can view the images instantly on a television, computer, or LCD (liquid crystal display). The LCD can be a separate monitor or a display attached to a digital camera.

In Simulations Computers are so good at manipulating images that they are often used to **simulate** (imitate) real life. People can create and test computer models and prototypes to decide whether a product will fit their needs before they spend a great deal of time or money building the real thing. For example, airplanes, cars, and buildings can be modeled as three-dimensional images on a computer. You can even test your performance on a product using computer simulations. Flight simulators, for example, can be used to train pilots to fly and to handle emergency situations. Race car drivers can make a practice high-speed run on a simulation to check their responses to different situations that might occur in a real race.

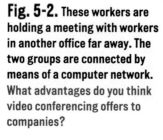

TechnoFact

HOW SMALL CAN IT BE? Making digital cameras as small as point-and-shoot film cameras is challenging enough. But camera makers have now made one that small with a zoom lens, thanks to a breakthrough in the development of the world's smallest all-glass spherical lens from Fuji. Another device, called an image builder, now makes downloading pictures from a digital camera to a computer even faster.

Fig. 5-2. These workers are holding a meeting with workers in another office far away. The two groups are connected by means of a computer network. What advantages do you think video conferencing offers to companies?

In Communications You can access high-quality pictures, animations, video clips, information, and computer simulations on the Internet (a worldwide network of computers). You can send and receive e-mail. Computers also are used in **video conferencing,** where video cameras hooked to the computers in a network allow people from all over the world to talk with each other at the same time. Fig. 5-2.

INFOLINK

You will learn more about the Internet in Chapter 15, "Finding and Using Information."

Throughout this book, you will have opportunities to use the computer in many activities. In this chapter, you will learn more about how computers are used in drawing, digital photography, and animation.

TechnoFact

VIRTUAL AIRSPACE! The virtual retinal display developed for military aviators uses a laser, monocle-size optics, and tiny scanners to project an image directly onto a pilot's retina. The device lets the pilot see the surrounding airspace while also getting digital images that appear in front of him or her.

SECTION 1
TechCHECK

1. Name three ways computers are used beyond the basic applications of word processing, spreadsheets, and databases.
2. What does *digitizing* mean?
3. What is a simulation?
4. **Apply Your Knowledge.** Find out what kinds of simulation programs are available in your school.

Drawing with a Computer

THINGS TO EXPLORE

- Describe the advantages in using computers to design and draw.
- Explain how **CAD** and other graphics software programs are used in drawing and designing.
- Create a graphic using **CAD** or other graphics software.

The old saying "a picture is worth a thousand words" is really true. Adding pictures, or **graphics,** to your words makes a document more meaningful and exciting. Fig. 5-3.

Using Graphics

Drawing is an important form of communication. Engineers use drawings to show the circuits in computers or television sets. Architects make drawings to show contractors how to build structures and what materials to use. Technical illustrators make drawings to illustrate how to hook a video game to a television.

Traditionally, most drawings and sketches were done on paper. The work was often time consuming and difficult. Now the computer, along with graphics software, makes it easy to draw and design.

Fig. 5-3. This teen is creating her own graphics using special computer software. Try designing a holiday greeting card using a graphics program.

Graphics software programs allow you to make drawings using a mouse, a keyboard, or other input devices. You don't have to be an expert artist to use graphics software. There are many levels of graphics programs, from ones that let you use **clip art** (pre-drawn images) to very complicated **computer-aided design (CAD)** software.

The best part about drawing with a computer is that it is very easy to modify (change) your drawing. Not only can you erase easily, but you can also change the size or shape of things you have drawn.

Fig. 5-4. CAD software provides users with many time-saving and other useful features. Here it is being used to create architectural details.

Using CAD

Today, many designers use computer-aided design, or CAD, software for product designs, technical illustrations, or architecture plans. There are many types of CAD software available for all users, from beginners to those who have had lots of specialized training. Fig. 5-4.

CAD has several advantages over traditional drawing or drafting. You can make drawings more quickly. For instance, if you are designing a home that needs the same door in several places, you can design the door once. Then you can copy the door graphic to any part of your design. You can also make your CAD drawing more accurate and neater than you can a hand-drawn design. CAD drawings can be accurate to more than 1/10,000 of an inch. That would be tough for a person to match!

CAD software also lets you see an object in three dimensions (height, width, and depth). This is an advantage to designers, architects, and engineers. For example, architects may design a floor plan for a house. A floor plan shows the house from above, as if you were looking down on a house that had no roof. It is difficult to picture in your mind how the house will look when you have only a flat drawing of it. With CAD software, you can see the rooms on the computer monitor as if you were standing inside or outside them. Some programs actually let you "walk" through the house! Fig. 5-5.

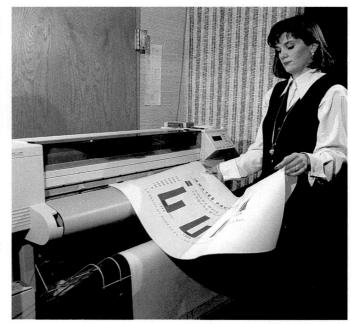

Fig. 5-5. Architects help clients visualize how a house will look by creating a CAD simulation. In what ways do you think this benefits the architects themselves?

Fig. 5-6. This plotter is used to produce large drawings. What kinds of drawings might need to be this large?

CAD drawings and other graphics can be saved to a computer file, added to other documents, and printed. Fig. 5-6. When a CAD drawing is sent directly from the computer to automated machines that make parts, the process is called **computer-aided manufacturing (CAM)**. The combination of CAD and CAM makes it possible to produce products quickly and inexpensively.

SECTION 2
TechCHECK

1. What is one advantage in using a computer for drawing?
2. What is CAD?
3. How can CAD help a designer?
4. Apply Your Knowledge. Ask a local architect who uses CAD if he or she will talk to your class.

Computer-Aided Design (CAD)

Real World Connection

Even if you can't draw a straight line with a ruler and a pencil, you can make very complicated drawings easily with a computer. Computer-aided design (CAD) or other graphics software makes it easy to draw lines and shapes. In this activity you will make a drawing with CAD or another graphics software program.

Design Brief

Make a drawing using CAD or other graphics software and a computer. You might choose one of the following ideas for your drawing or think of your own:

* Design and draw a school logo. A *logo* is a symbol that represents a product or company or, in this case, your school.
* Design and draw a floor plan for a house or school of the future.
* Design and draw a sign that reminds people to recycle waste materials such as aluminum, glass, or plastics.
* Design and draw a cover for a report. This cover might include a drawing of something related to science, social studies, mathematics, or art.
* Design and draw a cover for a CD or tape.
* Or better yet, think of your own!

Materials/Equipment

* computer with CAD or other graphics software
* printer or plotter

Be sure to fill out your TechNotes and place them in your portfolio.

SAFETY **FIRST**
* **Follow the safety rules listed on pages 42-43 and the specific rules provided by your teacher for tools and machines.**
* **Avoid eye strain. Look away from the screen often. Refocus your eyes on distant objects.**
* **Avoid strain on your neck, back, arms, or wrists. Stretch frequently. Get up and walk around at least once an hour.**

(Continued on next page)

Procedure

1. Boot a computer with CAD or other graphics software.
2. Start your design using one of the suggested topics, or get your teacher's permission to do your own.
3. Experiment with erasing, moving, and resizing part of your design until you are happy with the way it looks.
4. Ask a friend to check your drawing. Look for things that will make the drawing better. Watch for overlapping corners, rough "freehand" curves, and so on.
5. Make any final corrections to your drawing.
6. Save your drawing.
7. Print or plot your design.

Evaluation

1. If you used clip art for your project, explain why.
2. Is it easier to change a drawing on paper or on a computer? Explain.
3. When you get used to drawing with CAD software, do you think you will go back to traditional methods? Explain.
4. **Going Beyond.** Use the documentation (instructions) for the software to find out how to make a scale drawing, duplicate objects, and use clip art images with your design.
5. **Going Beyond.** Do you think that great works of art will ever be made on a computer? Explain.
6. **Going Beyond.** Ask your teacher to show you the traditional tools used for drafting. Fig. A. Try to compare tools like T-squares, triangles, and compasses with their CAD counterparts. Which are easier and faster to use?

Fig. A.

Using Digital Photography

THINGS TO EXPLORE

- Explain how digital photography has changed the way newspapers are produced.
- Tell how digital photography works and how you can use it.
- Take digital photographs.

TechnoTerms
cropped
download
negative
print

Today's news media want instant high-quality photos that can be **downloaded** (sent electronically) over phone lines using a modem or direct Internet connection. Computer technology has made it possible for still pictures to be transmitted electronically at incredible speeds. Thanks to digital photography, we can have "on-the-spot" pictures in today's newspaper from events happening around the world right now. Fig. 5-7.

Graphic Arts versus Desktop Publishing

Before digital photography, newspapers used a traditional graphic arts process to lay out each page of a paper. It took a lot of time. Photographers would use cameras to expose a light-sensitive film. The film had to be developed into a **negative** (reverse image) in a series of chemicals and then dried. **Prints** were then made by shining light through the negative onto light-sensitive paper. The photographic paper was then developed in another series of chemicals. Fig. 5-8. The finished print was dried and pasted onto a page along with the words for news stories. Another photograph was

Fig. 5-8. Traditional photographic prints are made by placing the exposed paper in a series of chemicals. (Note the trays in this photo.) The finished print is then hung up to dry. Find out which chemicals are used and how they work. Report your findings to the class.

Fig. 5-7. Digital photography makes it possible to download photos into a computer and print them out. Why do you think newspaper photographers are in a hurry to get their photos to the paper?

Chapter 5, Section 3 · 115

Fig. 5-9. The digital camera on the right can be hooked up to the computer. The photos can be downloaded directly into a page layout. Using special software, such as Photoshop, make changes in a sample photo so it can be used in a greeting card.

taken of the "camera-ready" page by a process camera and developed into a negative. Finally the negative was used to expose a printing plate for use on a printing press.

Compare that process with today's methods. Digital photography is all electronic. Instead of having light expose film, the light is focused on a light-sensitive chip called a CCD (charged coupled device). Digital photographs can be **cropped** ("cut" to remove unwanted background) on the computer. The finished digital photo can be electronically pasted into the page along with the news stories created by word processing. Finished printing plates are made directly from the computer file. This entire process, called desktop publishing, takes much less time. Fig. 5-9.

INFOLINK

See Chapter 4, "Introducing Computers," for more about desktop publishing.

Digital Photography and You

If you have a digital camera, you can send photographs through electronic mail or post them to a website for others to enjoy. You can easily change them, and you can add them to cards or letters. Your yearbook team can even make a digital yearbook. Yearbook photographers can provide the school pictures on CDs. The digital pictures can then be put into a yearbook using desktop publishing software.

TechnoFact

DIGITAL NEWS News photographers are always in a race to take pictures and get them published in newspapers or magazines or seen on TV as quickly as possible. You may hear news broadcasters use the phrase "exclusive coverage." That means they won the race and beat other news programs to the story.

SECTION 3
TechCHECK

1. How has the use of digital photography changed the way newspapers are produced?
2. What is a CCD?
3. How can you use digital photography?
4. **Apply Your Knowledge.** Check with your local newspapers to see how they use digital photography.

ACTION ACTIVITY

Digital Photography

Real World Connection

Digital photography is changing the speed with which still images and video are recorded and edited. This is important to the news media who need information and photos as quickly as possible. In this activity, you will learn to take digital photographs. Fig. A.

> Be sure to fill out your TechNotes and place them in your portfolio.

Design Brief

Use a digital camera to take photographs. Download the digital photos to a computer. Select the best one and adjust the photo using computer software. Print the finished photo.

Materials/Equipment

- digital camera/cables
- computer
- graphics software

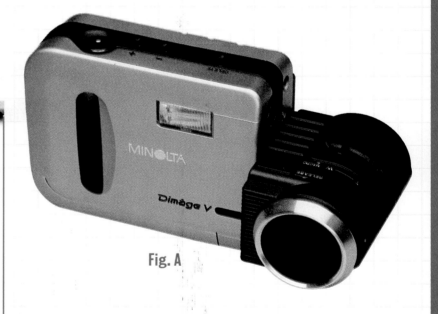

Fig. A

SAFETY FIRST

- Follow the safety rules listed on pages 42-43 and the specific rules provided by your teacher for tools and machines.
- Using a digital camera isn't dangerous, but you need to be careful to prevent damaging the expensive equipment. Use the neck strap to avoid dropping the camera accidentally. Use a tripod to steady the camera. Be careful not to trip over the tripod legs.

(Continued on next page)

ACTION ACTIVITY

Procedure

1. Ask your teacher to demonstrate the controls on your school's digital camera. Be sure you know how to do the following:
- Turn the camera on and off
- Focus, use flash, and zoom
- Preview the photographs you have taken
- Delete (erase) unwanted photos
- Download to a computer

2. Take a series of 5 to 10 photos that demonstrate your ability to compose a good photo. Remember to fill the frame with the subject. Use a tripod to steady the camera.

3. Download your photos to a computer. Using graphics software designed to work with pictures, do the following to your best shot:
- Crop unnecessary background, or outline the subject of the photo
- Adjust the brightness and contrast of the photo to increase its quality
- Use special effects to create a more interesting photo
- Correct the color if needed. Fig. B.
- Print the finished photo

Evaluation

1. What does it mean to download a photo?

2. What does it mean to crop a photo?

3. Can you use all the functions of the camera your teacher wants you to use?

4. Going Beyond. Use graphics software to cut and paste people from one photo to another.

5. Going Beyond. Find out how your school yearbook is made. Could digital photos be used in your yearbook?

Fig. B

Color Balance

Original

More Green

More Yellow

More Cyan   **Current Choice**   **More Red**

More Blue

More Magenta

Animating with Computers

THINGS TO EXPLORE

- Define persistence of vision and explain how it is used in animation.
- Describe the difference between making a traditional animated movie and making a computer-animated movie.
- Create framegrabs to make a claymation.

TechnoTerms
cell
claymation
frame
in-betweening
persistence of vision
tweening

Did you know that when you watch TV you are looking at thirty still pictures each second? The individual still pictures in live TV or on videotape are called **frames**. Each frame is a complete picture that is scanned onto the inside of a cathode ray tube (CRT)—sometimes called a picture tube—in computer monitors as well as TVs.

Use a stopwatch to try to count to thirty in one second. Can't do it? Don't feel bad; nobody can. Your brain is unable to process the information fast enough. When you watch TV, you can't see the individual still pictures because they are changing so fast. Your brain thinks you are watching moving pictures. This effect is called **persistence of vision**. Persistence of vision is also what makes animation possible.

In animation, the characters are drawn a little differently in each frame, or **cell.** When the cells are displayed quickly, the characters appear to move.

Animation Today

Animations, like Disney's *Snow White*, used to be handmade by many artists drawing thousands of high-quality images. To make the process go faster, artists drew the characters with larger than normal movements. Other artists, called "in-betweeners," filled in the steps needed to make smooth motion. The finished paintings, called cells, were photographed one at a time using movie film.

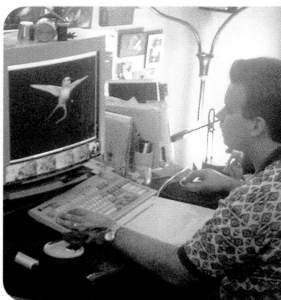

▼ This artist is working on a computer-animated figure of a bird. Most animation today is done with computers.

"Bunny" © 1998 Blue Sky Studios, Inc.

Fig. 5-10. This scene was not created using real figures. It is completely computer generated. Have you seen a completely computer-generated film? What elements made it look real?

Today, we can use the power of computers to do the same process much faster. The basic cells are drawn on the computer by the artist. Then software is used to automatically create the cells needed in between. This process is called **in-betweening,** or just **tweening**. Artists who use animation software can transform drawings into images that look incredibly real. Fig. 5-10.

COMMUNICATION CONNECTION

Talking Tech

Computer technology is sometimes confusing because many acronyms and abbreviations are used. An acronym is a word made from the beginning letters of other words. Here is a list to help you.

bit	=	binary digit
byte	=	eight bits
CAD	=	computer-aided design
CAM	=	computer-aided manufacturing
CD-ROM	=	compact disk read only memory
CD-RW	=	compact disk read and write
CIM	=	computer-integrated manufacturing
CPU	=	central processing unit
DTP	=	desktop publishing
DVD	=	digital video disk
Gb	=	gigabyte (1024 megabytes)

What Is Claymation?

One specialized form of animation called **claymation** uses clay figures that seem to come alive. You have probably seen claymation used in commercials for TV. It takes many hours for artists to adjust the scenes and characters by small increments (amounts). A computer can be used to store and quickly play back the individual pictures, called *framegrabs*.

SECTION 4
TechCHECK

1. What is persistence of vision?
2. How is computer animation different from traditional animation?
3. What is claymation?
4. **Apply Your Knowledge.** Watch for television commercials that use claymation. Record them and share them with your class.

IC	=	integrated circuit
K	=	kilobyte (1024, or 2^{10}, bytes)
Mb	=	megabyte (1024 kilobytes)
MIDI	=	musical instrument digital interface
modem	=	modulator/demodulator
PC	=	printed circuit (also, personal computer)
RAM	=	random access memory
ROM	=	read only memory
WYSIWYG	=	what you see is what you get

ACTIVITY

Make your own acronyms for things or events around school. Share them with the class.

ACTION ACTIVITY

Computer Animation

Be sure to fill out your TechNotes and place them in your portfolio.

Real World Connection

You have seen animated cartoons, but have you ever thought about the technology behind making them? In this activity you will make your own claymation sequence. Fig A.

Design Brief

Plan and produce a claymation animation using computer and video technology. Your claymation should last about ten seconds and have a backdrop and movable characters.

Materials/Equipment

- computer with VIDEO IN and OUT
- animation software
- colored paper rolls
- modeling clay or plastic models
- markers
- lights
- camcorder and tripod

SAFETY FIRST
Follow the safety rules listed on pages 42-43 and the specific rules provided by your teacher for tools and machines.

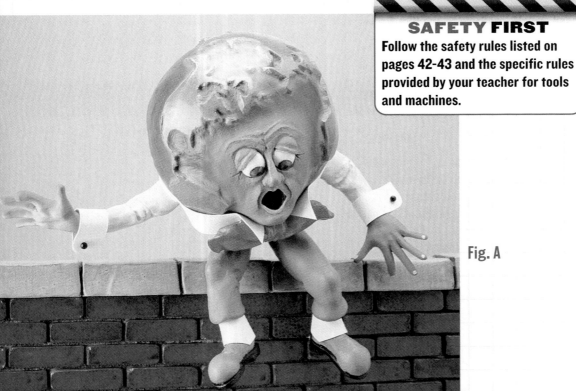

Fig. A

Procedure

1. Work in small groups of two to four students. Plan your animation to last about ten seconds.

2. Use colored paper or other props to make a backdrop for your characters.

3. Make your characters out of modeling clay, or use plastic action figures that have moveable arms and legs.

4. Set up a camcorder and a tripod. Connect the camcorder's VIDEO OUT to VIDEO IN on your computer.

5. Adjust the lights to make realistic shadows. Check with your teacher to find out which lights to use and how to set them up.

> ### SAFETY **FIRST**
> You will need to use bright lights to make your animation. Lights can get very hot and could cause burns. Be careful not to touch the hot lights or trip on power cords.

6. Create the first scene. Use software to take each picture, or framegrab. Save all of your framegrabs in one folder or directory. Ask your teacher how many framegrabs you can save. You can save time by using the in-betweening feature of animation software.

7. Use animation software to assemble all of your framegrabs into a movie. Play your movie and make adjustments as needed.

Evaluation

1. Why is it easier to animate using a computer than using hand-drawn cells? Fig. B.

2. How many framegrabs did you record when making your claymation sequence?

3. If you were going to do this activity again, what would you change?

4. **Going Beyond.** Add sound to your animation.

5. **Going Beyond.** Design, plan, and animate a science fiction story using plastic models and black paper for a background.

6. **Going Beyond.** Use computer graphics software to add special effects to your animation.

Fig. B

REVIEW &

CHAPTER SUMMARY

SECTION 1

- Computers can be used for many jobs beyond the basic applications of word processing, spreadsheets, and databases. Digital photography, simulations, and video conferencing are some of these other uses.

SECTION 2

- Graphics software makes it easy to draw and design.
- Computer-aided design (CAD) is an important tool used by engineers, architects, technical illustrators, and others.

SECTION 3

- Digital photography uses a CCD and is all electronic.
- In desktop publishing, a finished digital photo can be electronically pasted into the page along with the news stories created by word processing.

SECTION 4

- Persistence of vision is what makes still pictures appear to move.
- Animations, like Disney's *Snow White,* used to be handmade by many artists drawing thousands of high-quality images. Today, computers make the animation process much faster.

REVIEW QUESTIONS

1. How do computer simulations help in testing products?
2. Why is a computer a good tool to use in digital photography?
3. Why might an architect use CAD?
4. Why does the traditional graphic arts process take longer than desktop publishing?
5. What are the individual still pictures in live TV or on videotape called?

CRITICAL THINKING

1. If it is so easy to change photos, how can you be sure that what you're seeing in any photo is real? What effect do you think this technology has had on photographic evidence used in court?
2. Describe the process of in-betweening in animation.
3. List as many uses for computers as you can.
4. Why do you think using digital cameras makes it easier to do commercials and news shows?
5. Why do newspapers need photographs and information faster now than they used to?

ACTIVITIES

CROSS-CURRICULAR EXTENSIONS

1. SCIENCE Take a series of digital pictures that illustrates the growth of a plant.

2. MATHEMATICS Use CAD or other graphics software to draw common geometric designs, such as a triangle, square, rhombus, parallelogram, or hexagon. Make a cover for your math book or folder.

3. COMMUNICATION Use desktop publishing to make an announcement for a play or sports event at your school.

EXPLORING CAREERS

Some workers use computers to create information, such as drawings. Others use them to send information over a network or to program a machine. Here are two more of the many jobs that require the use of computers.

Network Administrator In many companies, if the computer network goes down, work grinds to a halt. The network administrator is the person who keeps everything up and running. Network administrators maintain hardware and software and analyze problems. They need good communication and troubleshooting skills.

Automotive Specialty Technician You have probably noticed that many auto service centers work on specific parts of a car, such as tires or mufflers. Automotive specialty technicians are trained to repair one system or component, such as brakes, suspension, or radiators. Understanding the mechanics of an automobile is important in this job, but so is using computerized diagnostic equipment.

ACTIVITY

Look through the employment ads in your Sunday newspaper and make a list of all of the jobs that require computer skills. Which jobs would you be interested in?

Inventing Things

What Is Innovation?

THINGS TO EXPLORE

- Explain what innovation is.
- Explore how serendipity affects innovation.
- Identify ways innovations affect our lives.

TechnoTerms
innovate
serendipity

To **innovate** is to create a new idea, device, or way of doing something. Innovation can take place in many ways. Usually it is the result of creative thinking.

How We Innovate

Sometimes innovation happens when people or companies work toward a goal using a combination of skill, creativity, and knowledge. Fig. 6-1. At other times, valuable innovations may result from **serendipity**—having a lucky accident. Teflon (a nonstick plastic), safety glass for car windshields, and even the process of making breakfast cereals like corn flakes were all lucky accidents. Inventors were looking for something else and accidentally came up with a new product.

(INFOLINK)

See Chapter 1 for more about creative thinking.

Fig. 6-1. This Ford *Probe* concept car is a result of innovation. How does its appearance differ from cars sold today?

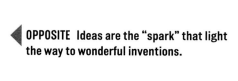

OPPOSITE Ideas are the "spark" that light the way to wonderful inventions.

Fig. 6-2. Many of the parts in these Boeing F/A-18 Hornets and other modern planes are held together with adhesives. Look around your home and find at least five other items that are glued together. What other fasteners would work as well?

Many innovative ideas go unnoticed. For example, today many parts of cars are made of plastics to reduce weight and to prevent rusting. Many of your clothes are made of new kinds of plastics that don't feel or look any different from the fabrics that were used before. In the aerospace industry, adhesives (glues) are used instead of rivets to hold airplane wings together. Fig. 6-2.

Why We Need Change

Why do we need to continually make changes? The main reason is that what worked for us yesterday does not work for us today. Sometimes we simply want variety. We want new car styles, different exercise equipment, or new foods. At other times, we have a specific need. For example, today we don't know what to do with our nuclear waste products. We haven't found a way to store them safely for long periods. If someone like you could think of a way to use or recycle the materials, it would really benefit our world.

Innovation is necessary in every field from aerospace to education. For example, aerospace engineers working for NASA (National Aeronautics and Space Administration) designed a space vehicle in the 1960s that could hold only one person. The astronaut could not move very much and could not stay in space for long periods of time. Innovations were needed. The result is the Space Shuttle, in which several astronauts can exercise, eat, and move around freely. Fig. 6-3. What do you think the next innovations in space travel will be?

TechnoFact

A USEFUL INVENTION Did you know that Teflon, a trade name for a special plastic, is used for many purposes besides nonstick frying pans? It is used in artificial heart valves and bone replacements for the human body, as well as for the outer layer of space suits.

Fig. 6-3. Need often plays a part in innovation. Soon the cramped quarters of early spacecraft (left) were abandoned in favor of the roomier Space Shuttle (right). Have you ever spent a long time in a confined space? Describe your experience.

Innovation can make our lives easier. It can also save lives. In many cities and towns in the United States, you simply dial 911 to get emergency help. That is a recent innovation. You might even use a cellular or cordless phone to make that call—two more innovations.

SECTION 1
TechCHECK

1. What is meant by innovation?
2. Explain what serendipity is. Give two examples.
3. What are some innovations that have made your life easier?
4. **Apply Your Knowledge.** Research innovations that came about by serendipity and compare how they were discovered.

Getting Ideas

THINGS TO EXPLORE

- Explore ways inventors get ideas.
- Explain what trends and fads are.
- Practice visualization to help you invent a product.

Techno Terms
fad
trend
visualizing

Many discoveries and inventions that have been improved on over hundreds of years can't be credited to any one inventor. Sometimes they occurred at a time when no records were kept. For instance, we don't know who discovered the uses of fire or invented the wheel. But we do know something about how inventors think.

What Makes Someone an Innovator?

Being innovative means you can do something new with your knowledge and experience. It also means you can recognize when something useful happens accidentally, even if you're not exactly sure what it is! Because innovation involves "new" things, it is usually tied closely to change. People who are innovators enjoy making changes.

Another thing that all innovators have in common is making good use of their brains. Inventors are good at **visualizing**, or picturing ideas in their minds. Often they can see a different way to do something that everyone else has missed. They can look at an everyday object and imagine new uses for it. Can you look at a regular pencil and visualize it being used for something other than writing?

Thomas A. Edison was one of the greatest American inventors. Do some research on Edison's life. What was the first device he patented?

Fig. 6-4. Somebody has to be the first to come up with an innovative idea like a video game. Can you be innovative? Make a list of at least 10 new uses for an ordinary tennis ball.

Noland Bushnell is an example of a really innovative person. In the 1970s he created "Pong," the first interactive video game, because he was bored with just "watching" television. He wanted to play with the television and have the television respond. His table tennis game was the start of the video games you play today. Fig. 6-4.

Some people, like the cartoonist Rube Goldberg, like innovation just for the pure fun of it. Goldberg invented complicated, funny ways to do simple jobs. Innovative ideas like his usually don't become an actual product or service.

Thousands of inventions, such as television, have had important effects on society. Many others have not. You might not have heard of a device called an automatic hat-tipper, for example. It was invented at a time when most men wore hats and tipped them politely to every lady they passed. The people who helped invent the television set are remembered. The inventor of the hat-tipper is not. However, because we keep good records today, we can research hundreds and even thousands of inventions to find out who had the original idea. Fig. 6-5.

Fig. 6-5

Inventors and Inventions

Lewis Howard Latimer received several patents throughout his career. He is best known for his work on lightbulbs and lamps. In 1882, he received a patent for long-lasting carbon filaments that glowed when heated. He later worked with Thomas Edison.

Madame C. J. Walker was an inventor and entrepreneur. In 1906 she started a business that made cosmetics and hair care products. By 1919, her products were sold nationwide and her business was making nearly half a million dollars a year.

Beulah Louise Henry was a businesswoman and inventor. She invented over 100 devices and received 52 patents. Her first patent was for an ice-cream freezer. In 1924 she invented an umbrella with a set of snap-on cloth covers in different colors.

Garrett A. Morgan
received a patent in 1914 for a gas mask for firefighters. In 1916, he used it to rescue workers trapped in a smoke-filled tunnel under Lake Erie. He also invented a traffic light (patented in 1923) and many other useful products.

Katherine Blodgett
received a patent in 1938 for a process that makes glass nonreflecting, which means it won't have a glare. As a result, things like eyeglasses, telescopes, and camera lenses could allow images to pass through clearly.

Tuan Vo-Dinh
has invented several life-saving devices. His first patent was a small badge that detects exposure to poisonous chemicals. In 1996, he invented a laser technique to detect cancer without requiring surgery. By 2000, he had received 20 patents.

Ellen Ochoa
is an astronaut and inventor. She and her co-inventors have three patents related to optical systems. One of them can identify and "recognize" objects. In 1990, she was selected by NASA to be an astronaut and three years later became the first Hispanic woman in space.

Where Do Ideas Come From?

Where do ideas for inventions start? There really isn't a specific set of steps to follow. Lots of innovations are the result of looking at old ideas or things in a new way. For example, Johann Gutenberg combined two unconnected ideas, the coin punch and the wine press, to create a new product—a printing press that used movable type.

Other inventors set out to solve a specific problem. The search for cures for diseases like AIDS is ongoing in medical research laboratories. People also come up with ideas for inventions by serendipity—just being in the right place at the right time. Edward Jenner's smallpox vaccine was a result of serendipity. It has since saved millions of people from a horrible disease.

Fig. 6-6. Nothing is as trendy as clothing styles. Do you think clothes like these will ever catch on again? Give reasons for your answer.

SCIENCE CONNECTION

The Divided Brain

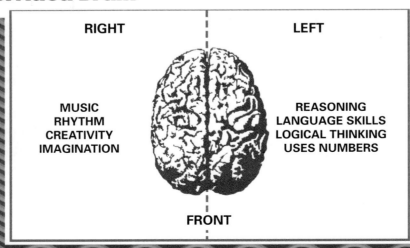

RIGHT

MUSIC
RHYTHM
CREATIVITY
IMAGINATION

LEFT

REASONING
LANGUAGE SKILLS
LOGICAL THINKING
USES NUMBERS

FRONT

As recently as 500 years ago, most people did not even know what the brain did. Many people thought thinking and emotions were centered in the heart or abdomen because that's where you feel pain.

In manufacturing, innovation might involve a new product or a different way to make the same product. **Trends**, or current preferences that people have, help determine what changes companies will make in their products. Fig. 6-6.

Many companies have research and development (R & D) departments whose job is to develop new product ideas. How practical an idea is, how economical it is, and how it is marketed often determine whether you and I ever see it. Fig. 6-7.

Fig. 6-7. The research and development department of this manufacturing company tries to find new uses for plastics.

After years of research into how people think, there is still much to learn about the brain. However, scientists have discovered that the brain has two halves. The halves are linked together by a complex network of nerves called the corpus callosum.

The left and right halves of the brain work in different ways. In most people, activities such as thinking logically, dealing with numbers, reasoning, and using language skills are handled by the left brain. The right brain deals with imagination and creativity and activities such as music, art, and daydreaming. It has been found that great inventors, such as Leonardo da Vinci and Albert Einstein, use more of their brains than most of the rest of us.

ACTIVITY

Read *Drawing on the Right Side of the Brain* by Betty Edwards. Report what you learned to the class.

For example, one product under development is the X-33. this is an experimental vehicle being developed by Lockheed Martin. It will fly 13 times the speed of sound. That's about 9800 miles per hour!

The X-33 is part of the Reusable Launch Vehicle (RLV) Technology Program, a partnership between industry and NASA. The program's goal is to develop a vehicle that can climb into orbit without booster rockets and can land like an ordinary airplane. Unlike the Space Shuttle, an RLV would be fully reusable and therefore less costly than the Shuttle.

Fads

Some innovations result in **fads**—things that are temporarily popular. Because there is no real need for the product or idea, many fads disappear quickly. What are some fads of today?

Can you imagine a world where nothing changed? Innovations make your world exciting. Inventors have to try new ideas, change old ideas, and look for new ways to solve today's problems.

SECTION 2
✓ TechCHECK

1. What is a fad?
2. How does visualizing help an inventor?
3. What is a trend?
4. **Apply Your Knowledge.** Make a drawing of a new eating utensil.

ACTION ACTIVITY

Inventing New Tools

Real World Connection

New products and inventions are often a combination of old ideas put together in a new way. In this activity, you will invent new tools that are a combination of tools you are already familiar with.

Inventors must be able to communicate their ideas in words and drawings. In this activity, you will make sketches and finished drawings that get your ideas across.

Design Brief

Invent four new tools that are a combination of two or more existing tools. You may combine some of the tools shown in Fig. A or think of others, such as a spatula, egg beater, or spoon.

Materials/Equipment

* pencil
* paper

Optional
* computer
* graphics or CAD software
* clip art
* word-processing software
* printer
* tape recorder
* drawing tools
* sample tools

> **Be sure to fill out your TechNotes and place them in your portfolio.**

SAFETY FIRST
Follow the safety rules listed on pages 42-43 and the specific rules provided by your teacher for tools and machines.

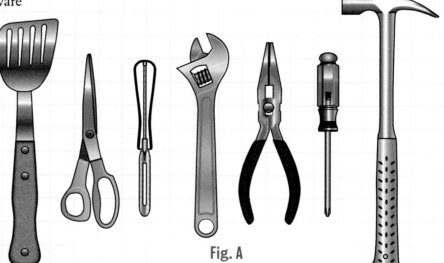

Fig. A

(Continued on next page)

ACTION ACTIVITY

Procedure

1. Work with a partner. Brainstorm some possible ideas for a new tool.
2. Choose four of your ideas that you think have the most ability to succeed.
3. Make some rough sketches of your four tools. *Optional:* Use a computer and graphics software to make your drawings.
4. Refine your ideas, and show them to your parents, relatives, teachers, and friends. Listen to the reactions of others. This is called *feedback*.
5. Redesign your ideas with the feedback in mind.
6. Make finished drawings of all four of your ideas. *Optional:* Use a computer and graphics or CAD software to make your finished drawings.
7. Choose the idea that you like best. Write a script for a radio commercial to sell your product. If possible, use a computer and word-processing software. Include in your script your product's name, cost, possible uses, and where to buy it.
8. *Optional:* Record your commercial using an audio tape recorder. Play your commercial for the class. Ask the class to sketch what they think your tool looks like without showing them your design.

Evaluation

1. How many different tool ideas were you able to brainstorm? List them.
2. What was the hardest part of this assignment? What was the easiest? Explain your answers.
3. Did you consider whether or not your tool design was safe? Does it have dangerous pinch points or sharp edges?
4. **Going Beyond.** Produce a video commercial for your product.
5. **Going Beyond.** Find someone in your community who has designed a product. Ask that person to talk to your class, or interview the person on the phone.
6. **Going Beyond.** Think of other possible combinations of existing products. Some you might consider include athletic and sports equipment, camping gear, food products, and home appliances. Fig. B.

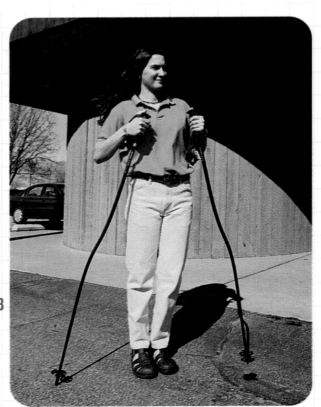

Fig. B

Protecting New Ideas

THINGS TO EXPLORE

- Explain what a patent is.
- Identify the five types of patents.
- Patent one of your ideas.

TechnoTerms
patent
prototype

A **patent** is a special government license that protects an invention from being copied for 20 years. When you patent an invention, anyone who wants to use your idea must get your permission and perhaps pay you. Fig. 6-8.

Applying for a Patent

Anyone can patent an invention, but getting a patent takes time and money. You must first prove your invention is new or the first of its kind. This means going back through the thousands of patents filed at the United States Patent and Trademark Office to be sure no one else has an existing patent for the same invention. Then you have to provide written plans and sketches that show how your invention works. You may have to make a **prototype**, or model. Fig. 6-9.

TechnoFact

TIMING IS EVERYTHING Alexander Graham Bell filed his patent application for the telephone on February 14, 1876. A few hours later, Elisha Gray also filed a document with the U.S. Patent Office. It stated that *he* was inventing a telephone. Lawsuits followed, but eventually the Supreme Court decided Bell was the sole inventor of the telephone.

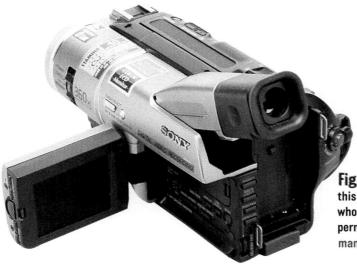

Fig. 6-8. The person or company that developed this digital camera holds the patent on it. Someone who wanted to copy the design would have to get permission to use it. Use the Internet to find out how many patents are registered with the Patent Office.

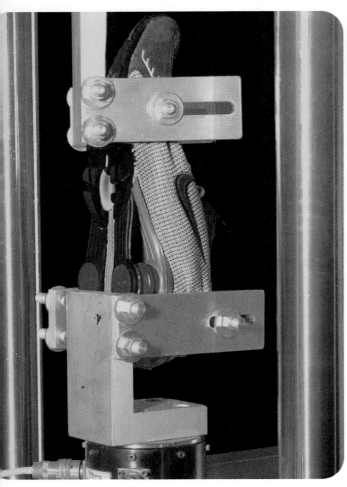

Photo courtesy of Nike, Inc.

Fig. 6-9. Running shoe prototypes are tested for performance. What qualities do you think this shoe is tested for?

Not every idea or device can be patented. Things that can be patented include processes or machines, manufacturing methods, and a new material or life form. Things that cannot be patented include naturally occurring materials, ways of doing business, and newly discovered scientific principles.

Types of Patents

There are three types of patents.
- **Utility patent.** The most common type of patent is a utility patent. This type of patent protects inventions considered to be "new and useful." The invention must meet certain requirements to qualify.
- **Plant patent.** A plant patent protects the invention or discovery of new varieties of plants. This includes cultivated spores, mutants, hybrids, and newly found seedlings. The new plant must have been reproduced asexually, which means that the process did not involve the union of nuclei, sex cells or sex organs.
- **Design patent.** A design patent protects the invention of a new design for an item that will be manufactured. It protects the invention's general appearance but not its structural or functional features.

Fig. 6-10. Every patent is given a number. How many patents were taken out on this product?

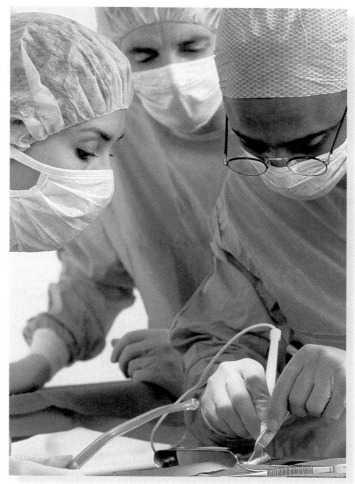

Fig. 6-11. Developers of lasers used in surgery protected their ideas with patents. What standards would you use in deciding whether or not to spend money developing a product fully?

A patent number appears on every patented product. Fig. 6-10. If you see the words "Patent applied for" or "Patent pending" on a product, it means the inventor has applied for a patent and is waiting to receive it. This is done because sometimes companies want to start making a product before the patent is given. It gives them some protection until the patent is approved.

It is important to protect inventions with a patent. You can imagine how disappointed you would be if you spent a great deal of time and money on an idea only to find out someone else already had a patent on something similar. The key is knowing when, or if, to move from a prototype to the real thing! Fig. 6-11.

SECTION 3
TechCHECK

1. What is a patent? List the three types.
2. Why might a product have the words "Patent pending" printed on it?
3. What is a prototype?
4. **Apply Your Knowledge.** Find someone in your community who has designed a product and applied for a patent. Ask that person to talk to your class, or arrange a phone interview.

ACTION ACTIVITY

Patenting Your Ideas

Be sure to fill out your TechNotes and place them in your portfolio.

Real World Connection

Your special ideas may be worthy of patents. In this activity, you will simulate applying for a patent to protect your ideas. After you have a better idea of how the process works, you may want to consider applying for a real patent on one of your inventions.

Design Brief

Design, test, and patent a package that will protect a raw egg from breaking. Your package or method will be tested by dropping the package with the egg in it from a 15-foot height to the ground. At least one-half of the egg must be visible at all times during the test.

Materials/Equipment

You may use any materials you like for the package, but they must be safe to drop from a second-story window. For example, you might use recycled packaging materials such as plastic, foam, or paper. Other supplies you will need might include string, tape, rubber bands, and scissors.

SAFETY FIRST

Follow the safety rules on pages 42-43 and the specific rules provided by your teacher for tools and machines. Ask your teacher for help on how to test your egg packages.

Procedure

1. Work in invention teams of two to four students. Your teacher will assign one group to represent the Patent Office.
2. Brainstorm ideas on how to meet the design brief requirements.
3. The students working in the Patent Office should write down the requirements for the design brief. They should then design a patent application form. A sample is shown in Fig. A.
4. With your team, decide on a design and complete the patent application.
5. If another team had an idea similar to yours, your patent will be denied. If this happens, your team must choose a different design. If your team is granted a patent, it will be given a patent number.
6. Make your egg-protection package according to your design. Be sure to put your patent number on your package.
7. When all the teams have finished, test each egg package by dropping it from a 15-foot height. Clean up any mess!

Evaluation

1. Have you ever found a product that was broken in the package? Explain how that package could have been made better.

2. Did the Patent Office evaluate your patent application fairly? Explain.

3. List five different packaging materials, and rate them from 1 to 5 (best is 1).

4. Going Beyond. If your egg package design succeeds in the first test, try dropping the package from a greater height. Then try tossing it down instead of just dropping it. What did it take to crack the egg?

Official Patent Application
Form 1-a

• Patent Office •

INSTRUCTIONS:

1. Complete this form.
2. Submit the form in person to the Patent Office.
3. You will be notified by a patent officer if your patent was approved or denied.

Patent Office Use Only:
Patent application date:__/__/__
Patent application time:____ a.m.-p.m.
Patent approval date:__/__/__
Patent approval time:____ a.m.-p.m.
Patent number_____

1. Complete the following information for each of the invention team members.

Invention Team Member 1:
•PLEASE PRINT•

NAME: _____ ___ _____
       LAST   M.I.   FIRST

Invention Team Member 2:
•PLEASE PRINT•

NAME: _____ ___ _____
       LAST   M.I.   FIRST

Invention Team Member 3:
•PLEASE PRINT•

NAME: _____ ___ _____
       LAST   M.I.   FIRST

Invention Team Member 4:
•PLEASE PRINT•

NAME: _____ ___ _____
       LAST   M.I.   FIRST

2. Describe your invention in the space provided below. Be specific.

_____
_____
_____

3. List the quantities and types of materials used in your invention

QUAN	MATERIAL	QUAN	MATERIAL
____	_____	____	_____
____	_____	____	_____
____	_____	____	_____

4. Make a detailed drawing of your invention. Make the drawing on a separate sheet of paper and attach it to this form. Be sure to clearly draw and label each part.

Fig. A

R E V I E W &

CHAPTER SUMMARY

SECTION 1

- To innovate is to use a new idea or approach in doing something.
- Having a lucky accident is called serendipity.
- Innovations may come from combining two existing ideas, looking at things in a new way, or from trying to solve a problem.

SECTION 2

- Research and development departments are always trying to create innovations or identify trends.
- How practical an idea is, how economical it is, and how it is marketed often determine whether most people ever see it.
- Some innovations are popular only temporarily; these are called fads.

SECTION 3

- Inventors may make a prototype, or model, of their invention.
- Inventors apply for patents to protect their inventions from anyone using them without their permission.
- Some companies produce a product before they have a patent. The product is then usually marked "Patent pending" or "Patent applied for."
- All patented inventions are assigned a special patent number.

REVIEW QUESTIONS

1. What effects do trends have on the products that companies produce?
2. Why do we need innovations?
3. What is the job of research and development departments in companies?
4. What is the purpose of a patent?
5. Why don't we always know who some inventors were?

CRITICAL THINKING

1. What could you do with a newspaper besides read it? Make a list of innovative ideas.
2. Why do NASA engineers have to continually make changes in spacecraft designs or in the equipment astronauts use?
3. Choose an innovation and tell how it has made your life easier.
4. Why do you think some inventions become only fads?
5. What do you think is the world's most important invention? Explain your answer.

ACTIVITIES

CROSS-CURRICULAR EXTENSIONS

1. **SCIENCE** Make a working Rube Goldberg-type invention that uses simple machines. (If you haven't seen one of Rube Goldberg's inventions, do some research!)

2. **MATHEMATICS** Make a drawing and build a scale model of an invention.

3. **COMMUNICATION** Write an article for your school newspaper on current fads in your school or community.

EXPLORING CAREERS

As you've read, many companies have research and development departments that develop new product ideas. Here are two careers involved in research and development.

Electrical Engineer Almost any piece of equipment that plugs into electrical current or runs on batteries was designed in part by an electrical engineer. These workers design products and test them to be sure that they work. If not, it's back to the drawing board. Electrical engineers have excellent mathematics skills and are good problem solvers.

Computer Microchip Designer Microchip designers create the "brains" of computers and other electronic devices. Also known as semiconductor processors, microchips can be found in everything from garage-door openers to portable CD players. Designing involves teamwork, so designers must like working closely with others and sharing ideas. Good problem-solving skills are also a must.

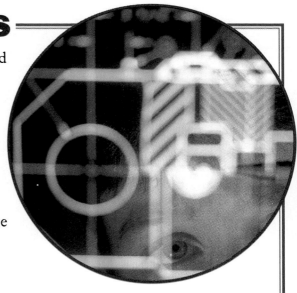

ACTIVITY

Take a vote in your class as to the most useful invention in the classroom. What do you think you would miss the most if it were not there?

Making Things

What Resources Do We Need?

THINGS TO EXPLORE

- Tell what a resource is.
- Identify different categories of resources needed to make a product.
- Tell the difference between renewable and nonrenewable resources.

TechnoTerms

fossil fuel
nonrenewable
profit
renewable
resource

Before a product can be made, resources are needed. A **resource** is anything that is used in the production of the product. What are some of the resources used to make products, and where do you find them? Fig. 7-1.

Resource Groups

Resources fall into seven basic groups that include the following:

- **People.** Technology is created by people. People have used their ideas and knowledge to invent and build products that meet their needs. Companies hire people for their ideas and skills.

Fig. 7-1. Many resources are used to make a product. Name examples of these resources that can be found in your community.

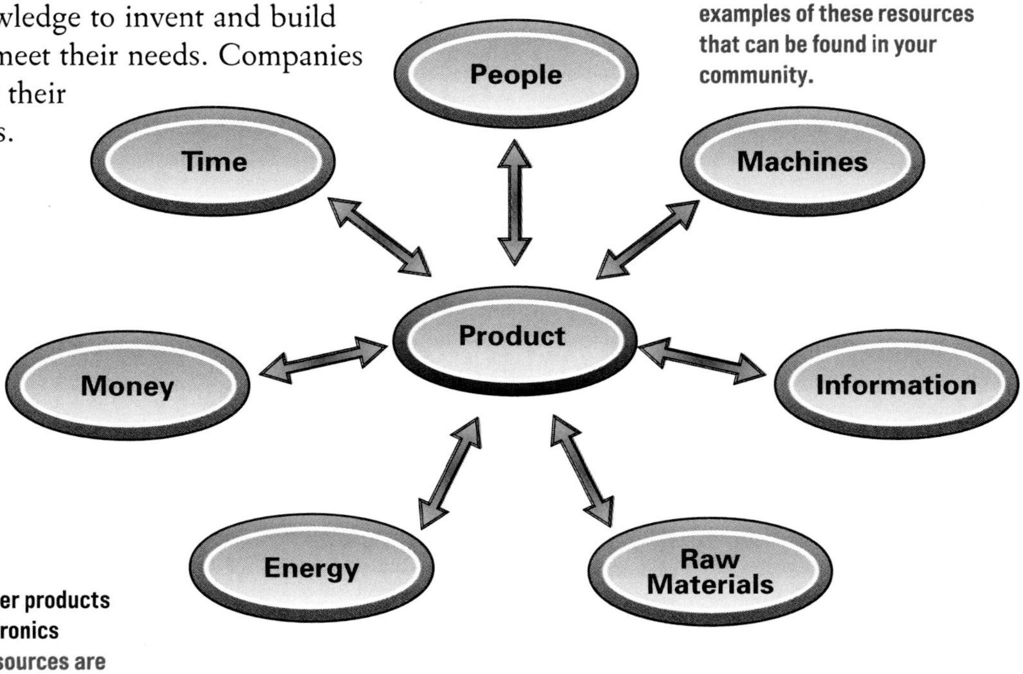

OPPOSITE TVs and other products are made at this electronics factory. How many resources are being used in this picture?

Fig. 7-2. Electricity produced by this hydroelectric plant on the Columbia River in the Pacific Northwest is renewable. As long as the river flows, its energy can be gathered and used. Many people are concerned about the effects of dams on the environment. Research the pros and cons of dams and hold a class debate on the issue.

(INFOLINK)

See Chapter 13 for more information on energy.

- **Machines.** We rely on machines as a resource to help us do work. By means of computer technology, machines often run automatically.
- **Information.** People use information to design, produce, and sell products. Because our knowledge base is growing so quickly, there are companies that gather, organize, and sell information.
- **Raw materials.** You probably know that natural resources include water, land, minerals, fuels, and timber. But did you know that some of these resources, such as **fossil fuels** (oil, coal, natural gas), are **nonrenewable**? Nonrenewable means that once the resource has been used up, it is gone forever!
- **Energy.** Energy is used to make things and to transport products. It is also used to heat, cool, and light the buildings that we live and work in. Some sources of energy are limited. **Renewable** resources can be replaced. Renewable energy sources include plant and animal matter (biomass), geothermal heat (heat from the earth), and the energy of moving water. Fig. 7-2. The sun is the source of nearly all forms of energy on Earth.
- **Money.** You need money to start and maintain a business. Once a company is making a **profit** (money left over after all bills are paid), it uses much of that money to expand the business.
- **Time.** Time is a resource because it takes time to make a product. In many cases, time helps determine which other resources you can use and how you can use them.

SECTION 1
TechCHECK

1. What is a resource?
2. Name six different resources you might use to make things.
3. What does *nonrenewable* mean?
4. **Apply Your Knowledge.** Research renewable and nonrenewable resources. Make a newsletter using desktop publishing to share this information with others.

THINGS TO EXPLORE

- Explain why resources are selected to make certain products.
- Tell how you can conserve resources and protect the environment.

TechnoTerms
conserved
decompose

Early people relied on muscle power to survive and to make the things they needed. As technology grew and changed, people were able to extend that power by using machines. Fig. 7-3. The energy of moving water, steam, oil, and the atom have been added to the list of resources that people can use to make things.

Choosing Resources

Whether or not a resource is used to make a product often depends on its availability. Suppose you were going to start a company that needed to use a great deal of electricity. You would probably try to find an area in the country where electricity is always available and low in cost.

As you design products, you should pick materials that fit the product. For example, paper clips made of pure silver might look good, but they would be too heavy and too expensive to be practical.

INFOLINK

See Chapter 1 for more information about the development of technology.

Fig. 7-3. Machines help us multiply our human powers—even for playing soccer! What games would you like to see robots play?

Fig. 7-4. You can save fuel resources by carpooling with a friend to school. Name some other methods for saving energy.

You must also use resources wisely and not waste them. For example, the cost of energy used to make and transport products is increasing. If energy can be **conserved** (saved), that may help reduce the cost of products. Fig. 7-4.

Some nonrenewable resources, such as aluminum, can be recycled (used again). Aluminum requires a great deal of electricity to produce the first time. By recycling, you can save not only the mineral resource but the energy used to produce it as well.

Is Our World Disposable?

Are companies producing products without thinking about the appropriate use of resources? These are hard questions for you to answer, but you need to ask them.

For example, you are surrounded by products that companies claim are "disposable." Think about all the paper and plastic cups, food containers, and packaging materials that your family throws in the garbage each week. As you learned in Chapter 3, these materials end up being thrown into landfills where they take many years to **decompose** (break down). Today, you must use resources carefully and be aware of how they will affect the environment after they are no longer in use.

INFOLINK

See Chapter 3 for more information about landfills and recycling.

Be Part of the Solution, Not Part of the Problem

The problems facing us and our environment did not happen overnight. They cannot be solved overnight either. Some of the suggestions below might not sound as if they would be much help, but they will. Think about the results over your entire lifetime! We can all do things that will make a difference.

- **Recycle materials.** Aluminum, newspaper, glass, and other materials can be recycled. Recycling helps in two ways. It saves the energy that would be needed to make new materials. It also keeps the materials from taking up space in landfills. In addition, recycling can save manufacturers and consumers money.

- **Conserve energy.** Using less energy means burning less fossil fuel or producing less nuclear waste. If everyone used less energy, it would help them save money and help the environment, too. Fig. 7-5.

Fig. 7-5. The solar car (top) has photovoltaic cells in its roof. The sun provides the energy. The state of Iowa helps conserve energy by using flexible fuel vehicles (bottom) that run on blends of gasoline and up to 85 percent ethanol (an alcohol made from grains). Which service stations in your community sell ethanol blends?

Fig. 7-6. People can have an effect on technology by the way they cast their votes. Ask a family member how he or she voted regarding a recent technology-related issue. How did the voting turn out?

- **Get involved.** Be a part of public service organizations in your school, your community, your state, or even at the national level. Help these groups to clean up the streets, build parks, fix up old houses, or do other things to conserve resources and improve the environment.
- **Be technologically literate.** Studying technology can help you make informed choices about appropriate uses of resources and how they affect the environment. You will also be able to make informed choices when you vote for people who will represent you in government. Fig. 7-6. That is where many decisions about handling our resources are made.

As you design systems and products in the next chapters, be aware of all the resources you use. Most important, be a good consumer of these resources.

SECTION 2
TechCHECK

1. Why might a manufacturer choose one resource over another?
2. What does it mean to use a resource carefully?
3. Name four things you can do to conserve resources and help the environment.
4. **Apply Your Knowledge.** Organize a group of students to collect the litter around your school for a week. Recycle as many materials as you can. Keep track of what things you find and create a news bulletin or newscast to share your findings.

Setting Up a Recycling Center

Be sure to fill out your TechNotes and place them in your portfolio.

Real World Connection

Just think of all the paper and aluminum cans discarded every day. Fig. A. If your class can help to make recycling easy and convenient, more people might pitch in and help. In this activity, you'll set up a recycling center.

Design Brief

Design and make a recycling center for your school or classroom. The center should provide containers that are clearly marked for the type of materials to be put in them. Your design should also include posters that explain the importance of recycling and how to use the recycling center.

Materials/Equipment

* posterboard
* markers
* garbage cans
* plastic bags
* computer with graphics software (optional)

SAFETY FIRST

Follow the safety rules listed on pages 42-43 and the specific rules provided by your teacher for tools and machines.

Fig. A

(Continued on next page)

Procedure

1. Work as a large group. Brainstorm ideas for the recycling center. Ask the principal of your school to give your class advice on the best way to set up your recycling center.

2. Decide how you would like to have your center work and choose its location. If your school is very large, you might consider having two or more centers or a small center for each classroom.

3. Make a plan for who will be responsible for each detail of your center. For example:
   - Who will empty the containers when they are full?
   - Where will the materials be taken for bulk recycling?
   - How will you safely handle broken glass or heavy paper?
   - Which teachers can you get to help?
   - Who will make the posters to encourage people to use the recycling center?

4. After your class has built the center, advertise it in the student newspaper, on bulletin boards, and in any other place where students and teachers will notice.

5. Start your recycling project with a special announcement from the principal or a cooperating teacher. Encourage other students to help with the project.

Evaluation

1. Which recyclable material is most common in your school? Which material is least used?

2. What would have happened to the recycled material if your project had not been started?

3. **Going Beyond.** Make a large poster that illustrates in a graph the amount of material your school is recycling.

4. **Going Beyond.** Research materials used commonly in your school, such as copier and computer paper. Find out if those materials are available in recycled form. Compare the prices of new paper and recycled paper.

THINGS TO EXPLORE

- Define what a process is.
- List four different processes used to make things.
- Use manufacturing processes to make a product.

TechnoTerms
manufactured
processes
quality control

Everyday products, such as the pen or pencil you write with, the paper you use, and the bus that may have brought you to school, are made, or **manufactured**, in factories. The raw materials are changed into products by means of **processes**. Fig. 7-7.

Manufacturing Processes

The major manufacturing processes and the machines used for them can be put into the following groups:

- **Forming**: Molding or changing the shape of materials
- **Machining**: Changing the shape of materials by cutting away pieces, or "chips"
- **Fastening:** Holding materials together with nuts and bolts, welding, or adhesives
- **Finishing:** Painting, varnishing, or coating products with plastic

Fig. 7-7. An automated machine does precision drilling. Drilling is which of the four processes?

INFOLINK

See Chapter 11 for more information about quality control.

Fig. 7-8. A laser is an intensely focused beam of light. Here a laser is used to cut metal. Name some other uses for lasers.

MATHEMATICS CONNECTION

The Builders' Favorite!

Did you know that concrete is the most commonly used building material on Earth? Think about the fact that buildings, roads, dams, house foundations, and even sidewalks are often made of concrete.

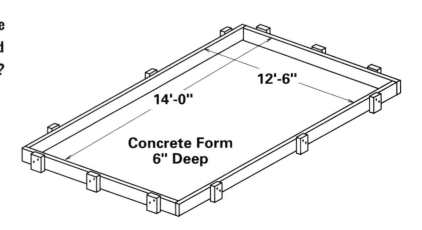

12'-6"

14'-0"

**Concrete Form
6" Deep**

In all processes, accurate measurements are important. Measurement can be done by hand or automatically by a machine. As products are made, they are often checked to see if they are the right size or shape. This inspection of parts is called **quality control**.

People have been developing new processes and perfecting old processes since the Stone Age. You have learned about a few of them. Today the development of manufacturing processes is especially rapid. For example, technology has made it possible to use lasers and high-pressure jets of water to cut through tough materials. Fig. 7-8.

INFOLINK

See Chapter 9 for more information about measurement.

TechnoFact

OSHA Many of the activities in this book include safety reminders. Your school also has rules about safety. In the workplace, the Occupational Safety and Health Administration (OSHA) sets safety standards. OSHA was established by the federal government and is part of the U.S. Department of Labor. Representatives of OSHA visit workplaces to make sure that safety standards are met.

SECTION 3
TechCHECK

1. What is a process?
2. Describe four different manufacturing processes.
3. What process involves painting or varnishing?
4. **Apply Your Knowledge.** List and describe the processes you think would be needed to build a car.

Concrete is a mixture of sand and gravel held together by a "paste" of cement and water. The mixture has to be held in shape in a *form* while a chemical reaction called *hydration* makes the concrete harden, or *cure*.

The amount of concrete needed to fill a form is both a mathematics and a technology problem. The example here shows you how much concrete it takes to make a patio 14'-0" x 12'-6" x 6".

Step 1. Calculate the volume of the form. (Hint: Change inches to decimal feet.)

14 x 12.5 x .5 = 87.5 cubic feet

Step 2. Change cubic feet to cubic yards. Concrete is sold by the cubic yard. A cubic yard is 27 cubic feet (3' x 3' x 3').

Step 3. Divide cubic feet by 27 (one cubic yard).

$$\frac{87.5}{27} = 3.24 \text{ cubic yards}$$

ACTIVITY

How much concrete would you need to make a sidewalk 6" thick, 4' wide, and 50' long? Ask a local concrete supplier how much one cubic yard of concrete costs. Calculate the cost of your sidewalk.

ACTION ACTIVITY

Making a Can Crusher

Be sure to fill out your TechNotes and place them in your portfolio.

Real World Connection

One of the problems with recycling cans is that they take up a lot of space. By crushing them, many more cans can be transported easily to a recycling center. In this activity, you will design a product that can be used to crush aluminum cans.

Design Brief

Design and make an aluminum can crusher that will help people save space when recycling. The can crusher must be safe to operate, easy to use, and effective. The cans must be crushed to less than one-half their original volume. The materials used to build the can crusher must be some that have been recycled.

Materials/Equipment

* wood (plywood, etc.), scrap pieces 2" x 4", 2" x 6"
* steel or aluminum scrap, 1" x 1" x 1/8"
* miscellaneous fasteners
* abrasive paper
* band or scroll saw
* belt or disk sander
* drill press and drill bit set
* hacksaw
* screwdriver
* wrenches
* file
* computer with CAD software (optional)

SAFETY FIRST
* Your teacher will show you how to use the machines in your technology lab. Ask questions about anything that is not clear to you.
* Follow the safety rules listed on pages 42-43 and the specific rules provided by your teacher for tools and machines.

Procedure

1. In this activity, you will build your own can crusher. First you must decide on a design. Think about how you would like your crusher to work—by stepping on it? by using your hands?

2. Your design must consider safety. You should avoid pinch points, where fingers could be caught during use. Some possible designs are shown in Fig. A.

3. As you put the finishing touches on your design, list the materials that you will need. Your list might look like this:

Quantity	Part Name	Material
2	linkage	1" x 1" x 1/8" steel

4. After you have finished your design and list of materials, have your teacher check your plans.

5. Follow the safety rules for each machine you use in making your crusher. Remember to ask the teacher for help if you are not sure how to do something.

6. Assemble your crusher, and check it for splinters or sharp edges that could cause injury. Use abrasive paper or a file to remove any sharp edges.

7. Test your crusher with an empty soda can.

Evaluation

1. How can your design be improved? Discuss the changes with your teacher.

2. Write a set of instructions for the safe use of your product. Attach the instructions to your crusher before you take it home.

3. **Going Beyond.** Design a machine that could safely crush hundreds of cans per hour.

4. **Going Beyond.** Research machines that are designed to crush materials such as rock or coal. Make a sketch that illustrates how the machines operate.

SAFETY FIRST

It is important that you wear safety glasses and follow the general safety rules for proper operation of any hand tools or machines.

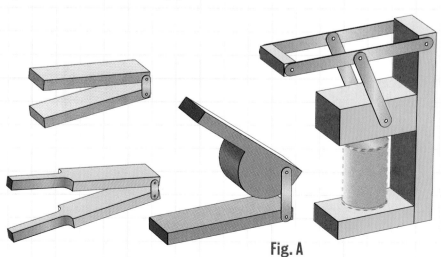

Fig. A

CHAPTER SUMMARY

SECTION 1

- Resources can be anything used in the production of a product.
- People, machines, information, raw materials, energy, money, and time are categories of resources used in manufacturing.
- Renewable resources can be replaced; nonrenewable resources cannot be replaced.

SECTION 2

- Choosing a resource depends on its availability, usefulness, and cost.
- Recycling materials and conserving energy are just two ways you can help save resources and protect the environment.

SECTION 3

- Everyday products are manufactured in factories using various materials and machines.
- Manufacturing processes are operations that help change a raw material into a finished product.
- Major manufacturing processes include machining, forming, fastening, and finishing.

REVIEW QUESTIONS

1. Name some advantages to conserving resources.
2. Name the seven basic groups of resources.
3. Name the four categories of manufacturing processes.
4. What is quality control?
5. Name one example of a renewable resource.

CRITICAL THINKING

1. If you had to produce a product in a short time, what kinds of resources would you try to use?
2. Contact an environmental protection agency near you. Find out what special problems affect your town and how you can help solve them.
3. Interview a business person. Find out how that person is a resource person for you or others.
4. Create your own logo for recycling.
5. Create a commercial with video and sound effects or create a flyer to use in your school.

ACTIVITIES

CROSS-CURRICULAR EXTENSIONS

1. SCIENCE Contact the company that recycles aluminum in your community. Ask them to describe what happens to the aluminum cans that you recycle.

2. MATHEMATICS Design a method for keeping track of the number of aluminum cans recycled at your school.

3. COMMUNICATION Make safety posters or ads for the machines you use in class. *Optional:* Use a video camera, computers, or darkroom processes for special effects.

EXPLORING CAREERS

To produce high-tech products requires the input and skills of people in a variety of jobs. These people are on the leading edge of technology and are constantly striving to create new products or improve existing ones. Here are two of the careers involved in making things.

Computer Engineer Computer engineers work as part of a team to develop new hardware and software. Most specialize in the design and testing of computer hardware or the research and design of software programs. They must have strong communication skills and be willing to keep up with advances in technology.

Computer Chip Technician Manufacturing a computer chip involves several hundred steps that are controlled by computer chip manufacturing technicians. They make sure that the expensive equipment used in manufacturing chips continues to run. Most technicians have training in electronics, with a solid background in mathematics and science.

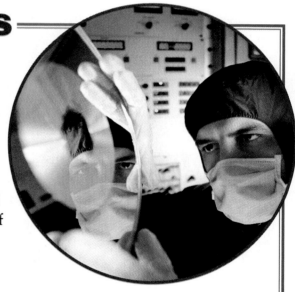

ACTIVITY

List the things you took to school today, including clothing, backpacks, and purses. What materials (cotton, plastic, wood, etc.) were used to make each product?

How Things Work

What Is a System?

THINGS TO EXPLORE

- Tell what a system and a subsystem are and give examples.
- List and describe the four parts of a general systems model.
- Identify five systems used in technology.
- Explain how to troubleshoot problems.

TechnoTerms
feedback
input
output
process
subsystem
system

Most of us don't know how the machines we commonly use work. If a stereo, television, or bicycle breaks down, we often have no idea about what could be wrong or what to do about it. Fig. 8-1. The fact is, you don't have to be a rocket scientist to understand the basic operation of most machines and other devices. Even the most complex machine can be learned about as a system.

The Systems Model

A **system** is a combination of parts that work together as a whole. To understand systems, a general model can be used. It has four parts: input, process, output, and feedback. Fig. 8-2.

INFOLINK

See Chapter 4 for more information about computers.

Inputs are things that are *put into* a system. For example, in a fast-food restaurant, first you place your order. This is the input. The next step is the **process**—what is done with inputs. In a restaurant, the workers prepare the food you ordered. The final result is called the **output**—what comes *out of* the process. In our restaurant example, the output is your meal. If the system is working right, the output will be what you ordered. **Feedback** is information about the output. If the meal you received was not what you ordered, you might send it back. That's feedback.

◀ OPPOSITE Interlocking gears help transfer power in many machines.

Fig. 8-1. Technology can be frustrating when it doesn't work. Has a technological device ever given you trouble? Describe your experience.

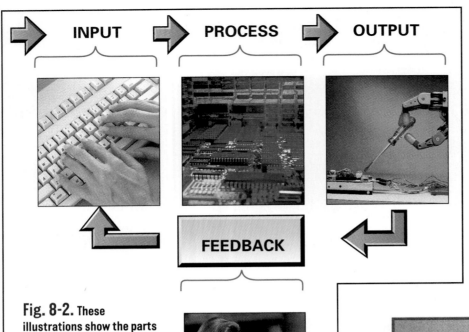

INPUT → PROCESS → OUTPUT

FEEDBACK

Fig. 8-2. These illustrations show the parts of a system. What is going on in each picture?

Fig. 8-3. This technician is using an oscilloscope to find problems in computer components.

Complex systems can be broken down even further into smaller **subsystems**. A bicycle, for example, is made of subsystems such as brakes and steering. Knowing about subsystems makes it even easier to understand how things work.

INFOLINK

The problem-solving strategy introduced in Chapter 1 showed you how to look at things *systematically*. Systematically means simply that you are using a system.

Troubleshooting System Problems

Do you know what *troubleshooting* is? It is trying to find the problem in a system. Fig. 8-3. Did you know that doctors troubleshoot? When you are sick, doctors investigate your different body systems to locate the problem. They look for *symptoms*, or signs, that might give them a clue to what's wrong.

Fig. 8-4. Autotechnicians often use electronic tools to diagnose problems. What kinds of symptoms do you think they look for?

Part of your job in troubleshooting technology system problems will be to look for symptoms. Instructions that come with some products often include troubleshooting charts to help you find and fix problems.

Systems That Make Things Work

Technology usually depends upon five basic systems. They are mechanical, electrical, fluid, thermal (heat), and chemical systems. These five basic systems can be used independently or in combination to make something work. A car, for example, is a complex machine made up of all five types of systems.

- **Mechanical:** Door latches, fan belts, pulleys, gears
- **Fluid:** Water pump, shock absorbers, hydraulic brakes
- **Electrical:** Battery, lights, radio, ignition
- **Thermal:** Radiator, air conditioner, heater
- **Chemical:** Fuel, battery fluid, antifreeze

When a car isn't working right, the problem is found by carefully checking each system related to the problem. Fig. 8-4.

SECTION 1
TechCHECK

1. Define *system* and *subsystem*.
2. List the five basic systems used in technology.
3. What is *troubleshooting*?
4. What are the four parts of the general systems model?
5. **Apply Your Knowledge.** Dissect a ballpoint pen and identify the systems used to make it work. Identify the inputs and outputs.

Exploring Mechanical Systems

THINGS TO EXPLORE

- Identify mechanical parts used in everyday machines.
- Give examples of mechanical subsystems.
- Use levers and links to build a mechanical device for a robot.

TechnoTerms
force
resistance
standard size

Most machines and products contain at least a few mechanical parts. Can you identify some of the mechanical parts in the bicycle in Fig. 8-5?

Standard Sizes

Many mechanical parts, such as screws and other fasteners, are made in **standard sizes**. That means they are interchangeable and easy to replace. You can look in a parts catalog and find the right combination of parts needed to do a certain job.

Fig. 8-5. A bicycle has several mechanical parts. Identify as many as you can.

Mechanical Subsystems

Levers, springs, nuts and bolts, screws, belts and pulleys are just a few of the many mechanical parts that are used in everyday machines. In designing mechanical systems, you will need to know about mechanical subsystems such as levers and linkages, gears, and chain and belt drives. Fig. 8-6.

> ### Fig. 8-6. MECHANICAL SUBSYSTEMS

- **Levers and linkages.** Levers help people to multiply their muscle strength and move heavy loads. How do levers work? When you use a lever to push or pull something, you are applying a **force**. The pivot point of a lever is called the *fulcrum*. The load you are trying to move is called the **resistance.** Machines sometimes use levers called *linkages* or *cranks*.

- **Gears.** Gears transmit forces from one part to another. They can be used to change the speed or direction of spinning parts. The speed at which a gear turns is measured in *RPMs,* or *revolutions per minute*. Gears are made in different shapes depending on how they are to be used. One type used in a car's steering mechanism is called a *rack and pinion gear*.

- **Chain and belt drives.** When forces have to be transmitted over a longer distance than gears can handle easily, a chain or belt is often used. Look at a bicycle chain. It rides on a toothed wheel called a *sprocket*. The chain transmits the force applied to the pedals to the back wheel through this chain and sprocket. Belts can do the same thing. You may have seen a belt drive on a washing machine or on a cooling fan in a car.

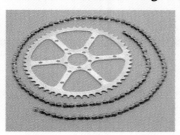

Pulleys grip the sides of the belt just as a chain's holes mesh with the teeth in a sprocket. Belts are lighter and run more quietly than chains, but they can slip more easily.

(Continued on next page)

Fig. 8-6. MECHANICAL SUBSYSTEMS (Cont'd.)

- **Cams.** A mechanical part that changes rotational (turning) motion into reciprocating (up-and-down) motion is called a *cam*. Cams are used in automobile engines.

- **Flywheels.** A flywheel is a metal wheel that is heavy enough to keep spinning once it is set into motion. A flywheel is used to keep engines running in such devices as lawn mowers.

- **Springs.** Mechanical energy can be stored in springs. Springs come in many sizes, strengths, and shapes. Take a look at the spring inside a ballpoint pen. Compare it with the spring in a car's suspension system. What do they have in common?

SECTION 2
TechCHECK

1. Name some mechanical parts that you can buy in standard sizes.
2. Why do machines use springs?
3. Give three examples of mechanical subsystems.
4. Apply Your Knowledge. Check out the mechanical parts in an old computer keyboard or typewriter. Describe the parts in the system.

Dissecting a Machine

Be sure to fill out your **TechNotes** and place them in your portfolio.

Real World Connection

Most machines are a combination of many parts from mechanical, electrical, fluid, thermal, and chemical energy systems. The parts are designed to work together to make a complete system. Even the most complex machines can be divided into subsystems. In this activity you will dissect a mechanical system. Fig. A.

Design Brief

Dissect a "junk" machine and group the parts into the five energy systems. Ask your teacher to approve your junk item before you begin. You will need to use hand tools to disassemble (take apart) your machine. Identify and save all of the parts for possible future use.

Materials/Equipment

- "junk" machine, such as a toaster, alarm clock, TV, video game player, toy, lawn mower, mixer, drill, blender, VCR, lamp, heater, stereo, record player, exercise machine, or bicycle
- small bags or paper cups
- various hand tools such as screwdrivers, nut drivers, Allen wrenches, adjustable wrenches, socket sets

SAFETY FIRST
Follow the safety rules listed on pages 42-43 and the specific rules provided by your teacher for tools and machines.

Fig. A

Procedure

1. In this activity, you will be working individually. Ask your parents, friends, or relatives if they have a "junk" machine that you could have. Please make it clear that you will not be repairing the item, and it will not be returned to them. Ask your teacher about storing any large junk items before you bring them to school.

2. Plan how you will dissect your machine so subsystems can be saved for future use. You should look for some of the following parts that can be reused: speakers, motors, batteries, pulleys, belts, gears, and fasteners such as nuts and bolts.

3. Take apart your machine using the right tools for each part. Ask your teacher for help if you can't loosen or remove a part.

4. Put the pieces into separate bags or cups according to their system. Identify as many of the parts as you can by name. Ask your teacher for help with parts that you can't identify.

5. Save all of your parts for future use. Explain to the class how you took the machine apart. Name all of the parts for them.

Evaluation

1. Into which of the five systems did the most parts of your junk item fall?

2. Make a list of all of the parts you found in your junk item.

3. **Going Beyond.** Make a display showing several parts from a dissected machine. Fig. B.

SAFETY FIRST

Ask your instructor to inspect your junk item and warn you of any dangerous parts you should avoid. Specifically, you should be careful with old TVs or computer monitors. They might have an electronic capacitor that can hold a charge even if the machine has been unplugged for many hours. Be careful! Ask your teacher for help.

Fig. B

Leapin' Links 'n' Levers

Real World Connection

Levers have been used throughout history to help move heavy objects. When you pull out a nail with a hammer, crack open a nut with a nutcracker, or tighten a bolt with a pair of pliers, you're using levers.

In this activity, you will make a lever-and-linkage device that will work like a mechanical hand on the end of a robot arm. Fig. A.

Design Brief

Build a robot gripper that can be used to pick up a pencil from a desktop. The gripper will be used in the last activity for Section 4 of this chapter. To make it you can use any type of lever or linkage you can think of.

Materials/Equipment

- acrylic plastic
- wood
- wood screws
- drill press or power drill and drill bits
- scroll saw or band saw
- screwdriver
- ruler

Be sure to fill out your TechNotes and place them in your portfolio.

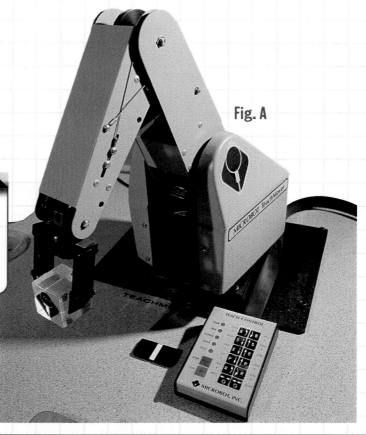

Fig. A

SAFETY FIRST
Follow the safety rules listed on pages 42-43. Follow safety rules for using saws and drills as directed by your teacher.

(Continued on next page)

Procedure

1. Work in groups of four. Each group will work together to make one robot gripper. Each member of the group should brainstorm at least four different design ideas. Each member should make four sketches on paper or use graphics or CAD software on a computer.

2. As a group, choose one design. Refine the design and have it approved by your teacher. If your idea is original and workable, your teacher will issue a patent to your group. Then no one else in the class will be able to use your idea without your permission.

INFOLINK

See Chapter 6 for more information about patents.

3. Safely and carefully cut the materials you will need to make your gripper. Have each person in the group make a different part to save time.

4. Assemble and test your gripper. Later you will attach your gripper to the end of a robot arm.

Evaluation

1. Name a lever that is a part of most cars.

2. What mechanical parts (or simple machines) are of help in making a robot gripper?

3. How could you design a robot to pick up small delicate parts such as a watch crystal?

4. **Going Beyond.** As you use tools and machines to make your gripper, make a list of all the levers and linkages you notice.

5. **Going Beyond.** Find out about the Space Shuttle's Remote Manipulator System (RMS). How do astronauts use this "arm"? Fig. B.

INFOLINK

See Chapter 18 for more information about the RMS.

Fig. B

Exploring Electrical & Electronic Systems

THINGS TO EXPLORE

- Define electricity and electronics.
- Identify conductors, insulators, semiconductors, and superconductors and give examples.
- Explain how to properly set up series, parallel, and series-parallel circuits using electronic components.
- Make an electrical circuit.

TechnoTerms
circuit
component
electricity
electronics

Electricity is the flow of electrons (small, negatively charged parts of an atom) through a material. **Electronics** is a part of technology concerned with the movement of these electrons through conductors, insulators, semiconductors, and superconductors. Fig. 8-7.

- **Conductors.** A conductor is a material that lets electrons pass easily from one atom to another. Most metals are good conductors. Can you think of other materials that are good conductors?

- **Insulators.** Materials that do not allow electrons to flow easily are called insulators. Examples of insulators include plastic, rubber, and glass. Can you think of other materials that are good insulators?

- **Semiconductors.** Semiconductors are important to electronics because they conduct electricity only under certain conditions. Silicon is a commonly used semiconductor. It is used to make integrated circuits (ICs).

- **Superconductors.** Superconductors can conduct electricity perfectly. Even some of the best ordinary conductors, such as copper, *resist* (hold back) the flow of electrons a little. Superconductors have no resistance at all.

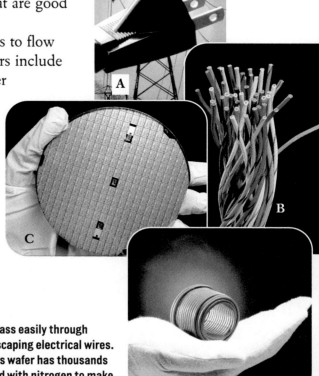

Fig. 8-7. (A) Conductors allow electricity to pass easily through them. (B) Insulators prevent electricity from escaping electrical wires. (C) Semiconductors are used in electronics. This wafer has thousands of semiconductors on it. (D) A wire can be cooled with nitrogen to make it a superconductor.

TechnoFact

FLOATING TRAINS
Experiments are being conducted to build trains that float on a magnetic field instead of rolling on a track. These *magnetic levitation*, or *maglev*, systems would allow trains to travel over 300 miles per hour! What makes maglevs possible? Superconductors.

INFOLINK

See Chapter 20 for more information about maglev trains.

Electrical Circuits

To be useful, conductors, insulators, semiconductors, and even superconductors must be connected in some way into a circuit. A **circuit** is the complete path along which electrons flow. Electronic parts, or **components**, are commonly connected in three basic types of circuits: series, parallel, and series-parallel. Fig. 8-8.

When components are connected in line, one after the other, they are in series. *Series* circuits are very common. One disadvantage to them is that if one part in the series fails, the entire circuit fails. Some holiday lights are wired in series, so if one light burns out the whole string goes out.

Parallel circuits are arranged so that other parts continue to work even if one part fails. Parallel circuits are common in all electronic products such as televisions, radios, and stereos.

The third type of circuit consists of a combination of series and parallel circuits. This type is called a *series-parallel* circuit.

In a *short circuit*, electrons bypass the proper path. For example, two uninsulated wires may touch in a way that causes the electrons to pass through the wires rather than the components. This causes too much current to flow in the circuit, which can be dangerous. If the excess current is high enough, it can cause a fire or even death.

SCIENCE CONNECTION

Superconductivity

What is superconductivity? You might think it is something brand new, but scientists have known about superconductivity since 1911.

Superconductivity was discovered by Dutch physicist Heike Kamerlingh Onnes. He was researching the effects of extremely cold temperatures on different metals. He discovered that mercury lost all resistance to the flow of electricity when cooled to about 4 Kelvin (K) (about -452° F or -269° C).

To understand how important his discovery is you need to think about how electricity works. Even the best conductors are not perfect because some electrical energy is lost to resistance. Before Onnes' discovery, there was no way to eliminate resistance. Superconductors were the answer. They let electricity flow with no resistance at all.

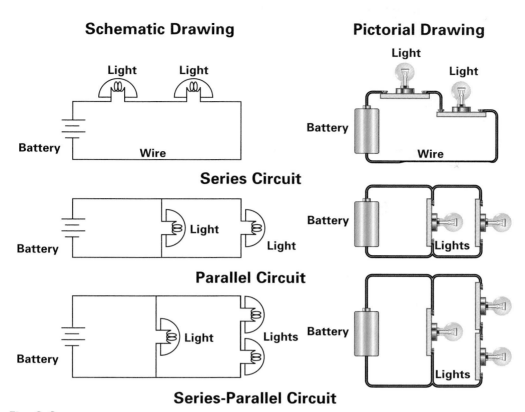

Schematic Drawing

Series Circuit

Parallel Circuit

Series-Parallel Circuit

Pictorial Drawing

Fig. 8-8. Components are connected in a circuit in one of three basic methods. What would happen to the second light in the series circuit if the first light went out?

Imagine all the uses for such a discovery! Superconductors can be used to save energy and money in power systems. Generators wound with superconductor wire instead of copper wire can generate the same amount of electricity with smaller equipment and less work.

The magnet to the left is floating above a superconductor that was cooled with liquid nitrogen. The magnetic field of the magnet produces an opposing magnetic field in the superconductor. The result is magnetic levitation.

Scientists see superconductivity as a very promising, exciting application of technology that will help us save energy and provide better ways to make things work.

ACTIVITY

Design a futuristic device based on superconductivity or write a story about how superconductivity could benefit your life.

Electronic Components

The components that are put into circuits are designed to control the flow of electrons. Some you might use to make electrical circuits include resistors, capacitors, diodes, and transistors. Fig. 8-9.

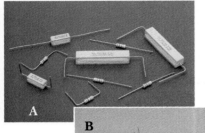

- **Resistors.** Resistors resist the flow of electricity. They come in many sizes and shapes. The most common types are coded with colored stripes to show their resistance level.
- **Capacitors.** Components that temporarily hold an electrical charge are called capacitors. Televisions and computer monitors, for example, have high-voltage capacitors that can hold a charge for many hours after the power is turned off. However, they can be dangerous because, even though the power is off, the capacitor holds enough charge to cause injury.
- **Diodes.** In a diode, electrons flow in only one direction. Diodes are marked in some way to show the direction of flow. They come in a variety of sizes, depending on the amount of current that will flow through the circuit.
- **Transistors.** Transistors are made with a semiconducting material. They have three wires called *leads*. A very small current or voltage applied to one lead can control a large amount of electric current at the other two leads. Transistors are commonly used as switches or to *amplify* (make larger) electrical circuits. Transistors started a revolution in technology because they made it possible for electronic products to be smaller, lighter, more reliable, and less expensive.

Fig. 8-9. Electronic components include (A) resistors, (B) capacitors, (C) diodes, and (D) transistors.

SECTION 3
TechCHECK

1. Define *electricity* and *electronics.*
2. What is the difference between a semiconductor and a superconductor?
3. How is a parallel circuit different from a series circuit?
4. Apply Your Knowledge. Make a display of electronic components found when you dissect an electronic device such as a radio.

Motor Motion Magic

Real World Connection

A simple battery-operated motor can be used in many ways. Small motors that run on 1 1/2 to 12 DC volts can be found in many products. Fig. A. Portable tape players, toys, and even full-size cars use battery-operated motors. You can probably think of many more examples.

In this activity you will learn to reverse a small motor using a type of switch called a double-pole, double-throw (DPDT) switch. This is one of many different types of switches used to control the flow of electrons.

Design Brief

Make an electric circuit using a DPDT switch that will reverse the direction of a motor.

Materials/Equipment

- small DC motor (1 1/2-12 volts), such as from the machine dissection activity for Section 2
- hookup wire
- power supply (0-12 volts DC)
- wire strippers

Be sure to fill out your TechNotes and place them in your portfolio.

SAFETY FIRST

- Follow the safety rules listed on pages 42-43 and the specific rules provided by your teacher for tools and machines.
- When you are experimenting with electricity, use low-voltage batteries to avoid the possibility of a painful or life-threatening electrical shock.

Fig. A

Procedure

1. Work in pairs. Use a small electric motor from the junk machine dissection activity, or get one from your teacher. You will be connecting your motor in a series circuit with a switch to control its direction.

2. Electronic circuits are drawn using a set of symbols to represent real parts. These drawings are called *schematic diagrams*, or simply *schematics*. Connect the circuit according to the schematic in Fig. B.

3. Be sure to set the power supply to zero before you connect the motor. Have your teacher check your circuit.

4. Turn on the power supply. Check your circuit to see if it will reverse the direction of the motor when you flip the switch.

Evaluation

1. Did your circuit work the first time you tried it? If not, what was wrong?

2. How could you use a circuit with a DPDT switch and motor in a crane or winch?

3. **Going Beyond.** Design an electrical maze using wires, batteries, and a buzzer.

Schematic Drawing

Fig. B. The symbols on the schematic represent the items pictured below.

Exploring Fluid Systems

THINGS TO EXPLORE

- Tell what a fluid is and give examples.
- Explain how hydraulic and pneumatic systems operate.
- Design and build a robotic arm powered by hydraulics.

TechnoTerms
air compressor
hydraulic pump
hydraulic system
piston
pneumatic system

When someone says "fluid," what do you think of first? Most people think of water or some other liquid. The fact is that fluids can be either liquids or gases. Both air and water are examples of fluids.

Fluid Systems

Fluid systems are one of two types: hydraulic or pneumatic. **Hydraulic systems** operate using a liquid, usually oil. **Pneumatic systems** operate with a gas, usually compressed air. Fig. 8-10. These fluid systems apply pressure on the fluid to do work. Since it is harder to compress a liquid than a gas, hydraulic systems apply more pressure and can lift heavier loads or stop heavier objects.

Fluid systems need a source of power. In hydraulic systems, a **hydraulic pump** is used. Pneumatic systems use an **air compressor** (a machine that squeezes, or compresses, air) as a power source.

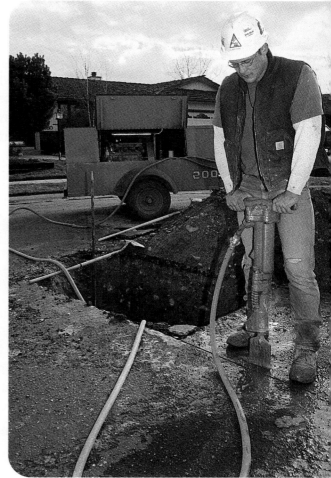

Fig. 8-10. The compressed air that powers this pneumatic hammer is created in the generator in the background. How does the air travel to the hammer?

Fig. 8-11. Hydraulic pistons are used in locomotives. Why do you think they are so large?

Cylinders and Pistons

A common component in both hydraulic and pneumatic systems is a cylinder containing a **piston**. As air or oil is pumped under pressure into the cylinder, it makes the piston move. Fig. 8-11.

Cylinders commonly come in two types, single-acting and double-acting. In single-acting cylinders, pressure is applied to the piston in one direction only. Automobile shock absorbers and hydraulic door closers contain single-acting cylinders. In double-acting cylinders, pressure is applied in either direction. This type of cylinder can push as well as pull. You may have seen hydraulic cylinders working on a backhoe or a dumptruck. Which kind of cylinder do you think is at work in the backhoe? Which is used in the dumptruck?

SECTION 4
✓ TechCHECK

1. What is a fluid?
2. Explain how pneumatic and hydraulic systems operate.
3. Which fluid system would you use to lift heavy loads? Why?
4. **Apply Your Knowledge.** Research fluid systems used in airplanes. Make a chart to show your findings.

ACTION ACTIVITY

Making a Fluid-Powered Robot

Be sure to fill out your TechNotes and place them in your portfolio.

Real World Connection

Real robots can do many different jobs. Many are operated using hydraulic or pneumatic power. In this activity, you will have a chance to put fluid systems together to make a hydraulic robot. Real robots use computers to control their actions. In this activity, you will act as the computer-controller for your robot.

Design Brief

Design, build, test, and refine a robot arm that can pick up a pencil from the surface of a table by means of fluid power. The robot must be able to move the pencil to another part of the table and release it. During testing, you may touch only the controls of your robot, not the gripper or the arm. The controls must be at least 12 inches from the gripper.

Materials/Equipment

- acrylic plastic or foam core poster board
- wood, wood screws
- syringes
- plastic tubing
- water
- food coloring
- drill press or power drill, drill bits
- scroll saw or band saw
- screwdriver
- ruler

- usable parts from the machine dissection activity in Section 2 (optional)
- gripper from the links and levers activity in Section 2

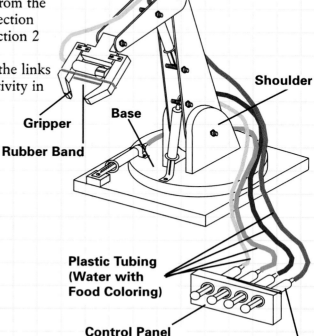

Elbow

Shoulder

Gripper

Base

Rubber Band

Plastic Tubing
(Water with
Food Coloring)

Control Panel

Plastic Syringes

Fig. A

SAFETY FIRST

Follow the safety rules listed on pages 42-43 and the specific rules provided by your teacher for tools and machines. Do not use too much force to push the plunger in the syringes. If you do, the tube might come loose and the colored water might leak out.

(Continued on next page)

ACTION ACTIVITY

Procedure

1. In this activity you need to remember how to solve a problem systematically. Do you remember the problem-solving steps?

- Identify or define the problem.
- Gather ideas or solutions.
- Use your best judgment to pick the best solution.
- Test your idea.
- Evaluate your idea, and refine it until it is the best solution to the problem.

2. Work in groups of four or five. Design your hydraulic robot to work with plastic syringes and tubing containing colored water. Fig. A.

3. Brainstorm different ideas. Make sketches or use graphics or CAD software on a computer.

4. As a group, discuss how the robot will operate and what materials will be best to use.

5. Make a list of the materials needed, their size, and quantity. This list is called a *bill of materials*. Fig. B.

6. Following your teacher's instructions, carefully cut the materials for each part of your robot. Have everyone in the group work on a different part to save time.

7. Assemble and test your robot arm. Remember that very few new ideas work perfectly the first time. The last step in solving a problem is to evaluate and refine the solution as needed.

Evaluation

1. List the steps in problem solving. Next to each step, write a description of what you did to solve the robot problem.

2. What is a bill of materials?

3. How are real robots controlled?

4. **Going Beyond.** Try using your robot arm with just air in the syringes and tubing. Now you have a pneumatic or air-powered robot. How does it work? Which system works better? Why?

5. **Going Beyond.** Try to animate your robot using animation software on a computer.

6. **Going Beyond.** Measure the maximum range of your robot's movements. This is called a robot's *work envelope*. Make a sketch with dimensions showing the work envelope.

Quantity	Item Name	Description
3	Syringe	50 cc
2	Wood Screw	1 1/2" #8 Flat Head
3 ft.	Tubing, plastic	1/8" I.D. (Inside diameter)

Fig. B

Exploring Chemical and Thermal Systems

THINGS TO EXPLORE

- Identify various chemical systems.
- Explain the difference between dry-cell and wet-cell batteries.
- Tell what a thermal system is and how it works.
- Design and build a hot wire cutter.

TechnoTerms
electrolyte
petrochemical
petroleum
refining
thermocouple

Chemical systems are those based on—you guessed it!—chemicals. Thermal systems have to do with heat. Both are important to technology.

Chemical Systems

Have you ever heard of sodium tallowate, stearic acid, ammonium chloride, or methylchloroisothiazolinone? Read the label on everyday products such as toothpaste, soap, or shampoo and you might find these names as well as some that sound even stranger. Chemicals play an important part in technology.

Batteries Did you know that a chemical system enables you to use a portable stereo? Chemical reactions produce electricity in batteries, providing a portable electrical source. The chemicals used to produce voltage are called **electrolytes**. Small batteries such as those used in a flashlight require a paste of chemicals (*dry cell*). Fig. 8-12. Larger batteries, such as those used to start a car, use a liquid (*wet cell*).

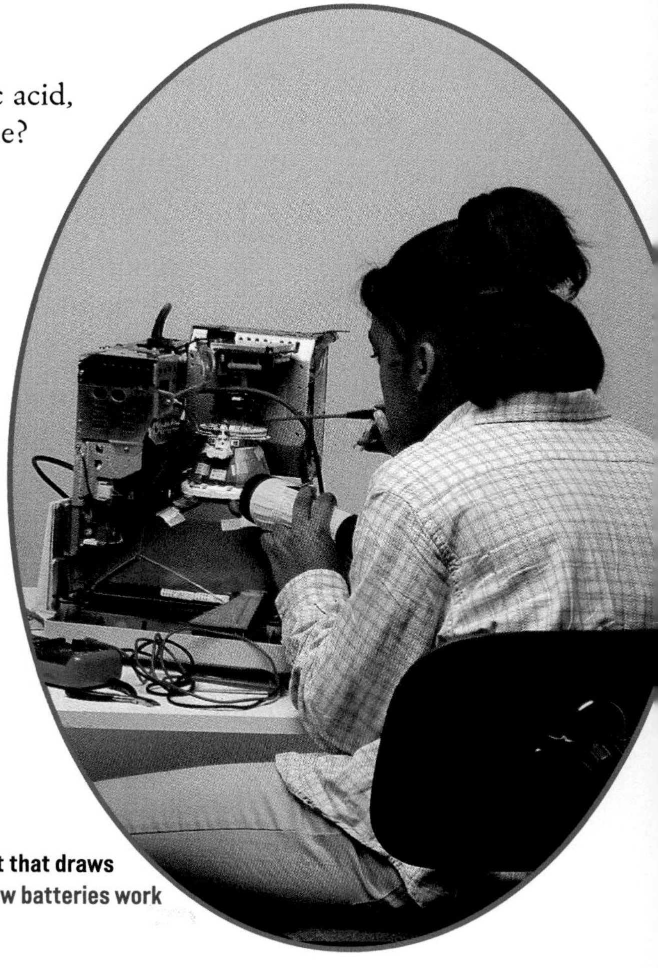

Fig. 8-12. This teen is using a flashlight that draws power from dry-cell batteries. Find out how batteries work and report your findings to the class.

Fig. 8-13. Petroleum comes from wells like this one off the coast of California. Why do you suppose a small boat is used to travel between the rig and any visiting large ships?

Batteries are only one example of a chemical system used for energy.

Chemical systems produce most of the electrical energy we use at home, in school, and in factories. For example, the chemical energy stored in coal is changed into heat (thermal) energy and then into electrical energy.

Petroleum Products The gasoline we use in cars or trucks is a product of very specialized chemical systems based on **petroleum**, or oil. Fig. 8-13. The oil found in the ground is called *crude oil*. Crude oil is not usable in its natural form and must be changed into other products by **refining**.

Did you know that some plastics and medicines come from oil? Products produced from oil are called **petrochemicals**.

Thermal Systems

What do thermostats, thermal underwear, and thermos bottles have in common? If you said "heat," you're right! Thermal systems control heat—the temperature of your toaster as well as the temperature of your automobile engine.

Devices that control temperature often contain a bimetallic strip. *Bimetallic* means it is made of two different metals. As metals heat up, they expand (get larger), but at different rates. One metal in the bimetallic strip expands more than the other. This difference causes the bimetallic strip to bend to one side. As it bends, it triggers an electric circuit to start or stop heating or cooling equipment.

Another interesting device in thermal systems is a **thermocouple**. A thermocouple is like an electric thermometer. Thermocouples are connected to gauges in control rooms away from areas affected by extreme heat or cold. Some are used to read very high temperatures in dangerous areas, such as the core of a nuclear reactor.

SECTION 5
TechCHECK

1. Name two chemical systems that affect your life.
2. What is the difference between a wet-cell and a dry-cell battery?
3. What is a thermal system?
4. **Apply Your Knowledge.** Write down the contents of several household products, such as toothpaste and shampoo.

ACTION ACTIVITY

Putting a Thermal System to Work

Be sure to fill out your TechNotes and place them in your portfolio.

Real World Connection

In long-distance power lines, resistance to electron flow is not something we want. However, resistance is not always bad. Sometimes it is used to create thermal energy. In this activity, you will use resistance to create thermal energy to cut plastic foam.

The hot wire cutter you will make is an example of a simple series circuit. Electrons flowing through *Nichrome wire* make it hot. Electric heaters and toasters use this same type of wire to heat your home or to make toast.

Design Brief

Design and build an electric hot wire cutter that you can use safely and accurately to cut shapes out of plastic foam. Fig. A.

Materials/Equipment

- Nichrome wire (28 gauge x 12" long)
- prefinished particleboard, 3/4" x 12" x 24"
- steel rod, 1/4" diameter x 18"
- DC power supply, adjustable 0-12 volts
- hookup wire
- wood glue
- Styrofoam plastic foam
- wood block, 1 1/2" x 3 1/2" x 3 1/2"
- stick-on plastic "feet"
- drill press or power drill with 1/4" drill bit
- wire cutter or stripper
- hacksaw
- band saw
- scroll saw
- 6" C-clamp

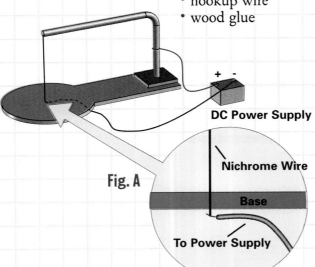

+ -

DC Power Supply

Nichrome Wire

Base

To Power Supply

Fig. A

SAFETY FIRST
Follow the safety rules listed on pages 42-43 and the specific rules provided by your teacher for tools and machines.

Procedure

Part 1 · Designing the Foam Cutter

1. Work in groups of five or six. Design a base for your hot wire cutter. Your base should be sturdy and have a wide area for holding the plastic foam flat. You might design your base using graphics or CAD software on a computer.

2. Lay out your design on particleboard. With your teacher's help, carefully cut the particleboard using a scroll saw for sharp curves and a band saw for straight lines or wide curves.

3. Cut a 2" x 4" to the 3 1/2" length needed, using a hand saw or a band saw.

4. Glue and clamp the wood block to the back of the base to help support the steel rod. Let the glue set.

5. With your teacher's help, use a drill press to drill a 1/4"-diameter hole in the center of the block and press through the base. If you are using a portable power drill, be sure to hold the drill in a vertical position.

6. Place an 18"-long piece of 1/4" steel rod in a vise. Carefully bend the rod to almost a 90° angle. You may have trouble bending the rod. Try to think of a safe way to make it easier for you. Ask your teacher for help with your idea. (When you finish, the Nichrome wire will pull the rod into a 90° angle. The steel rod works like a spring to keep the Nichrome wire tight.)

7. Locate a point in the center of your cutter base for the Nichrome wire to pass through. Drill a 1/4"-diameter hole.

8. Cut and strip the ends of the hookup wire needed to attach your cutter to the power supply.

9. Assemble the cutter so the Nichrome wire is pulled tight enough to bend the steel rod to 90°.

10. Finish your cutter by putting on the plastic "feet" to help steady it and to protect countertops.

Part 2 · Testing the Hot Wire Cutter

1. Ask your teacher to inspect your work. Be sure to start with the power supply turned off and adjusted to 0 volts. Connect the hookup wires to the positive (+) and negative (-) terminals. It doesn't matter which wire is positive or negative.

SAFETY FIRST

The Nichrome wire will become very hot. Do not touch it or let it touch anything other than the plastic foam you are cutting. Adequate ventilation must be provided to remove the fumes produced when plastic foam is cut. Do not use the hot wire cutter in closed spaces, without teacher supervision, or to cut anything but foam.

2. Test your hot wire cutter by holding a piece of Styrofoam plastic foam against the wire. Slowly turn up the voltage until the Nichrome wire can melt the foam. Try cutting various shapes.

Part 3 · Improving Your Hot Wire Cutter

1. Design and test a method to accurately cut the following shapes: long, thin rectangles; perfect circles; cones; other geometric shapes.

2. Sketch your methods on paper.

Evaluation

1. What happens if you start to cut into Styrofoam plastic foam and stop before the cut is finished?

2. Why is it important to watch what you are doing when using the hot wire cutter?

3. **Going Beyond.** List three electrical appliances you might have at home that have Nichrome wires in them that get hot.

8 REVIEW &

CHAPTER SUMMARY

SECTION 1

• A system is a combination of parts that work together as a whole.

• The systems model includes input, process, output, and feedback.

• The five basic systems used in technology are mechanical, electrical, fluid, thermal, and chemical.

• Trying to find the problem in a system is called troubleshooting.

SECTION 2

• Mechanical systems often include levers, gears, chains, cams, flywheels, springs, and other parts.

SECTION 3

• Electronics involves the movement of electrons through conductors, insulators, semiconductors, and superconductors.

• Electronic components such as diodes, capacitors, transistors, and resistors are connected in three basic circuits: series, parallel, and series-parallel.

SECTION 4

• Fluid systems are either hydraulic or pneumatic.

SECTION 5

• Chemical systems include batteries and petroleum products.

• Thermal systems control the temperature of things.

REVIEW QUESTIONS

1. What systems can you find in a flashlight? Do they work together or independently?

2. What is the difference between a single-acting cylinder and a double-acting cylinder?

3. What is a superconductor?

4. List three materials that are insulators and three that are conductors.

5. Why is it important that mechanical parts be made in standard sizes?

6. What parts of a car would be considered thermal systems?

CRITICAL THINKING

1. Name a situation in which pneumatic systems would be better to use than hydraulic systems. Explain why.

2. Research ways to use a computer to control a part of your robot.

3. Make a chart identifying different kinds of mechanical fasteners by name.

4. Design a hand-held foam cutter. Sketch your design on paper, and discuss it with your teacher. Make and test your hand-held cutter with the help of your teacher.

5. Design a flywheel-powered car.

ACTIVITIES

CROSS-CURRICULAR EXTENSIONS

1. MATHEMATICS Figure out the gear ratios for a bicycle.

2. SCIENCE Design and build a test circuit to check for materials that are conductors or insulators.

3. COMMUNICATION Write or e-mail an oil-refining company and ask for information on oil refining and petrochemicals.

EXPLORING CAREERS

Have you ever wanted to take something apart to find out what makes it work? Following are two careers that require that you ask how a product works.

Data Processing Equipment Repairer When a computer crashes, an equipment repairer determines the cause of the problem. These workers install and repair computers and peripheral equipment, such as printers. They use a variety of hand tools to adjust the mechanical parts. Equipment repairers have computer knowledge and a strong interest in fixing things. They must also have good customer-service skills.

Data Retrieval Specialist Many companies have large computer databases where they keep information (data) that is often needed for reports. Data retrieval specialists spend hours and sometimes days searching for a particular piece of information, such as sales figures from a specific day three years ago. They often write computer programs to assist them in locating the information, so programming skills are also needed. This is a good career for someone who is persistent when faced with solving a problem.

ACTIVITY

Find instructions that came with a product that include a troubleshooting chart. Would the chart be helpful in fixing the product? Explain.

CHAPTER 9
Designing Things

Why We Need Measuring Tools

THINGS TO EXPLORE

- Explain why precise measuring tools were developed.
- Compare old ways of measuring with today's measurement tools.
- Tell what a standard is and why standards are used.

TechnoTerms
precision
standard

When you design and make things, measurement tools can help you be as accurate as possible. You've used rulers before, but there are many other measurement tools used in technology. Tools such as stopwatches, thermometers, multimeters (meters that measure electricity), and meter sticks are devices that you will use in designing, building, and testing things. Fig. 9-1.

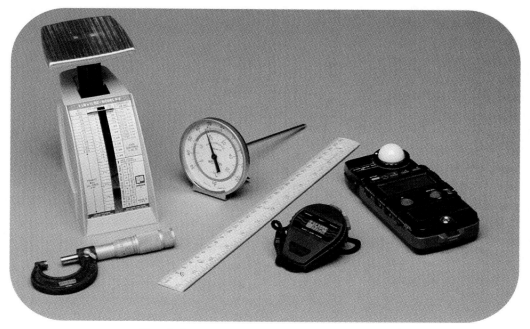

Fig. 9-1. These measuring instruments include (from left to right) a micrometer, scale, cooking thermometer, ruler, stopwatch, and photographic light meter.

OPPOSITE Design is where all products start. Design requires imagination!

Early Measurements

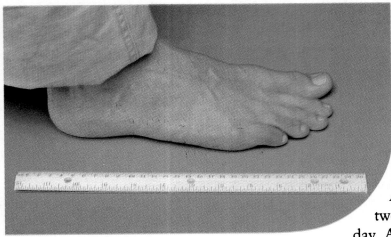

Fig. 9-2. Compare this man's foot to a one-foot ruler. How does your own foot compare?

Early measurements were often based on human dimensions. For instance, a foot was the length of an average man's foot. Fig. 9-2. Other measurements, such as an acre and a furlong, were built around practical activities like plowing. An acre was the amount of land two yoked oxen could plow in one day. A furlong was the distance a horse could pull a plow without stopping to rest. Can you see some problems with these early systems of measuring? They were not very *precise*, or accurate.

Standard Measurements

Scientists in particular needed more precise **standards** (exact units used by everyone) of measurement so they could build on one another's findings. This need for **precision** and standards produced the measurement systems we use today. We use standard measurements for such things as the speed of light and the amount of electricity used by an appliance.

SECTION 1
TechCHECK

1. Why do we need to make precise measurements?
2. How did people measure things in times past?
3. What are standards?
4. Apply Your Knowledge. Measure 50 feet using a rule or yardstick. Then measure the same distance using your foot as the measuring tool. How close is the measurement taken with your foot to that taken with the standard tool?

Using Measurement

THINGS TO EXPLORE

- Explain the difference between the English system of measurement and the metric (SI) system.
- Use the English and metric systems to measure accurately and quickly.
- Tell what measuring to scale is and how it is useful.

TechnoTerms

drawing to scale
estimate
International System
of Units (SI)
metric system

Designing things often means using measurement in different ways. In this section you'll learn about metric measurement, drawing to scale, and estimating.

Metric Measurements

You've probably learned about the **metric system** (base 10) of measurement in mathematics classes. Parts of the metric system make up the **International System of Units** (called **SI**, for the French name, Système International) which is used internationally for trade. SI makes it easier for scientists, engineers, and construction industries all over the world to work with materials and parts that are interchangeable.

Commonly used metric base units are meters, liters, and grams. Fig. 9-3. All metric measures are based on multiples of ten, which makes them easy to calculate with.

Fig. 9-3. This chart shows English and metric equivalents. Using information shown here, convert 6 ounces to grams.

	English Unit	Abbrev.	SI Equivalent	Abbrev.
Distance 1 Inch	Inch Foot Yard	In. or " Ft. or ' Yd.	25.4 Millimeters 304.8 Millimeters .914 Meter	mm mm m
Area 1 Square Inch	Square Inch Square Foot Square Yard	Sq. In. or In.² Sq. Ft. or Ft.² Sq. Yd. or Yd.²	645 Square Millimeters .0929 Square Meter .836 Square Meter	mm² m² m²
Volume 1 Cubic Inch	Cubic Inch Cubic Foot Cubic Yard	Cu. In. or In.³ Cu. Ft. or Ft.³ Cu. Yd. or Yd.³	16,387 Cubic Millimeter .0283 Cubic Meter .7646 Cubic Meter	mm³ m³ m³
Mass 1 Pound	Ounce Pound	Oz. Lb.	28.35 Grams 453.6 Grams	g g

COMPUTER SPEEDS What are nanoseconds, MIPS, and flops? These are standard measurement terms used in computer technology. A nanosecond (one-billionth of a second) is a unit of measurement for computer speed. MIPS stands for *m*illions of *i*nstructions *p*er *s*econd. Flops stands for *fl*oating point *op*erations *p*er *s*econd. Both MIPS and flops are used as a measurement of how fast supercomputers can process information.

At one time, the United States planned to switch completely to the metric system, but it never did. We still use the English system, which is based on the foot, the pound, and the quart. However, many products now show both English and metric measurements.

When you are designing and building, you will find it easier if you stick to one system rather than mixing metric and English units.

Drawing and Making Things to Scale

Sometimes things are too large or too small to draw in their actual size. For instance, you couldn't make a full-sized drawing of a house because paper isn't made large enough! Fig. 9-4. You also would have a difficult time drawing the parts of an integrated circuit in their actual sizes because they are too small. Can you imagine drawing more than 1,000 circuits in a space smaller than a pencil eraser?

To solve this problem, objects are often drawn to scale. **Drawing to scale** means that the object is drawn larger or smaller than it really is, but all its parts are still in the correct proportion.

MATHEMATICS CONNECTION

Measuring Electricity

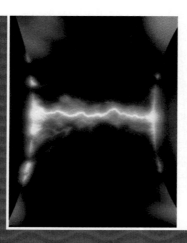

Measuring electricity is important in designing the electrical circuits found in many products. The three basic units used to measure electricity are the volt, the amp, and the ohm.

• **Volt.** Electricity flowing through wires in a circuit is very similar to water flowing through a hose. The voltage that pushes electrons through the wire is similar to the pressure that pushes the water. *Voltage* is a unit of electrical pressure.

• **Amp.** The amount of water flowing through a hose might be measured in gallons. In an electrical system,

Estimating

To **estimate** a measurement means to figure closely but not exactly. For example, you have probably learned about how long an inch or a centimeter is in your mathematics classes. Now you can estimate about how long something is in inches or centimeters. It is important in technology to be able to estimate the size of objects without actually measuring them.

Fig. 9-4. The structure being built on this site requires many drawings. All of them are drawn to scale. Draw your desk at a scale of 1" = 1/4".

SECTION 2
TechCHECK

1. What is the difference between the English and metric systems of measurement?
2. Name three base units used in the metric system.
3. Why do architects draw to scale?
4. **Apply Your Knowledge.** Design and build a scale model of a car.

the amount of electrons flowing through the wire is measured in amperes (amps). *Amperage* is a measure of quantity.

• **Ohm.** A nozzle at the end of a hose resists (holds back) the flow of water. The higher the resistance in an electric circuit, the more electrons are held back. The resistance in electrical systems is measured in *ohms*.

There is a simple mathematical relationship between volts, amps, and ohms. This special formula is known as *Ohm's Law*.

amps = volts ÷ ohms

ACTIVITY

All three units—volts, amps, and ohms—can be measured with a multimeter. Use a digital multimeter to measure the voltage in different batteries.

ACTION ACTIVITY

Measurement Mania

Be sure to fill out your TechNotes and place them in your portfolio.

Real World Connection

The ability to estimate and measure quickly and accurately is an important skill in technology. Fig. A. In this activity, you will first exercise your ability to use English and metric measurement. After you are able to measure accurately, you will be challenged to estimate and measure objects quickly and accurately.

Design Brief

Demonstrate your ability to measure accurately and quickly using both English and metric measurement. Put your measurement skills to use in a measurement test.

Materials/Equipment

- string
- miscellaneous materials for measurement game
- rulers, tape measures
- calculator (optional)

SAFETY FIRST
Follow the safety rules listed on pages 42-43 and the specific rules provided by your teacher for tools and machines.

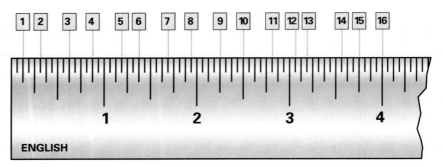

ENGLISH

Fig. A

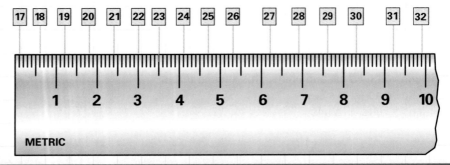

METRIC

Procedure

Part 1 · Measurement Test

1. Complete the measurements for 1 through 32 shown in Fig. A. Write your answers on a separate sheet of paper. You must reduce all fractions and use decimals where needed.

2. Ask your teacher to check your test. If you missed even one of the answers, you must take the test again.

3. If you missed some answers, your teacher will help you understand your mistakes.

4. When you are sure you know how to measure in both English and metric (SI) units, go on to the next part of this activity.

Part 2 · Estimating and Measuring

1. Using a separate sheet of paper, write your estimate of both the English and metric size of each object in the list shown in Fig. B.

2. Using the proper measurement tools, make accurate measurements of each object and record them on your sheet.

3. Find the difference between your estimate and the actual size by subtraction. You may use a calculator.

4. Figure the total of the differences between your estimates and the actual sizes of the objects in both the English and metric (SI) sections. Check with your teacher on how to make an English total.

Evaluation

1. Which measurement system is easier for you to use? Why?

2. List three occupations that require fast and accurate measurement.

3. How is measurement important in sports such as volleyball, football, basketball, and baseball?

4. **Going Beyond.** Do other countries use the English measurement system? Do some research to find out.

5. **Going Beyond.** Find out the meaning of the following units: newton, furlong, joule, light-year.

Fig. B

	Name of Object	Estimate	Actual Size	Difference
English	Width of sheet of paper			
	Diameter of a globe			
	Thickness of a pencil			
	Height of a desk			
	One Meter			
			TOTAL	
Metric (SI)	Width of classroom door			
	Height of this book			
	One Inch			
	Circumference of globe			
	Width of computer disk			
			TOTAL	
			ENGLISH TOTAL	
			GRAND TOTAL	

Designing Products for People

THINGS TO EXPLORE

- Define ergonomics and tell how it affects the way things are designed.
- Explain how anthropometric data is used in design.
- Design a space helmet based on anthropometric data.

Have you ever tried on gloves and found the sizes either too large or too small? Have you ever sat in a chair that hurt your back? Simple everyday things like water faucets and door knobs can sometimes be hard to use because they weren't designed with people in mind.

Look around you. Is the room you're in designed to fit people your size? Can you reach all the shelves? Are the chalkboards at a proper height so you can easily see them from your desk? Can you easily reach all the materials you need to do your work?

Tradition

Many times products are designed just for looks. Tradition also plays a part. Scissors are an example. Originally, they were designed for right-handed people. That design has not changed much, although special versions are made for left-handed people.

The products you use and the places where you live, work, and play are safer, easier to use, and more comfortable if they are designed based on how a real human body is made and works.

Fig. 9-5. People come in all shapes and sizes. Gather anthropometric data for people in your family. What is the average height and weight of your family members? How about arm reach?

Human Sizes

Have you ever wondered how designers decide on what size things should be? When designers made the chair you are sitting in, for example, they used anthropometric data. **Anthropometric data** is size information collected from many people. Fig. 9-5. Designers use this data to determine the dimensions of products such as clothing, furniture, sporting goods, car interiors, and even spacesuits. In most cases, they create a size that "fits" about 90 percent of a product's users. That means for five percent of the people it will probably be too large and for another five percent it will be too small.

Ergonomics

Ergonomics is the study of how the human body relates to things around it, such as furniture and clothing. It is also called "human engineering."

Ergonomics plays an important part in today's high-tech workplace. For example, many people spend long hours using computer keyboards. But nerves in the wrist can be damaged by repeating a simple movement like pressing the keys over and over again. The result is **carpal tunnel syndrome**. To help prevent this problem, designers have come up with wrist braces, adjustable chairs, and even specialized keyboards. Fig. 9-6.

Part of ergonomic design is to make products safer. For instance, special dashboard and ceiling padding, seatbelts, harnesses, and other safety features are built into today's automobiles. Even the seats are designed so you will sit in a proper driving position.

Fig. 9-6. Ergonomic design makes products, such as computer keyboards, easier to use.

SECTION 3
TechCHECK

1. What is ergonomics?
2. What kinds of anthropometric data can be gathered to help designers?
3. How are anthropometric data and ergonomics used to design things?
4. **Apply Your Knowledge.** Design a chair that best fits all the students in your classroom.

ACTION ACTIVITY

Designing a Space Helmet

Be sure to fill out your TechNotes and place them in your portfolio.

Real World Connection

Scientists, technologists, designers, architects, and engineers all use anthropometric data to help them design and make products for people to use comfortably. Special attention must be given to high-tech equipment such as spacesuits. In this activity, you will gather anthropometric data and use it to design a space helmet.

Design Brief

Research and design a space helmet that can be used easily and comfortably. Your design must consider the following:

- Proper size. Your team will determine the best size by gathering anthropometric data.
- Appropriate materials. The helmet should be able to withstand impacts. The visor should protect astronauts from the blinding glare of the sun.
- Weight. Even though things are "weightless" in space, the helmet must be light enough to wear during training on Earth.
- Comfort and safety. Your helmet should be comfortable to work in for many hours. Ventilation and a microphone for communication should be provided.

Materials/Equipment

- graph paper, pencil
- large calipers
- ruler or tape measure
- computer with spreadsheet and graphics software (optional)

SAFETY FIRST
Follow the safety rules listed on pages 42-43 and the specific rules provided by your teacher for tools and machines.

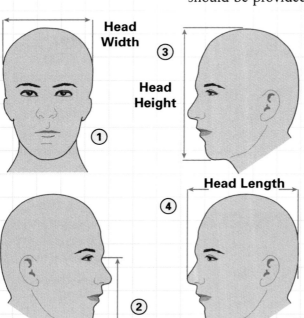

Head Width ①

Head Height ③

Head Length ④

Chin-to-Eye Height ②

Fig. A

Procedure

1. Work in groups of four or five. Read through the steps in the procedure and divide the tasks so that everyone in your group is helping to design the space helmet. Use a tape measure or calipers and a ruler to take the measurements shown in Fig. A. These measurements will be the anthropometric data you need for this activity.

2. Record the measurements for each member of your group on a chart similar to the one in Fig. B. You could use a computer and spreadsheet software to create your chart. Remember, however, that computers do not handle fractions very well. You might want to use the metric system to avoid problems.

3. Using graph paper, sketch a head to scale using the largest dimensions obtained in step 1. Then sketch your designs to scale. Make a front view and a side view of the helmet.

4. Label each part of your space helmet design. Give each part a name, such as visor, sealing neck-ring, microphone, air vents, padding, and antenna.

5. Choose the right materials for your helmet. Following is a list of materials and their properties that may be of help:

- *Fiberglass:* Fiberglass is made of very thin glass strands glued together with a liquid plastic that hardens. The glass fibers go in every direction and give fiberglass its strength.
- *Polycarbonate:* Polycarbonate is a clear, strong thermoplastic material that can be bent or shaped with heat. It is used to make safety glasses and windshields for snowmobiles and motorcycles.
- *Aluminum:* This is a very strong, lightweight metal that conducts heat and electricity easily. Aluminum can be formed into almost any shape, from very thin foil to thick castings.
- *Acrylic:* Acrylic plastic is clearer than glass. It is a thermoplastic that is easy to cut, bend, or shape with heat. It expands and contracts with temperature change but can be very brittle in thin sections.

Evaluation

1. How could your design be improved?
2. What is your estimate of how much your helmet design would weigh? What do you think it would cost to make? How could the cost and weight be reduced?
3. List two other examples in which anthropometric data are used to design products.
4. **Going Beyond.** Use a computer and CAD software to make a final design for your helmet.

Fig. B

ANTHROPOMETRIC DIMENSIONS					
	1	2	3	4	
Student 1					
Student 2					
Student 3					
Student 4					
Student 5					
Total					
Average					
Maximum					

INFOLINK

See Chapter 5 for more information about computer-aided design (CAD).

Choosing the Right Material

THINGS TO EXPLORE

- Explain why it is important to select the right material to make products.
- Distinguish between synthetic and natural materials and give examples of each.
- Identify different properties of materials that you can test.
- Test a material for fatigue strength.

TechnoTerms

composite
compression strength
fatigue strength
synthetic
tensile strength
thermoplastic
thermosetting plastic

Since the earliest times, people have been researching new materials and new uses for old materials. The materials chosen for a product can make it either useful and long-lasting or dangerous and short-lived.

Classifying Materials

Basically, materials can be divided into two major groups, synthetic and natural. Natural materials, such as copper and wood, can be found in nature. **Synthetic** refers to a material made by humans that cannot be found in nature. The many kinds of plastics are examples of synthetics.

Products often are combinations of both natural and synthetic materials. For example, a television set has a picture tube made of glass, a cabinet made of plastic or wood, and wires made of copper.

Did you know that materials can be further divided into categories based on their origin? In general, they fall into four main groups: woods, metals, plastics, and composites. Fig. 9-7.

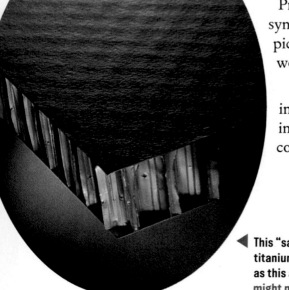

◀ This "sandwich" is made from a compound of the metals titanium and aluminum. Intermetallic compounds such as this are strong at high temperatures. What products might need a material with that property?

FIG. 9-7. CLASSIFYING MATERIALS

- **Woods.** Woods are either hardwoods or softwoods. Sounds simple, but the words *hard* and *soft* often have little to do with the hardness of the wood. The difference is in the tree that the wood came from. Hardwoods come from trees that have broad leaves, such as walnut and maple. Softwoods come from trees that have needles, such as pine and fir.

- **Metals.** Metals are either ferrous or nonferrous. *Ferrous* is a Latin word for iron. Ferrous metals contain iron and nonferrous metals do not. Ferrous metals include iron and the many types of steel. Nonferrous metals include copper, tin, lead, aluminum, gold, and silver.

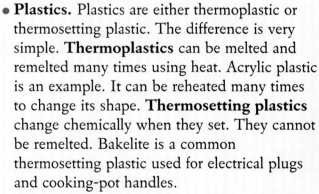

- **Plastics.** Plastics are either thermoplastic or thermosetting plastic. The difference is very simple. **Thermoplastics** can be melted and remelted many times using heat. Acrylic plastic is an example. It can be reheated many times to change its shape. **Thermosetting plastics** change chemically when they set. They cannot be remelted. Bakelite is a common thermosetting plastic used for electrical plugs and cooking-pot handles.

- **Composites.** By combining different materials, new and often better properties can be obtained. Composite materials, such as fiberglass and carbon graphite (graphite-epoxy), are often lightweight and strong. They are used to make high-performance aircraft wings and lightweight sporting goods such as tennis racquets.

Properties of Materials

Materials are chosen for products based on their properties, or characteristics. Fig. 9-8. Each material has special properties that make it useful for certain things. When you are designing products, you should consider using more than one kind of material to take advantage of their different properties. The following properties are found in many common materials:

- **Hardness:** The ability to resist dents
- **Tensile (tension) strength:** The ability to resist stretching or being pulled apart
- **Compression strength:** The ability to resist being squashed or smashed
- **Fatigue strength:** The ability to resist breakage after being bent back and forth

Testing Materials

Some materials have surprising uses. Would you ever consider building a boat out of cement or glass? Some very large boats are made of a cement mixture, called ferrocement, that is sprayed over wire mesh. And the boats float! Many boats are also made of fiberglass, which contains glass fibers.

Materials are first tested before they are used to make products. Fig. 9-9. This is done in order to learn if a material will be appropriate for the product. There are many different properties of materials that can be tested. Hardness, tensile strength, compression strength, and fatigue strength are some of the most common. Testing a material until it breaks or is destroyed is called *destructive testing.*

Fig. 9-8. The bend in this ergonomically designed golf club makes aiming easier. What properties must the materials used in golf clubs have?

Fig. 9-9. This engineered-wood beam will be tested for compression strength. The metal beam will press down on top of the wooden one. Why do you think a beam would need to withstand compression?

SECTION 4
TechCHECK

1. Why is it important to choose the proper materials to make a product?
2. What is the difference between a synthetic material and a natural material?
3. What kinds of tests can you do on materials?
4. **Apply Your Knowledge.** Test a rubber band, a pencil, and a soda can for the four properties. Make a chart showing the results of your test.

ACTION ACTIVITY

Have You Ever Felt Fatigue?

Be sure to fill out your TechNotes and place them in your portfolio.

Real World Connection

Fatigue strength can be a lifesaving property. Airplanes, for example, are sometimes damaged by metal fatigue. The metal in the wings flexes as the airplane flies. The metal fuselage (body) expands and contracts as the plane changes altitude. If they are not detected, small cracks caused by fatigue around rivets can cause the plane to break up. Fig. A.

In this activity, you will do a destructive test to determine the fatigue strength of a steel paper clip.

Design Brief

Design and perform a test that will determine the fatigue strength of a paper clip. Use mathematics to find the average fatigue strength of at least 10 paper clips.

Materials/Equipment

- paper clips
- pencil, paper
- computer with spreadsheet and word-processing software (optional)

SAFETY FIRST
Follow the safety rules listed on pages 42-43 and the specific rules provided by your teacher for tools and machines.

Fig. A. This photo (magnified 280 times) shows a welded joint between two steel plates in an aircraft wing. Metal fatigue caused a stress fracture in the joint (brown area).

Procedure

1. Work in groups of five or six. Discuss the possible ways a paper clip might be tested by bending it until it breaks. Experiment with a few paper clips to see if your idea will work.

2. As a group, decide which test method will be used. Write the procedure for your method using a computer. Your procedure should be detailed enough so that another group could follow your directions. Fig. B.

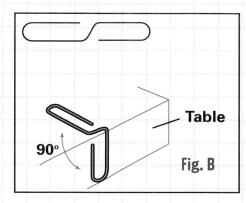

Table

90°

Fig. B

3. Have each person in your group test two or three paper clips. Each tester should write down the number of bending cycles the paper clip went through before it broke.

4. Each tester should add up the number of bending cycles for each paper clip and find the average number of bends required. (To find the average, divide the total number of bending cycles by the number of paper clips tested.) If possible, use a computer and spreadsheet software to help you find the average.

5. Compare the average number in your test with the average numbers of others in your group. Find an average number for the entire group by adding all the individual averages together and dividing by the number of people in your group.

Evaluation

1. Were the average numbers of bending cycles the same for every person in your group? If not, why do you think some people got different numbers?

2. What is the relationship between the size of the bending angle and the number of bending cycles that can be completed before breaking?

3. **Going Beyond.** Compare the average obtained by your group with the averages of other groups. Make a bar graph showing the results. You can use a computer to help make the graph.

4. **Going Beyond.** Design a test procedure to test other materials for fatigue strength.

SECTION 5

Making Models and Prototypes

THINGS TO EXPLORE

- Tell what a prototype is.
- Explain why designers, engineers, and architects often make prototypes or models.
- Make a prototype of a solar cooker.

TechnoTerm
wind tunnel

You know what a model is. Do you remember what a prototype is? A prototype is a model of a product being designed for production. Companies can't build thousands of products without being sure that all the defects or problems have been corrected. Without planning and testing by using models, expensive mistakes can be made. Instead, models and prototypes help them decide whether the product will fit their needs before they spend a great deal of time and money.

Using Prototypes

Prototypes are used by designers, engineers, and architects. When architects design a new skyscraper for a large city, for example, they often build models so people can visualize the shape of the building. Fig. 9-10A. The model lets people see what the building will look like better than a drawing can. Models are usually made to scale. Also, the model of a new building is sometimes placed in a model of the entire city to see how it fits in.

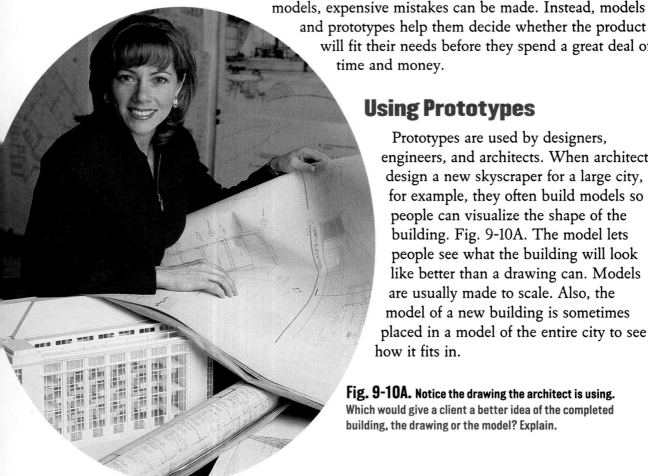

Fig. 9-10A. Notice the drawing the architect is using. Which would give a client a better idea of the completed building, the drawing or the model? Explain.

Fig. 9-10B. A true 3D model can be tested on the computer for strength and other properties.

In some cases, a structural model is placed in a **wind tunnel** to see how the building will be affected by wind currents. Some buildings have been built where wind currents were so strong that they sucked the windows out!

Some products, such as airplanes, buildings, and cars, can be modeled as three-dimensional images on a computer. Fig. 9-10B. In this way, you can save even more time because you don't have to build a real model. You can even test some products using computer simulations.

INFOLINK

See Chapter 5 for more information about computer simulations.

SECTION 5
TechCHECK

1. What is a prototype?
2. What is the purpose of making a prototype for a product?
3. Why do architects make models?
4. Apply Your Knowledge. Design and build a scale model of a house.

ACTION ACTIVITY

Cooking with Sunlight

Be sure to fill out your TechNotes and place them in your portfolio.

Real World Connection

Solar cookers use the energy of the sun to cook food. They are a much-needed product in developing countries. They would prevent people from cutting down so many trees for fuel and could even save lives. Many young children die each year because they do not have clean water to drink. A solar cooker could be used to *pasteurize* (purify) water and kill the harmful bacteria.

In this activity you will design and build a prototype of a solar cooker. Your design will be tested to see what temperature can be reached. Fig. A.

Design Brief

Design, build, and test a solar cooker that will cook food. You may use such materials as recycled cardboard boxes, newspapers, and aluminum foil to make your prototype.

Materials/Equipment

- recycled cardboard boxes, newspapers, foil, as needed
- clear acrylic or polycarbonate plastic (Plexiglas acrylic plastic or Lexan plastic)
- flat black spray paint
- hand and power tools, as needed
- one-quart container for water
- thermometer

SAFETY FIRST
- Follow the safety rules listed on pages 42-43 and the specific rules provided by your teacher for tools and machines.
- If you use food in your solar cooker, be sure to follow proper food safety and sanitation rules.
- Handle hot items with care.

Procedure

1. You will be working in groups of four or five. You should consider some of the following facts before starting your design:

- The cooking area should be insulated so the heat gathered is not lost.
- As more sunlight is reflected into the cooking area, a higher temperature will be produced.
- A flat black surface absorbs heat energy rather than reflects it.
- Clear plastic, like glass, helps to trap heat by the *greenhouse effect*. (The greenhouse effect is a natural buildup of heat trapped by atmospheric gases, mainly carbon dioxide. The gases let in visible light but keep some infrared radiation from leaving the Earth's surface.)

2. When your group has decided on a design, divide up the work needed to complete the cooker. Follow your teacher's safety rules when using any machines.

3. When you finish your prototype, test it by placing a one-quart container of water with a temperature of 72°F. inside your cooker.

4. Place a thermometer in the water, and point your cooker toward the sun.

5. After one hour, record the temperature of the water.

6. Continue to test your cooker for 4 hours. Move it each hour so it continues to point toward the sun.

Evaluation

1. What are the advantages of using solar energy to cook?

2. What are the disadvantages?

3. **Going Beyond.** Make a graph showing the temperatures and cooking times of your solar cooker.

4. **Going Beyond.** How could solar cookers be made more efficient using high-tech materials?

Fig. A

REVIEW &

CHAPTER SUMMARY

SECTION 1

• As you design and build things, you need to make accurate measurements.

• The need for more precision, as well as standard ways to measure things, produced the measurement systems we use today.

SECTION 2

• The SI (metric) and English systems are standard measurement systems used today.

• Drawings and models of things that are very large or very small are often made to scale.

SECTION 3

• Ergonomics involves designing products or places that fit human needs and sizes.

• Designing with ergonomics in mind means products will be more useful and safer for people.

SECTION 4

• Choosing the right material can determine whether a product will be useful and good for the environment.

• All materials can be classified into two major groups, synthetic and natural.

• Special properties of materials, such as hardness, tensile strength, compression strength, and fatigue strength can be tested.

SECTION 5

• Prototypes and models help people see what the actual products will look like.

REVIEW QUESTIONS

1. What is the purpose of measurement standards?

2. Why do engineers make prototypes?

3. Give an example of a synthetic material and a natural material.

4. What kind of anthropometric data would you use in designing a chair?

5. Name several materials that would be good choices for making a house.

CRITICAL THINKING

1. List some synthetic materials you might use to build a model car.

2. Research the composite materials used in making a space shuttle and jet fighters.

3. Design a nondestructive test method for a material.

4. What properties of aluminum make it a good choice for beverage cans?

5. Collect anthropometric data from students in your class to use in designing a student desk.

ACTIVITIES

CROSS-CURRICULAR EXTENSIONS

1. **SCIENCE** Compare the properties of steel, aluminum, and titanium. Which metal would be best for a bicycle frame? Why?

2. **MATHEMATICS** Collect anthropometric measurements such as height, head circumference, and arm reach for your class. Calculate an average for each measurement.

3. **COMMUNICATION** Write or call a business that makes a product. Find out what materials they use and how they choose those materials to make the product.

═ EXPLORING CAREERS ═

When the telephone was first developed, engineers measured hundreds of people's faces to ensure that the handsets would be comfortable for most people to use. How do you think human factors are involved in the following two design careers?

Commercial and Industrial Designer
These designers develop products like cars, home appliances, and toys. They combine their artistic talent with research findings to create the most useful and appealing design. Have you ever wondered why a shampoo bottle is shaped a certain way? Designers help create a product that will appeal to potential buyers.

Computer Game Animator
Animators create the lifelike characters, backgrounds, and 3-D graphics that make up computer games. This is a job that requires a great imagination and strong artistic and technical skills. Animators must have the ability to work well under pressure in order to meet tight deadlines.

ACTIVITY

Design an animated character for a computer game or develop a design for a liquid-soap dispenser. Share your results. Explain why you chose this character or design.

CHAPTER **10**

Exploring Automation

Producing Products and Moving Materials

THINGS TO EXPLORE

- Define automation and give examples.
- Relate how automation has affected the modern workplace.

TechnoTerms
automation
manually
robot

After you have solved design problems and picked the right resources and materials for your product, you need an efficient way to produce and move materials. That's where automation comes in.

Automation

Automation is the automatic control of a process by a machine. Today, **robots** (highly advanced, computer-controlled machines) do welding and other dangerous tasks automatically. Even soft drinks can be made using automated sensors that control the mixture of syrup and water.

Automation has really affected the modern workplace. Automation does away with jobs that are boring or dangerous. Fig. 10-1. It also creates more need for highly skilled people to control, repair, and program machines. Automation can speed production in many instances and can also save time and money.

Fig. 10-1. These workers are assembling electronic products. What parts of their job might be boring? Could a robot do them as well?

◄ OPPOSITE The movement of robots is only a blur in this auto assembly line.

Fig. 10-2. An automated conveyor speeds up the delivery of baggage at an airport. What do you think would happen if this equipment broke down?

Automation also has problems. It tends to take jobs that require less skill away from workers. Sometimes these workers cannot be retrained to do higher-skilled jobs. In addition, automated machines are expensive to buy and maintain.

It would be hard for us to live without automation now. Fig. 10-2. Think about the automatic clothes washer and dryer your family uses. Do they go through the cycles automatically, or do you have to run them **manually** (by hand)? What about the copy machine or the bell system used in your school?

SECTION 1
TechCHECK

1. What is automation?
2. How has automation affected the modern workplace?
3. Describe a problem created by automation.
4. **Apply Your Knowledge.** Working in small groups, make a list of things that are automated. Share your list with other groups.

Mass Production

THINGS TO EXPLORE

- Define mass production and give examples.
- Explain how mass production has changed manufacturing.
- Use mass production to produce a product.

TechnoTerms
assembly line
custom
     manufacturing
manufacturing
mass production

Before factories as you know them existed, people made products in their homes. Different households made different products and had different skills or crafts. Each item was produced by one person. Today this process of making items one at a time is called **custom manufacturing**. Fig. 10-3. It often takes a very long time to complete one item.

Fig. 10-3. Large airplanes like this Boeing 747 are made with custom production methods. What other types of products are still produced this way?

Fig. 10-4. Although Chrysler was first to use assembly lines, Ford improved them by moving the product past the workers. As a result, Ford cars could be made faster and cheaper. Do some research on other cars made at the time. How did their designs differ?

Mass Production

Later, groups of craftspeople began to work in factories **manufacturing** (making) products faster. In the 1700s, Eli Whitney and others developed the system of **mass production**. Mass production uses **assembly lines**, where products move past a worker who does a specific job. Henry Ford later improved this system. Fig. 10-4. Automation has helped increase the efficiency of mass production in today's factories.

SECTION 2
TechCHECK

1. What is custom manufacturing?
2. How were products made before factories became common?
3. What is mass production?
4. **Apply Your Knowledge.** Give examples of items you think are *not* made by mass production.

ACTION ACTIVITY

Mass Production vs. Custom Production

Real World Connection

Today, we rely on mass production to make products from cars to cookies faster, more accurately, and more economically. Is it really faster than custom production? In this activity, you will see for yourself!

Be sure to fill out your TechNotes and place them in your portfolio.

Design Brief

Determine the difference in efficiency between custom production and mass production. You will be assigned a specific job in either the custom production or mass production factory. Fig. A. Your job is to work as fast and as accurately as possible.

Materials/Equipment

- vanilla wafer cookies (plain)
- assorted flavors of frosting (squeeze tubes)
- M&M candies (plain)
- napkins or paper towels
- plastic gloves
- measuring tools

SAFETY FIRST
Follow the safety rules listed on pages 42-43 and the specific rules provided by your teacher for tools and machines.

Fig. A

(Continued on next page)

ACTION ACTIVITY

Procedure

1. Divide the class into two groups. Flip a coin to see which group will be the custom workers and which will be mass-production workers.

2. As a class, decide on a design for the finished cookies.

3. Divide the supplies of cookies, candy, and frosting equally among each group. Each of the craftspeople will need a supply of each item.

4. Your teacher will assign a job in the assembly line to each person in the mass-production group. The mass-production workers should sit at a long table or rearrange their desks so that each worker can easily pass the cookies on to the next worker.

5. All workers in both groups should wear plastic gloves to keep the cookies clean. When each group is ready, your teacher will tell you to start production. The custom workers will each make complete cookies. Each mass-production worker will do a specific job in the production process and pass the cookie down the assembly line until it is finished.

6. After a set time, your teacher will ask both groups to stop. At that time, you must stop where you are. Do not finish the cookie you were working on.

7. Each group should gather their completed cookies. All of the completed cookies will be inspected by your teacher and either accepted or rejected based on accuracy of design.

Evaluation

1. Which group produced the most cookies?

2. Which group had the most rejected cookies?

3. What do you think the advantages and disadvantages of custom production are?

4. What do you think the advantages and disadvantages of mass production are?

5. Did the production of cookies turn out the way you expected? Explain.

6. **Going Beyond.** Research to find out what products are custom made by craftspeople today.

7. **Going Beyond.** Try the production again with different products. Do you think you will get the same results?

8. **Going Beyond.** Try the production process again, but this time switch the groups. Do you think you will get the same results?

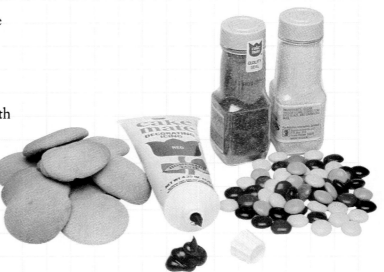

Moving Materials

THINGS TO EXPLORE

- Explain what materials-handling systems do.
- Identify basic types of materials-handling equipment.
- Build a conveyor belt system to move materials and parts.

TechnoTerms
automated storage
 and retrieval
 system (AS/RS)
automatic guided
 vehicle system
 (AGVS)
conveyor
crane
hoist
materials handling

During the Industrial Revolution, which you learned about in Chapter 1, automation and assembly lines became part of the factory setup. This marked the beginning of a need for efficient **materials handling** (moving and storing materials). Today, various kinds of equipment are used in materials handling. Most systems usually include several basic types. Fig. 10-5.

Fig. 10-5. Forklifts (top left) are used to carry pallets at this loading dock. A large crane (right) lifts a container onto the deck of a ship. Computers (bottom left) help schedule delivery of materials to a production area.

Equipment Used for Materials Handling

- **Conveyors.** Conveyors move materials and parts from one place to another. They always travel over a fixed path. Examples include roller conveyors, skate-wheel conveyors, and belt conveyors. In automated assembly lines, parts are often moved by conveyor from one workstation to another.
- **Trucks.** Materials-handling trucks are different from the trucks you normally see on the highways. An example is a *forklift*. Forklifts have forks on the front and are used to carry heavy loads.
- **Containers.** Large parts are moved onto *pallets*, or special platforms. The pallets can then be stacked in racks. Small parts are moved in containers such as trays or tubs.
- **Hoists and cranes.** Hoists are used for lifting heavy loads. Cranes are really hoists that move in a limited area. For example, an overhead crane lifts large items from above. A monorail crane is a hoist that moves on one overhead rail.
- **Automated storage and retrieval system (AS/RS).** This system uses a computer-controlled crane. It travels between pallet racks and automatically loads and unloads them.

MATHEMATICS CONNECTION

Determining the Speed of a Conveyor

The speed of a conveyor is often measured in parts per minute or feet per second. You can measure these speeds by doing the following:

How to Determine Parts Per Minute

- Set the drill or motor to a constant speed.
- Have one or two classmates feed the input end of your conveyor with small parts such as washers or nuts.
- Space the products (washers or nuts) evenly and close together.
- Count the number of products as they start coming off the end of the conveyor for one minute.

- **Automatic guided vehicle system (AGVS).** In this system, a computer controls the movement of several driverless carts called AGVs. These carts follow wire paths that are built into the floor. The computer keeps track of the location of all the carts so they don't run into each other as they move materials around a factory. What do you think would happen if one cart got off track?

SECTION 3
TechCHECK

1. What is materials handling?
2. Name six different ways to move materials.
3. What is a pallet?
4. **Apply Your Knowledge.** If you had to move all the chairs in your school to another school, what type of materials-handling equipment would you use?

How to Determine Feet Per Second

- Set the drill or motor to a constant speed.

- Put a mark or piece of tape on the edge of the conveyor belt. This mark will be used to measure how fast the belt is moving.

- Make another mark on the side board as a point of reference.

- Have someone count the number of times the tape passes the marked point in a set period of time, such as 30 seconds.

- To calculate the speed in feet per second, measure the length of the belt and multiply by the number of passes. Divide by the number of seconds.

Example:

6 passes × 3 feet each = 18 feet

18 feet ÷ 30 seconds = .6 feet per second

ACTIVITY

What was the speed of your conveyor in feet per second?

ACTION ACTIVITY

Can You Handle It?

Be sure to fill out your TechNotes and place them in your portfolio.

Real World Connection

Materials handling is an important part of designing factories that can make high-quality products quickly. In this activity, you will work in groups to make a set of conveyor belts that can be used to simulate an assembly line. Fig. A.

Design Brief

Build and test a conveyor belt system that will transport parts from one point to another in your classroom. The conveyor system must be safe to operate and have adjustable speeds so it can meet the needs of future activities.

(INFOLINK)

See Chapter 11 for more information about mass production.

Materials/Equipment

- curtain pleating tape
- 1/8" tempered hardboard
- 1/4"-20 threaded rod
- 1/4"-20 nuts
- 1/8" acrylic sheet (Plexiglas)
- 1 1/2" PVC pipe
- 1/4" x 1 1/2" fender washers
- masking tape
- wood glue, screws
- scroll or band saw
- disk or belt sander
- hacksaw
- adjustable wrench
- power hand drill
- sewing machine (optional)

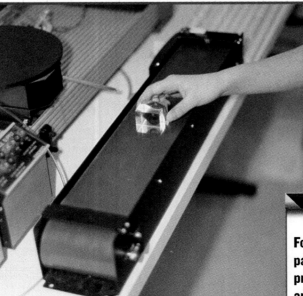

Fig. A

SAFETY FIRST
Follow the safety rules listed on pages 42-43 and the specific rules provided by your teacher for tools and machines.

Procedure

1. This activity will require you to work in large groups. There are many parts that need to fit together to make your conveyor belt. Your class will need two or three conveyors for the mass-production activity in Chapter 11.

2. Your group or your teacher might decide to change the design of your conveyor system to meet the needs of your classroom. The illustration in Fig. B will help you get a start on a design. Note that the length of the conveyor is up to you and your teacher. The width of the conveyor is determined by the material you use for the belt. The minimum size should be 4".

3. The design that you choose will be used by all the groups so the conveyors will match. Divide the tasks in building the conveyor among your group. Some jobs might require two students.

4. With your teacher's help, use a hacksaw to cut the 1/4"-20 threaded rod to length for the drive roller and the return roller. (The lengths will depend on the width of your conveyor belt.)

SAFETY FIRST
This activity requires all students to wear eye protection at all times.

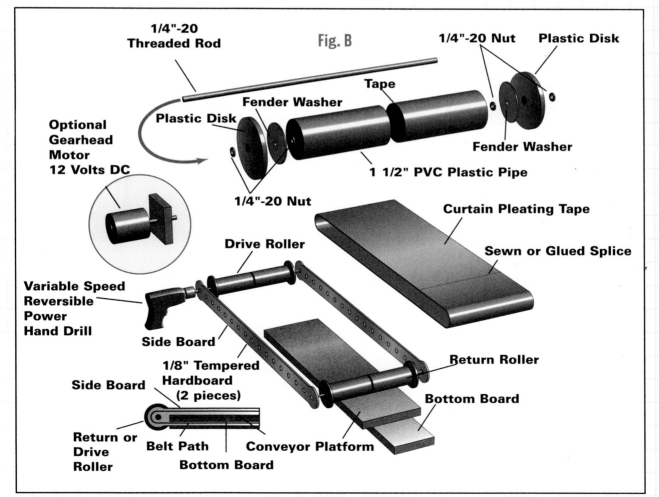

Fig. B

- 1/4"-20 Threaded Rod
- 1/4"-20 Nut
- Plastic Disk
- Tape
- Plastic Disk
- Fender Washer
- Optional Gearhead Motor 12 Volts DC
- Fender Washer
- 1 1/2" PVC Plastic Pipe
- 1/4"-20 Nut
- Curtain Pleating Tape
- Drive Roller
- Sewn or Glued Splice
- Variable Speed Reversible Power Hand Drill
- Side Board
- Return Roller
- Side Board
- 1/8" Tempered Hardboard (2 pieces)
- Bottom Board
- Return or Drive Roller
- Belt Path
- Conveyor Platform
- Bottom Board

(Continued on next page)

ACTION ACTIVITY

5. Cut the 1 1/2" PVC pipe to length with the hacksaw. Wrap and tape a piece of paper around the place where you are cutting to guide the saw for a square cut.

6. Wrap the center of each PVC pipe with two layers of masking tape. This will make the center of the pipe larger in diameter than the ends. This helps the belt track properly without going off to one side.

7. Cut four acrylic (Plexiglas) disks 3" in diameter using the scroll saw.

SAFETY FIRST

Ask your teacher to help you cut the disks on the scroll saw. Remember to keep your fingers away from the front of the blade.

8. Assemble the drive and return rollers as shown in the drawing. The fender washers are used inside the PVC pipe to keep the threaded rod centered. You might want to put a drop of glue on the threads near the nuts to keep them from coming loose.

9. Cut the conveyor platform and bottom board to length and width on the band saw. The length of the boards depends on your design.

10. Cut and sand the 1/8" tempered hardboard to make the sides of the conveyor. Round the ends to match the plastic disks on the rollers.

11. Temporarily assemble the rollers, sides, conveyor platform, and bottom board as illustrated. Check to see if the axles are parallel to each other. This is important for the belt to track properly. You might want to file the return roller holes into slots to make it adjustable.

12. Use a piece of string or a tape measure to find the proper length of the belt. Add 4" for the pleating tape to overlap at the splice. Cut the curtain pleating tape to the proper length.

13. Sew or glue the splice together at the proper distance for your conveyor. Be sure the edges of the pleating tape remain straight.

14. Glue and screw the completed conveyor system together.

15. Attach the power hand drill to the drive roller. (Your teacher may want to use a special gearhead motor for your conveyor instead of a drill.) Slowly start the drill or motor. Watch carefully to see if your conveyor belt stays on track. Make adjustments as needed.

Evaluation

1. Determine the speed of your conveyor belt system. How many parts per minute did it deliver?

2. List two other possible materials that could be used for the conveyor belt.

3. **Going Beyond.** Can you think of another way to measure the speed or capacity of a conveyor belt system? Explain.

4. **Going Beyond.** What would determine the speed of a gravity-fed roller conveyor system?

Robots, Robots Everywhere!

THINGS TO EXPLORE

- Explain the difference between a true robot and an automated machine.
- Identify jobs robots can do best and tell why.
- Relate how robots affect people's jobs.
- Program a robot to do a job.

TechnoTerms

cost-effective
industrial robot
pick-and-place
maneuver
programmed
teach pendant
work envelope

Robots, robots everywhere! Robots are one of the most exciting inventions in technology. They are entertaining you in the movies. They are doing jobs for you that you don't want to do or that are too dangerous for you. Fig. 10-6. Today's robots are highly advanced machines **programmed** (computer controlled) to do special jobs.

What Can Robots Do?

How would you like a robot that could do all your homework? You would have to program the robot with the correct answers. In other words, you would have to do your homework first and then program the robot to do it! That would be true for every different assignment. Doesn't sound like much help to you, does it? But robots cannot think for themselves yet. In fact, a true robot is controlled by a computer and must be reprogrammed to do different jobs.

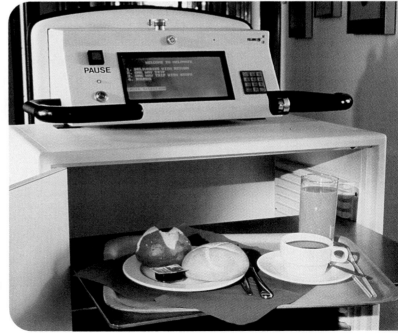

Fig. 10-6. This robot used to deliver food and other items in a hospital navigates by means of a map stored in its memory. In what ways would a robotic delivery system make your own life easier?

TRUE ROBOTS The word *robot* comes from a play by a Czech writer named Karel Capek. Robot means "to work" in Czech. But the idea of robots was around long before Capek used it in the 1920s. Originally, a robot was simply any machine that could do work without a person running it. By that definition, many devices such as clock radios, clothes washers, microwave ovens, and mechanical toys are robots. But they are really not. True robots can be reprogrammed over and over to do different jobs!

Robots can do many jobs that make life easier for us. They are used to help physically challenged people in hospitals. They are used in homes for entertainment as well as for doing specialized jobs such as walking the dog. They explore space and underwater locations in the oceans. In factories, robots are helping to produce goods more efficiently and **cost-effectively** (saving money). In one Japanese factory *only* robot workers are used! As technology advances, robots will be used in many new ways. Can you think of jobs robots can do that you wouldn't want to do?

Robots are like humans in that they can only move their arm or wrist a certain distance. The maximum distance that each part of a robot arm can move is called its **work envelope**. When a robot picks up a part and moves it somewhere else, the action is called a **pick-and-place maneuver**. Robots are sometimes programmed with a remote keypad called a **teach pendant.**

Robots Are Not All the Same

How robots look depends on the jobs they do. Robotic arms are used in manufacturing cars, for instance. These are called **industrial robots**. On an auto assembly line, a robotic arm can be programmed to pick up a door and place it in a certain spot. Another robotic arm fastens it while a third robotic arm spray paints or spot welds certain pieces. Fig. 10-7.

INFOLINK

See Chapters 17 and 18 about the use of robots in outer space.

The *Viking* landers that NASA sent to Mars are other types of robots. They moved around on the surface like huge bugs, picking up soil and doing experiments. Still other kinds of robots handle dangerous materials such as radioactive wastes.

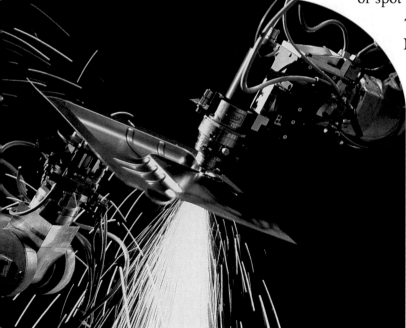

Fig. 10-7. Metal produces a shower of sparks as a robot cuts through it using a laser. Why do you think a robot would be ideal for this kind of job?

Fig. 10-8. Although robots do take some jobs from people, they also help create new jobs. These students are learning to be programmers. What other robot-related jobs might they learn?

TechnoFact

ROBOT POWER
Industrial robots can run on different power systems. Some move with electric motors, including special motors that turn a little bit at a time, called stepper motors. Pneumatic robots run on compressed air. Hydraulic robots use oil pumped under pressure to make the robot operate. The type of energy system used depends on the size of the parts being moved or lifted.

How Robots Affect People's Jobs

Have you heard that robots are taking jobs away from people? In many instances, robots do jobs that people don't want to do. These jobs are either too boring or too dangerous for humans. Sometimes robots can do a better job because they can be programmed to do the exact same thing over and over and still be *precise* (accurate) to one-thousandth of an inch.

Often, people replaced by robots can be retrained to become robot programmers. Fig. 10-8. For example, the best person to program or train a robot to weld is a welder.

SECTION 4
TechCHECK

1. What is the difference between a true robot and an automated machine?
2. What jobs can robots do?
3. How do robots affect people's jobs?
4. **Apply Your Knowledge.** List jobs you think could best be done by robots in a car production factory and tell why.

ACTION ACTIVITY

Programming a Robot

Be sure to fill out your TechNotes and place them in your portfolio.

Real World Connection

People have long dreamed of making machines that can follow instructions. Telling machines what to do is now possible, thanks to automation and robotics. In this activity, you will first control a simulated robot arm to move objects. When you understand the controls of the robot, you will learn to program it to work for you automatically.

Design Brief

Design a flowchart showing the movements the robot will make step by step. Determine the robot's work envelope. Program a simulated robot arm to perform a pick-and-place maneuver.

Materials/Equipment

- conveyor belt system developed for Section 3
- programmable robot arm
- wood or plastic blocks (sized for the robot to grasp)
- graph paper
- tape measure
- protractor

SAFETY FIRST

- **Always be careful with any electrical equipment. Keep liquids away, and place extension cords so that people will not trip.**
- **Most educational robots operate on low voltage, but always ask your teacher before changing any electrical connection.**
- **Do not attach or unplug equipment from computers while they are turned on. Computer circuits are easily damaged.**
- **Follow the safety rules listed on pages 42-43 and the specific rules provided by your teacher for tools and machines.**

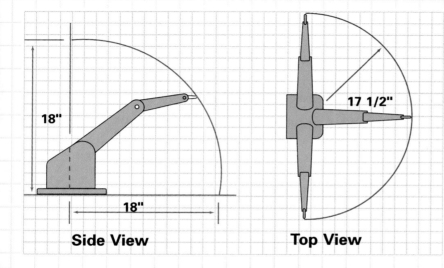

18"

18"

17 1/2"

Side View

Top View

Fig. A

Procedure

1. Work in groups of three or four. To program a robot, you must understand each of its movements. Ask your teacher to show you how to safely operate a simulated robot arm, or use the robot arm you made in Chapter 8. Each of the joints of a robot arm has a name such as base, shoulder, or wrist. Make a list of names of each joint in the robot arm you are using. Next to each name, write the maximum movement possible. Your list might look like this:

   - base, 180°
   - shoulder, 120°
   - elbow, 110°
   - wrist rotation, 360°
   - pitch, 180°
   - gripper, 0" to 3"

2. Your next job is to determine the work envelope of your robot. Use graph paper to make a sketch to scale of the top and side views of your robot arm. Choose a scale that will let you draw the robot and its maximum movements. Your sketch might look like the one in Fig. A.

3. Use a tape measure to find the maximum reach for the robot. Make a line on your sketch to show the work envelope. Show the work envelope in both the top and side views.

4. Now that you know the limits of your robot, it's time to put it to work. From your sketch, determine the best place to put your conveyor system. You will be using the robot arm to off-load (take parts off) your conveyor belt in a pick-and-place maneuver.

5. Set up the conveyor and robot according to your plan. Ask your teacher to show you how to program the robot arm. Each robot has a slightly different way of putting steps into memory. Your robot may be programmable directly, or it might require the use of a computer or teach pendant.

6. Make a flowchart of each step you are programming into the robot's memory.

7. When you have learned to program the robot arm, adjust your conveyor to a slow speed.

INFOLINK

See Chapter 3 for more information about flowcharts.

8. Have one student in your group place simulated products (wood or plastic blocks) on the conveyor belt at an even rate. Try to program the robot to grab the parts at the end of the conveyor and drop them into a box.

9. You might find it hard to place the parts on the conveyor belt in the exact spot the robot expects to find them. One solution might be to design and build an attachment for the conveyor that will direct the products to the center of the belt. You could also have a stop at the end of the conveyor that will give the robot time to grab each product.

Evaluation

1. What was the hardest part of this activity? How could it have been made easier?

2. What do you think would happen if the products on the conveyor were all different sizes?

3. What would be the advantage of having one computer control the robot and the conveyor belt?

4. **Going Beyond.** Design an electrical system that will start and stop the conveyor belt so that products don't pile up at the end of the belt.

Putting Robots to Work

THINGS TO EXPLORE

- Tell what an end effector is.
- Explain why special end effectors are necessary for certain jobs.
- Explain what robotic sensors are and give examples.
- Design and build an end effector for a robot.

TechnoTerms

end effector
robotic vision
sensors
universal gripper

Robots have changed manufacturing. Special equipment like end effectors and robotic sensors help them do many useful things.

The End Effector

The **end effector** of a robot is the device at the working end of a robot arm. Fig. 10-9. Designing a **universal gripper** that will work for every object gets tricky with some materials. Your fingers can pick up small pins and needles, catch a football, or grasp a tennis racket. But it is not easy for a single robot gripper to work as a pair of "hands" to pick up very small or large objects. If the robot has to pick up liquids, hot parts, radioactive materials, or glass sheets, the design must be changed.

Specialized end effectors can be made for robots to work with specific materials. An electromagnet might be used to attract metals such as steel. The electromagnet can be turned on and off by the robot's controlling computer. However, this method would not work for materials that are not attracted to magnets. Aluminum, copper, glass, plastics, or wood would require a different kind of end effector such as a suction cup. The controlling computer can start or stop a vacuum pump that creates the suction to pick up parts. Most robots are designed to do many jobs. The end effector can be changed depending on the job.

Fig. 10-9. This robot is used for testing. Its end effector can do things a human hand cannot.

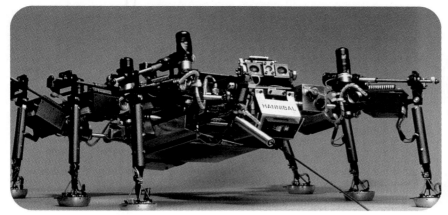

Fig. 10-10. Small microbots like this one can crawl across the floor. The blue "eyes" are sensors. Why do you think it has so many?

Robotic Sensors

Robots can be equipped with special **sensors** (electronic input devices) to help them do jobs. Some robots can hold fragile parts with just the right pressure so they do not drop them or squeeze them too tightly. Light sensors can guide robots along pathways. Fig. 10-10.

Some robots use television camera eyes to help them "see." This is called **robotic vision**. With eyes, robots can tell the difference between parts with different shapes. Vision systems also help robots perform tasks such as installing automobile windows or making sure packages are lined up in the right order to be labeled. Robots with vision do jobs that are often too detailed for humans to carry out. For example, they inspect hundreds of microscopic connections on semiconductor chips. Can you imagine doing that kind of close-up work over and over without making any mistakes? Robots can.

SECTION 5
TechCHECK

1. What is an end effector?
2. Why do robots need specialized end effectors?
3. What are sensors? Give examples.
4. What is meant by robotic vision?
5. Apply Your Knowledge. Research the types of sensors used by robots moving materials in a factory.

ACTION ACTIVITY

Designing a Robot's End Effector

Be sure to fill out your TechNotes and place them in your portfolio.

Real World Connection

Did you ever have trouble grabbing a bar of soap in the bathtub? Imagine that the thing you are trying to grab is worth hundreds or millions of dollars and the slightest wrong move would destroy it. This is part of the difficulty in using robots. Their end effectors must be sensitive and able to grab without causing damage. Satellites, for example, that need repair in orbit require careful handling by astronauts in the Space Shuttle. The grapple device on the shuttle's robot arm was designed for this purpose.

The robot arm is called the Remote Manipulator System (RMS). Eventually, it will find uses in manufacturing or in underwater robotic exploration. It will be used to grip large, delicate objects and materials.

In addition to a gripper with fingers, special cup-shaped devices can make it much easier to grab objects.

INFOLINK

See Chapter 18, "Living and Working in Space."

Design Brief

Design, build, and test a gripping device that might be used on a robot arm.

Materials/Equipment

- plastic cups
- sharp knife
- tape
- string

SAFETY FIRST

Follow the safety rules listed on pages 42-43 and the specific rules provided by your teacher for tools and machines. Use hand tools such as sharp knives with caution.

Procedure

1. Put two plastic cups together and cut them as shown in Fig. A.
2. Cut three pieces of string 5 inches long.
3. Tape the strings to the outside of one cup and the inside of the other cup as shown.
4. Put the cups together and adjust the lengths of the strings to cross the diameter of the cup.
5. Test your end effector by turning one cup in the opposite direction of the other. The strings should cross in the center.
6. Try to grab a pencil or other object using your grapple device. (The pencil simulates the grapple point on an object such as a satellite.)

Evaluation

1. What makes this device better than a gripper?
2. What does it mean to grapple something?
3. Why is it important for astronauts to be able to grab satellites in orbit?
4. **Going Beyond.** Design, build, and test a grapple device like the one in this activity that would fit on a robot arm in your school.
5. **Going Beyond.** Research other types of end effectors. Make a chart or a poster of various types and describe their uses.

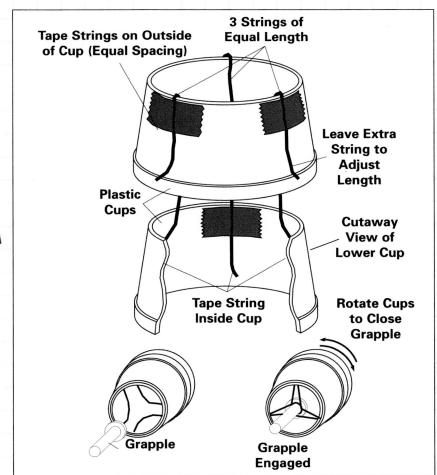

Tape Strings on Outside of Cup (Equal Spacing)

3 Strings of Equal Length

Leave Extra String to Adjust Length

Plastic Cups

Cutaway View of Lower Cup

Fig. A

Tape String Inside Cup

Rotate Cups to Close Grapple

Grapple

Grapple Engaged

CHAPTER SUMMARY

SECTION 1

• Automation is the automatic control of a process by a machine. We depend on automation in our homes, schools, and workplaces.

SECTION 2

• Mass production uses assembly lines, where products move past a worker who does a specific job.

• Automation has helped improve the efficiency of mass production in today's factories.

SECTION 3

• The Industrial Revolution marked the beginning of a need for efficient materials handling.

• Basic materials-handling equipment includes conveyors, trucks, containers, hoists, cranes, monorails, and computer-controlled systems.

SECTION 4

• Robots are highly advanced computer-controlled machines programmed to do special jobs that are often too dangerous or too boring for people to do.

• In factories, robots are helping to produce goods more efficiently and cost-effectively.

SECTION 5

• Specialized end effectors are made for robots to work with specific materials or to do certain jobs.

• Robots can be equipped with special sensors to help them have a sense of touch or hearing. Robotic vision helps them do jobs that are often too detailed for humans to carry out.

REVIEW QUESTIONS

1. Are any of the automated machines in your home or school robots? Explain.

2. How do assembly lines and automation affect the cost of many things you buy?

3. What kinds of materials-handling equipment might you see at a construction site?

4. What makes a robot different from an automated machine?

5. What kind of sensor might a robot have if its job was to pick up light bulbs and pack them?

CRITICAL THINKING

1. Design and build an easy method for adjusting the tension and tracking of a conveyor belt on the rollers.

2. Research how mass production in Japan is different from that in the United States.

3. Design a people-mover conveyor system that could help students move faster in your school.

4. Research the safety hazards of industrial robots.

5. Describe a factory that you think could run efficiently with only robot workers.

ACTIVITIES

CROSS-CURRICULAR EXTENSIONS

1. **SCIENCE** Design a robotic end effector that could be used to pick up samples of soil based on size and density.

2. **MATHEMATICS** If your company must pay $50,000 for a plastic mold for a pen that costs less than a dollar, how does the company make money?

3. **COMMUNICATION** Write a company such as Ford or General Motors for information on how it uses robots in the manufacture of cars.

EXPLORING CAREERS

Technology has made it easier and more efficient to perform many jobs. For example, cashiers in grocery stores no longer have to input by hand all of the prices for the products that you buy. Following are two jobs that have been affected by the increase of automation in the workplace.

Production Manager Production managers plan and organize the work necessary to manufacture products in a timely manner. They assign workers, keep projects on schedule, and monitor the quality of the work. Managers must be organized and flexible. They must also have good "people" skills in order to keep employees interested and productive.

Robotics Technician Robotics technicians program and monitor robotic equipment that is usually used for manufacturing. They must be able to set up, operate, and repair the robot if there is a mechanical problem. Technicians must have skills and knowledge in electronics and programming.

ACTIVITY

Divide into teams and select one fast food. Have each team produce enough small drawings of the item for every person in the class. Did you work well together? Were you efficient?

CHAPTER 11
How Business Works

What Is a Company?

THINGS TO EXPLORE

• Tell what a company is.

• Tell what the main goals of a company are.

• Describe three different ways to manage a company.

TechnoTerms

capital
company
corporation
dividend
entrepreneur
partnership
proprietorship

You've probably seen ads in magazines and on television sponsored by names such as Xerox, IBM, General Motors and Exxon. These are huge companies that hire thousands of workers to produce goods or services for you, the consumer. A **company** is an organized group of people doing business. An **entrepreneur** is someone who starts a company.

In the last chapter you learned that craftspeople used to make products in their homes. In this chapter you will investigate the business end of how factories and companies operate today.

Every company is in business to sell products and to make a profit. Remember, one resource needed to make things is money, or **capital**. In order to get a company started, someone has to *invest* (put in) money to start the company. Then, after the company is organized, someone has to *manage* (run) it.

Sole Proprietorship

Company Management

Different ways to manage a company include

• **Proprietorship.** A proprietorship is a business owned by just one person. Fig. 11-1. It is the easiest type of business to form because you as the owner have complete control over everything, including the profits.

Partnership

Corporation

◀ OPPOSITE There are many kinds of businesses throughout the world.

Fig. 11-1. A business may be a proprietorship, a partnership, or a corporation. Which kind would you like to try someday? Why?

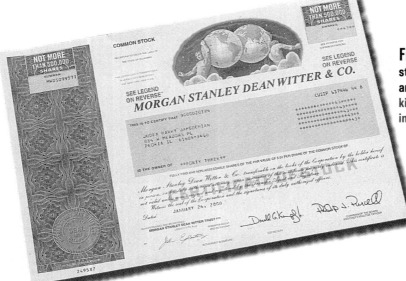

Fig. 11-2. People who own stocks, or shares, in a company are given a certificate. What kinds of companies would you be interested in investing in?

- **Partnership.** A business owned by two or more people is a partnership. It is also easy to form. The partners share the profits, the workload, and the responsibilities.
- **Corporation.** A corporation is a company organized and owned by stockholders. A *stockholder* is anyone who buys a share in a company. Fig. 11-2. Stockholders purchase stocks in companies hoping that the value of their stock will increase. While you own the stock, you also receive a dividend. A **dividend** is a payment you get as part-owner of the company. Forming a corporation is complicated. Corporations are very structured because there are so many people involved.

SECTION 1
TechCHECK

1. What is a company?
2. Describe three different ways to manage a company.
3. What is the main purpose of a company?
4. **Apply Your Knowledge.** List three companies and tell what their business is. Find out if they are corporations, proprietorships, or partnerships.

Starting a Corporation

THINGS TO EXPLORE

- Identify the positions within a corporate structure.
- Explain what a company's main departments do, including human resources, research and development, production, and marketing departments.
- Design a product prototype and conduct a consumer survey.

TechnoTerms
annual
stockholder

I f you wanted to start a corporation, you would first have to make out an application and get it approved by a government agency or department. After it is approved, you can sell stock to raise the capital needed to start and run the business.

Corporate Structure

After capital is available, you need to fill the positions within the corporate structure. Fig. 11-3.

- **Stockholders.** People who have bought shares in the company are the **stockholders**. They hold an **annual** (yearly) meeting. They elect the board of directors for the next year.
- **Board of directors.** Board members are elected. They set company policies and determine the main company goals. They report how the company is doing to the stockholders. The board also hires a company manager, or president, to run the company.
- **Management.** Managers run the company. They have to be good leaders and hard workers to make the company successful. Management picks the products the company will make, decides how to raise money to buy or rent buildings, locates raw materials, and determines worker salaries.
- **Administration.** The company president is the administrator in charge of the company and makes sure things are done right. The people in administration carry out the decisions made by the management department. Vice presidents are responsible for

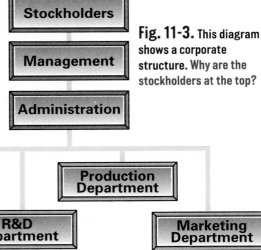

Fig. 11-3. This diagram shows a corporate structure. Why are the stockholders at the top?

certain areas in the company. In your school, for example, you might have a vice principal who has special duties to help the principal.

Main Departments in a Company

In a successful company, everyone works together to meet the company goal. Companies may have the right people and equipment to manufacture a product, but it takes even more than that to succeed. You also need to have a good production plan to produce a good product at the lowest cost.

Human Resources In some companies, the human resources department is called *personnel*. This department makes sure that the people who have the skills needed for certain jobs are hired. If extra training is needed, the human resources department makes sure workers get the training. They also make sure you are rewarded if you are a good worker.

Research and Development (R&D) The R&D department improves existing products or designs new products. Companies depend on R&D departments to find efficient ways to make products so they can save money and make a profit.

Production Department The production department is in charge of making the company's products. Production workers turn materials into parts for products and assemble the products.

Marketing Department The job of the marketing department is to sell the company's products. Sometimes marketing conducts consumer surveys to find out what people want, how much they are willing to pay for a product, and who would probably buy it. The marketing department must also have a *marketing plan* (strategy) for advertising and selling the product.

See Chapter 6 for more information about research and development.

SECTION 2

Tech*CHECK*

1. Who are stockholders?
2. What is the purpose of a marketing department?
3. What positions make up a corporate structure?
4. **Apply Your Knowledge.** Your school is a business. List the main departments and positions that make it run smoothly.

ACTION ACTIVITY

Starting a Company

Be sure to fill out your TechNotes and place them in your portfolio.

Real World Connection

Companies get started in order to make money providing a product or a service to you, the consumer. If the product is successful, the company may expand.

In the activities in this chapter, your class will organize into a company. Fig. A. You will sell stock, design products and advertisements, produce the product, package it, sell it, and make a profit (we hope). As in any business, there is no guarantee that there will be money left over after all the bills are paid.

Design Brief

Organize your class into a company, design a product prototype, and conduct a consumer survey. Your product should be something your class can produce in your technology lab for less than $5. It must be designed with safety in mind. Sharp edges, electrical shock hazards, or pinch points must be avoided.

Materials/Equipment

- materials needed to build a prototype
- computer with graphics and spreadsheet software
- machines and tools as needed to make a prototype

SAFETY FIRST
Follow the safety rules listed on pages 42-43 and the specific rules provided by your teacher for tools and machines.

Stockholders   **Board of Directors**   **Manager**   **Factory**

Fig. A

(Continued on next page)

ACTION ACTIVITY

Procedure

Part 1 · Designing the Product

1. Consider your market—the customers who will buy your product. If your product is going to be sold in school, you might think of the students as your market. What kinds of products would they like to buy?

2. Work with your teacher to brainstorm product ideas. Make a list of possible markets and products. Your list might look like the one shown in Fig. B.

3. Vote on each of the products on your list. Find the five products your class thinks have the best chance of selling. Work as a class to refine the ideas and make plans for building them.

4. Divide the work of building a prototype for each of the product ideas among the class. Your prototype does not need to be perfect. There will be time later to improve the design.

Part 2 · Making a Consumer Survey

1. You will need to get some information about what the people want in a product. By conducting a consumer survey (a series of questions), you can obtain feedback.

2. Create a survey form. All five of the products should be included on the form. The results of the survey will help you decide which of the prototypes will be produced by your company. Your survey might look like the one in Fig. C.

3. Give your survey to students, parents, or teachers—your market—and to other groups as well. The people you ask to complete your survey should be picked at random (by chance).

4. Evaluate the results of the survey. When you do this, you are compiling data. The data can be presented to the class in the form of a graph. It is important to evaluate the results fairly, without letting your feelings interfere with your judgment. Fig. D.

5. Make a list of the prototypes, ranking them from most popular to least popular.

Evaluation

1. Could your survey have been done by phone?

2. What questions would you change?

3. Why do companies make prototypes of products?

4. **Going Beyond.** Put the consumer survey results on a spreadsheet.

5. **Going Beyond.** Design another prototype where cost is not a consideration. Let your imagination work to come up with an idea that you think could be manufactured and sold.

Market	Elementary School	Middle School	Parents	Teachers
Product	pencil holder	clipboard	clipboard	clipboard
	locker shelf	locker shelf	desk tray	desk tray
	bookmark	book holder	can holder	key chain
	toy	game	game	game
	puzzle	CD or tape holder	CD or tape holder	CD or tape holder
	picture	school logo	school logo	school logo

Fig. B

CONSUMER SURVEY

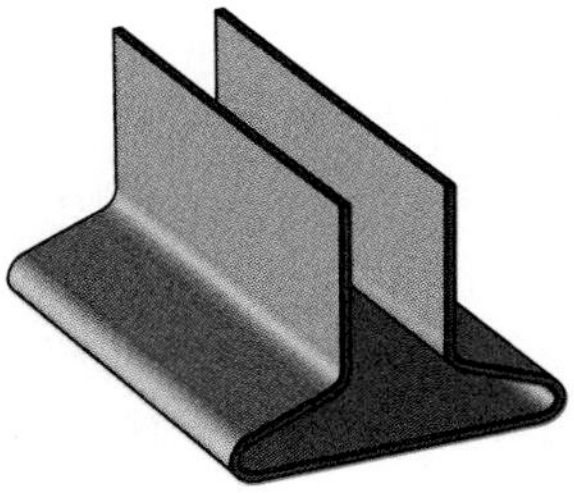

1. **Are you tired of a cluttered desk?**
   ☐ Yes   ☐ No
2. **Would you buy a plastic letter holder?**
   ☐ Yes   ☐ No
3. **Would a letter holder make a good gift?**
   ☐ Yes   ☐ No
4. **What color do you like best for a letter holder?**
   ☐ Clear ☐ Red ☐ Blue ☐ Yellow ☐ Other_____
5. **Can you think of other ways to use the letter holder?**

   _____

If you return this survey, you will be given a coupon for 20% off your letter holder purchase. Thank you for your help.

Fig. C

Fig. D

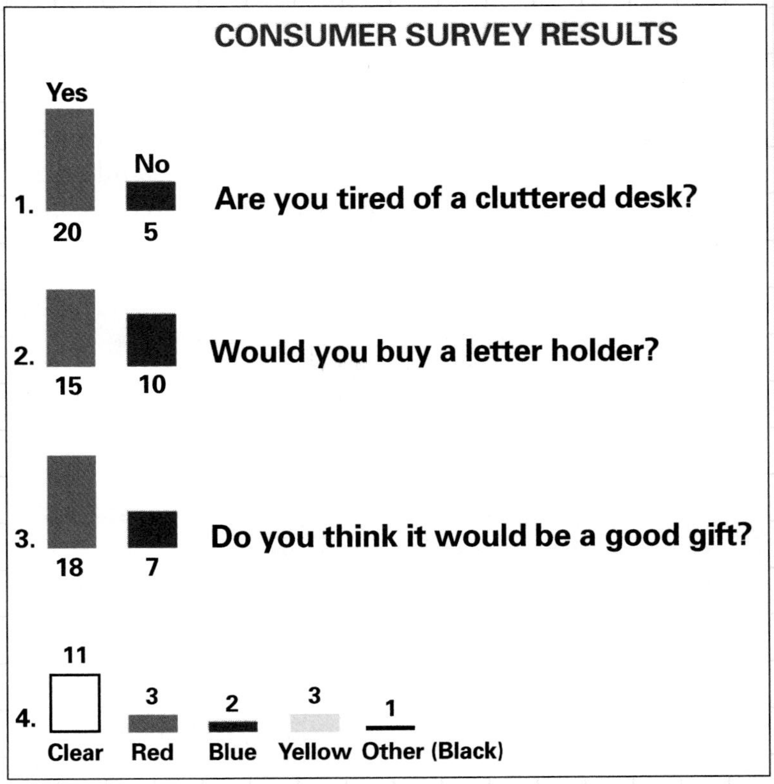

CONSUMER SURVEY RESULTS

1. Yes 20   No 5   **Are you tired of a cluttered desk?**

2. 15   10   **Would you buy a letter holder?**

3. 18   7   **Do you think it would be a good gift?**

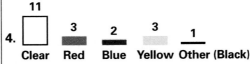

4. 11 Clear   3 Red   2 Blue   3 Yellow   1 Other (Black)

Mass Production

TechnoTerms

inventory
just-in-time
manufacturing
(JIT)
synchronized
production

THINGS TO EXPLORE

- Explain mass production and tell when and why it is used.
- Explain what just-in-time manufacturing is and why it is used.
- Apply and interview for a job in a school-organized company.

Mass production enables companies to produce large quantities of parts and products within a short time. To do this, each worker in a factory assembly line is assigned only one job. Each person does the same job over and over. Mass production also makes products less expensive to produce and therefore less expensive to buy.

Did you know that your shoes are probably mass produced by machines? Fig. 11-4. One machine cuts the material, another punches holes for laces, another glues the soles, and so on, until the final product is the shoe you recognize.

Fig. 11-4. Many parts of these shoes are made by machine. Here, a worker helps assemble them. Would you want to work on an assembly line? Why or why not?

Standardized Parts

Mass production works only if all the same parts are standardized (the same size and shape). The parts must be interchangeable.

Today's automobiles are made mostly with standardized, or interchangeable, parts. If one part, like a headlight or a door handle, breaks, you can buy another one just like it from an automobile dealer.

Just-in-Time Manufacturing

Where do companies store materials and purchased parts until they are needed on the production line? Many companies order large quantities (amounts) of materials ahead of time. They then must pay for storage space in a warehouse. People must be hired or robots

Fig. 11-5. The parts for this tractor arrived at the factory just before production began. What would happen if the parts were late?

TechnoFact

BUSINESS IN THE 21ST CENTURY. What will business be like in the 21st century? Some business specialists say that some current trends will continue. For example, product life cycles are becoming shorter. That means products are made that will not last long. People want more new products that can be developed in a shorter time. Managers and executives need more information faster to help them make decisions more quickly.

purchased to move the material to the production line. Once the finished product is made, it often spends time in a warehouse, waiting to be shipped.

One way to cut down on **inventory** (things in storage) and costs is to use a computer to schedule deliveries just in time. **Just-in-time manufacturing (JIT)**, also called **synchronized production**, is a method that many companies are turning to. It eliminates the need for storage space and workers to manage the inventory. All the materials and ordered parts get to the factory just in time to be used in production. Fig. 11-5. When the product is finished, it is not stored but is immediately shipped to the customer. Just-in-time manufacturing enables companies to cut back inventory as much as possible.

INFOLINK

See Chapter 4 for more information about computers.

SECTION 3
TechCHECK

1. Why do companies use mass production?
2. Why do you need standardized parts in mass production?
3. What is just-in-time manufacturing?
4. **Apply Your Knowledge.** List ten products you think are mass produced.

ACTION ACTIVITY

Organizing a Company

Be sure to fill out your TechNotes and place them in your portfolio.

Real World Connection

In the last activity, your class decided on a product to be manufactured by your student company. Now it's time to organize your company and get down to the business of doing business. Fig. A. In this activity, you will sell stock, organize into departments, and start producing your product.

Design Brief

Organize marketing, production, bookkeeping, and human resources departments. Design a stock certificate for your company. Sell stock in your company to raise capital. Apply and interview for a job in the company. Learn your job and start work.

Materials/Equipment

- materials needed to produce your product
- equipment needed to produce your product
- computer with graphics, database, word-processing, and spreadsheet software
- conveyor system created in Chapter 10 (optional)

SAFETY FIRST
Follow the safety rules listed on pages 42-43 and the specific rules provided by your teacher for tools and machines.

Fig. A

Procedure

1. Your company must be organized into departments. First, the company will need a dependable student to be president. The president will organize and supervise all of the company's operations. Put the names of the major departments on the board. You may volunteer for any of the departments.

2. Each department will need a vice president. The vice president of each department will report to the company president for assignments.

3. Each department must complete specific jobs before production can start. The assignments for each department are as follows:

   • **All departments:** Decide on a name for your product and a name for your company. Sell stock ($0.25 per share) to people in your company or outside of your company. Be sure to caution people that there is no guarantee of profit. Give the money and receipts to your teacher to help pay the bills for materials used to make your product.

   • **Bookkeeping:** Design a company stock certificate. The certificate must have the following information: company name, stockholder's name, date, value, receipt, and number. It might look like the example shown in Fig. B. Keep track of the stockholders on a spreadsheet.

   • **Production:** Refine the prototype so that the product is ready to be mass produced. Make a final drawing of the product that shows *dimensions* (sizes). Make a list of all of the parts of the product, their material, part name, and size. Make a flowchart of the production process. There should be a place on the flowchart for every operation that must be done. Your teacher will help you design and make jigs and fixtures for the machines. A *jig* holds an object and guides the tool during work. A *fixture* keeps the object in the proper place while it's being made. The jigs and fixtures will make your product easier to mass produce.

INFOLINK

See Chapter 3 for more information about flowcharts.

EGGSTRA SPECIAL CORPORATION

STOCK CERTIFICATE

This certifies that_____is the owner of_____shares of stock in the Eggstra Special Corporation. This stock is non-transferable and can be redeemed for $0.25 per share at any time or held until liquidation.

President_____V.P._____

Bookkeeper_____Stock Number_____

Fig. B

(Continued on next page)

- **Marketing**: An advertising campaign will help improve sales of your product. Your group should consider posters, flyers, announcements, demonstrations, video or audio commercials, and school newspaper ads to make your product more visible.

- **Human resources:** Your department will place all the workers in the jobs for which they are best qualified. First, design and make copies of an application form. Then get a list of the jobs from the production department. Post a list of positions available. Have each student (including those in the personnel department) complete an application for a job. The vice president of human resources and the company president will interview the job applicants and hire people for the jobs for which they are qualified.

4. When your company is ready for production to start, each worker must know his or her specific job. If possible, set up the conveyor belt systems you made in Chapter 10. The conveyors should be placed so the movement of materials follows the flowchart made by the production department.

5. As products are completed, they should be placed in storage for the next activity.

INFOLINK

See Chapter 10 for how to build a conveyor system.

Evaluation

1. What was your specific job in the assembly line? What did you like or dislike about your job?

2. How could your job have been changed to make the production of products more efficient?

3. Design your place on the assembly line with ergonomics in mind. How far do you have to reach for parts or tools? How would you feel after doing the job all day?

4. How could the conveyor systems be changed to make them better? Make a sketch of your ideas.

SAFETY FIRST

Safety on the job is important to millions of workers. Always keep safety in mind. If a worker is hurt on the job, it costs money and slows production. Be careful!

Total Quality Improvement

THINGS TO EXPLORE

- Define quality control and tell why it is important.
- Explain where inspections are done in the production cycle.
- Identify special tools and techniques used for quality control.
- Design a go-no-go gauge to use as a quality-control tool.

TechnoTerms

acceptance sampling
burn-in test
defect
quality assurance
quality control
statistical process control

Have you ever bought a new shirt only to find out after you got it home that a seam had come loose? Or maybe the zipper on your coat never worked right? Maybe you bought a compact disc (CD) or tape only to find a **defect** (something wrong) in it that ruins the way it sounds? Then you know one reason why we want quality control of products. How well it is made helps determine the *quality* of a product. Fig. 11-6. The reputation of a company can be hurt by just a few bad products.

Companies want to make high-quality products. To do this, they often set a quality standard before they make the product. **Quality assurance** means that a product is produced according to specific plans. Another name for quality assurance is **quality control** (**QC**).

Production Cycle Check Points

Inspectors examine parts, materials, and processes at all stages of production. For example, in manufacturing, parts are checked for strength and to meet standards set up in the production line. Inspections often are made during three key times during the production cycle.

- **Delivery of materials.** Materials are inspected as they arrive. If the materials don't meet the standards, they are rejected and sent back.
- **Work in process (WIP).** Inspectors check that work is being done in the right way and that the right parts are being used.

Fig. 11-6. At this BMW factory in Germany, inspectors check for possible problems. What kinds of things might they be looking for?

Fig. 11-7. Computers are tested to be sure they are working properly. Have you ever bought a defective product? Describe the experience.

- **Finished product.** This is the final inspection, where everything should work and look right! Fig. 11-7.

Quality Inspection Methods

Some special quality control techniques are used in manufacturing.

Measuring Parts A drawing is made for every part for a product. The drawing shows how to make the part and includes sizes in exact dimensions. Because most parts do not need to be perfect, a range of sizes is often given. For example, the range might be 1.5" ± 0.005". This range of acceptable sizes is called the *tolerance*. The part can be 0.005" over or .005" under and still be just fine.

To check materials, parts, and products, inspectors use some special tools, such as *gauges*. A go-no-go gauge, for example, is a simple tool that does not require workers to take time to read measurements. It is made to fit a particular size part. If the part fits, it passes.

Some inspection devices emit sound waves or X-rays. Laser beams are used to make very precise (exact) measurements. New quality control techniques even make it possible for you to check inside without damaging the product.

Acceptance Sampling Many products are made in large quantities. Can you imagine how hard it would be to check every M&M candy as it passed on a conveyor belt? When it is not always possible to check each product, inspectors use a procedure called **acceptance sampling.** Acceptance sampling means that you select a few samples and inspect them to see if they meet the standards. If the samples pass inspection, then the whole batch is approved. If the samples don't pass inspection, the batch is rejected.

Fig. 11-8. Because the circuit board is the heart of a computer, it must work right.

Statistical Process Control

Statistical process control (SPC) is used to make sure that a process is being done right. If the process is correct, then the product itself doesn't need to be inspected.

Suppose you have a certain machine that automatically fills empty boxes with crackers. Although the box label says it contains 12 ounces of crackers, not all the boxes are filled exactly the same. The machine is set to fill each box within certain limits—from 11.5 ounces to 12.5 ounces. A control chart on a computer keeps track of each box so if the machine goes over or under the limit, the workers can stop and adjust the machine.

Burn-in Tests The **burn-in test** is another quality assurance method. It is done mostly on electronic products like computers. Fig. 11-8. Usually if a computer is going to fail, it does so during the first few hours of operation. For this reason, manufacturers let every computer run for a couple of hours as the burn-in test. Computers that don't pass the burn-in test are repaired if possible.

SECTION 4
TechCHECK

1. What is meant by quality assurance?
2. When are inspections made during the production cycle?
3. Name some quality control tools and methods.
4. **Apply Your Knowledge.** Compare the number of cookies in two boxes of the same brand.

TechnoFact

CLOSE ENOUGH One of the problems with the production of a product may be that the parts are not the same size. There is no way to make every part exactly the same. Fortunately, most parts do not require extreme accuracy. Accuracy adds to production costs. Most machined parts can be a few thousandths of an inch off and still work. Wood or plastic parts may work just fine if they are as much as one-sixteenth inch off the desired size.

ACTION ACTIVITY

Making Quality Products

Be sure to fill out your TechNotes and place them in your portfolio.

Real World Connection

Now that you have some products made, it's time to try to sell them. It is important that every product sold be of the highest quality possible. For example, Fig. A shows two women inspecting french fries. In this activity, you will design ways to test the quality of your products to ensure customer satisfaction.

Design Brief

As a team, design and perfect a method of quality control that will help to make your products better.

Materials/Equipment

* wood or acrylic (Plexiglas)
* machines and tools, as needed

SAFETY FIRST
Follow the safety rules listed on pages 42-43 and the specific rules provided by your teacher for tools and machines.

Fig. A

Procedure

1. The accuracy of each part you make directly affects the quality of your product. Your task is to invent a way to ensure quality. Work in groups of four or five to analyze the jobs in the assembly line. Look for ways to improve quality.

2. Decide on the acceptable tolerance for each of your product's dimensions.

3. Design and make a go-no-go gauge for the parts of your product. Your gauge might look like the one in Fig. B.

4. To use the gauge in the assembly line, all you have to do is put the part next to the gauge. If the part is smaller than the "no-go" side or larger than the "go" side, it is not acceptable.

5. Your product may have parts that are difficult to measure with a go-no-go gauge. Your group should then design and build a testing device that will test your parts for accuracy.

6. Your president and the vice presidents of the departments should decide if some of the workers on the assembly line could be put into a new department called quality control.

Evaluation

1. Have you ever purchased a product and found that it was defective? What was the problem?

2. What happens to the quality of products when workers are tired, bored, or mad at their boss?

3. How could the quality of your product be improved?

4. Going Beyond. Visit a factory, and ask how the quality of the product is measured.

5. Going Beyond. How do you think modern automation and robotics in factories have changed product quality? Do you think a car built by robots is better than one built by people? Explain.

6. Going Beyond. Can you think of ways to use robots or automation to inspect the quality of products? Explain.

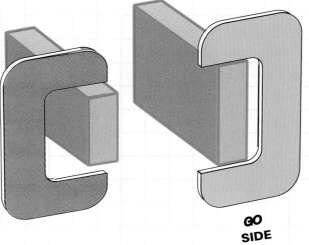

Fig. B. The part on the left is too small. The part on the right is too big.

GO SIDE

NO-GO SIDE

Packaging and Selling Products

TechnoTerms

direct sales
distribution
marketing plan
retail sales
sales forecast
wholesale sales

There are many reasons for packaging products. This section will explore several. As you read this section, think about how it applies to the product your class is making.

Why Are Products Packaged?

Have you ever ordered an item from a catalog and, when it arrived, it looked like a truck ran over it? One purpose of packaging is to ensure that a product gets from one place to another without a lot of damage. Packaging also protects the product from damage while it is on the store shelf.

Sometimes products need more than one package. Think of a bandage or a piece of chewing gum. To keep it clean and safe to use, each one is individually wrapped and packaged inside a larger package. Small parts are often packaged in larger containers to make them easier to keep track of in a store. If your product is a liquid, then the packaging has to contain the liquid safely. Packaging can be a way of making your product more attractive and appealing to customers. Packaging also helps prevent theft. Fig. 11-9.

Fig. 11-9. Because they're small, music CDs are targets for theft. How does packaging help prevent theft?

Packaging is done at the end of the assembly process. Most companies buy packages from a package manufacturer. The package might already be printed with the company symbol, or *logo*. You may often buy products, like Nike shoes, Apple computers, or Coca Cola and Pepsi soft drinks, just because you recognize their special logos. Fig. 11-10. Do you know what these company logos look like? Companies want to build an image that you recognize and can remember easily.

Marketing and Advertising Products

Before a company begins to sell a product, the marketing department makes a special **marketing plan**, or strategy. The plan might include a **sales forecast** (prediction of how many products the company will sell). If the market research is not accurate, the company could lose a lot of money. Can you imagine making thousands of products but selling only a few hundred?

Have you ever bought something just because an ad caught your attention? Companies count on that happening. The main goal of advertising is to make the product familiar to the public and to convince customers that they need the product. Trade names like McDonald's, Honda, Disney, and Hershey make it easier to sell products. People already know the name, so if something new comes out from that company you are quick to notice it.

Lots of companies try different advertising methods, such as television and radio commercials, billboards, and ads in magazines, newspapers, and on the Internet.

Fig. 11-10. Logos make products easy to recognize. Design a logo for a new soft drink.

Fig. 11-11. One form of direct sales is to sell items from a company website. Some websites can even help you locate a car to buy at a local dealership.

Selling Your Product

Selling the products to the buyers is part of a marketing strategy. There are three main ways to sell your product. Fig. 11-11.

- **Direct sales.** Direct sales means selling a product directly to the customer. Usually a company will have salespeople who make a *commission* (certain percentage) on the amount they sell.
- **Wholesale sales.** Wholesalers are people or companies that buy large quantities of products from manufacturers. Then they sell the products to other businesses in large quantities.
- **Retail sales.** Retailers buy products either from wholesalers or directly from the manufacturer. Then they sell them to you at discount stores or large department stores.

MATHEMATICS CONNECTION

Figuring the Profits

Companies need to find out if they are making a profit by calculating the *break-even point*. To calculate the break-even point, you need to make a graph that shows your sales and costs. It might look like the one shown in this feature.

After all the products are sold or everyone in the class has tried to sell the product, it is time for your company to come to an end. In business, closing out a company is called *liquidation*. The profit that your company makes will then be divided among the shareholders. Here is the formula for calculating the profit per share.

profit per share = total profit ÷ total number of shares sold

Distributing Your Product

In addition to selling their products, companies must have ways to get their goods to the buyers. This is called **distribution**. Sometimes the distribution path for a product can be short, as in direct sales. Sometimes products must be temporarily stored in warehouses until the right time for distribution.

At some point, transportation is involved in distributing the products. Depending on the product or the need, different kinds of transportation, such as air freight, trucks, trains, or ships, might be used. Most of the "18-wheelers" you see on the highways are carrying products to wholesalers or retail stores.

SECTION 5
TechCHECK

1. Name three ways to sell a product.
2. Give at least three reasons for packaging.
3. What is a marketing plan?
4. **Apply Your Knowledge.** Pick five products that you think have exciting or interesting ads. Why do you like the ads for these products?

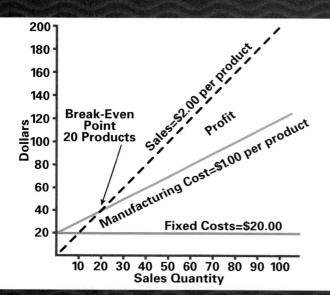

ACTIVITY

Your teacher will provide the cost of materials for your product. Calculate the break-even point for your company or for another product you sell at a school store.

ACTION ACTIVITY

Pack It Up and Sell It!

Be sure to fill out your TechNotes and place them in your portfolio.

Real World Connection

Now your company is well on its way to making a profit. Your products are being made with quality in mind, and your company is ready to roll. The next step is to package and sell your products. In this activity, you will design a package for your product.

Design Brief

Design a package for your product that will include the specifications listed below. Your package idea should

- Protect the product from being damaged
- Provide information about the cost of the product
- Include instructions on the safe use of the product
- Help prevent theft by shoplifting
- Be attractive and eye-catching
- Be environmentally safe and biodegradable

Materials/Equipment

- packaging materials
- packaging equipment such as a heat gun and plastic film sealer (optional)
- silk screen printing equipment (optional) Fig. A.
- computer with graphics software (optional)

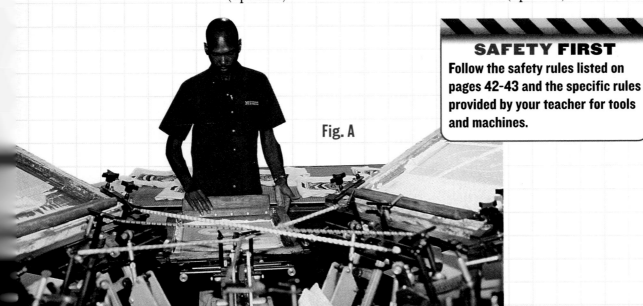

Fig. A

SAFETY FIRST
Follow the safety rules listed on pages 42-43 and the specific rules provided by your teacher for tools and machines.

Procedure

1. Work in groups of four or five to brainstorm ideas for possible package designs. Your design will depend on the product and the type of package desired. Your package might be in the form of a box, card, bag, label, or other device. Your group may decide to print graphics or extra instructions on your package. Most packages are thrown away after they are opened. You have a responsibility to design a package that will either decompose in a landfill, be recycled, or be reused.

2. Choose the idea that you think best meets the requirements listed in the design brief. Fig. B. The package you design must be produced in the same quantities as your product. Keep your design simple so that most of the company's effort can be directed toward making a quality product.

3. Sketch your design on paper, or make a preliminary, or first draft, computer drawing.

4. As a team, make a sample package.

5. As a group, present your package idea to the class.

6. Evaluate each of the packages on the basis of the design brief requirements. Use the most appropriate design for packaging your product.

Evaluation

1. How will your package protect your product?

2. When you walk through a grocery store aisle, which products catch your eye? Can you explain why?

3. What products have you purchased that seemed to be "overpackaged"?

4. **Going Beyond.** Evaluate five packages of products you have at home, using the design brief specifications.

5. **Going Beyond.** List five products that have packages that are heavier than the product.

6. **Going Beyond.** What do you think packages will look like in the future?

Fig. B

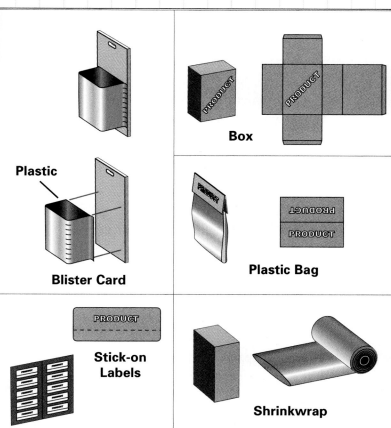

Blister Card

Box

Plastic Bag

Stick-on Labels

Shrinkwrap

11 REVIEW &

CHAPTER SUMMARY

SECTION 1

- A company is an organized group of people doing business.

- A company can be organized as a proprietorship, partnership, or corporation.

SECTION 2

- Stockholders, a board of directors, management, and administration are part of the structure of a corporation.

- Important company departments are human resources, research and development, production, and marketing.

SECTION 3

- Mass production enables companies to produce large quantities of products within a short time.

- Just-in-time manufacturing (JIT) eliminates the need for storage space and people to manage inventory.

SECTION 4

- Quality control makes sure products are made according to specific plans.

- Quality control inspections are made during the delivery of material, while work is in process, and when the product is finished.

SECTION 5

- Packaging keeps products from being damaged, protects products while on the shelf at a store, keeps small parts from being lost, and interests people in the product.

- Distribution departments get products to customers.

REVIEW QUESTIONS

1. Look at the ads for products in a magazine. What ads attract your attention? Why?

2. What quality control methods are you used to dealing with in school?

3. Why do you think products are always being tested and retested?

4. Was your product package design environmentally safe?

5. How have products such as milk and soda containers changed since your grandparents were your age?

CRITICAL THINKING

1. Design a consumer survey that people could fill out easily.

2. Why do people sometimes offer food samples in grocery stores?

3. What other techniques can you think of to get people to try a new product?

4. How would you feel if you were fired and your job on an assembly line was taken over by a robot? Explain.

5. Visit a local factory that makes products using the assembly-line process. Ask some of the workers what they like or dislike about their jobs.

6. Use computer graphics software to draw a floor plan (view from above) of an assembly line. How could you rearrange the equipment or conveyors to make the flow of materials more efficient?

ACTIVITIES

CROSS-CURRICULAR EXTENSIONS

1. COMMUNICATION Choose a product that you think is packaged without concern for the environment. Redesign the package. Make a sketch of your design, and send it with a letter to the company that made the product.

2. MATHEMATICS Use a spreadsheet to calculate the break-even point for your company. What was the biggest expense in your company?

3. SCIENCE Research bio-related technology companies on the Internet and find out what they do. See Chapter 19 for information on bio-related technology.

EXPLORING CAREERS

To be successful in business today, people need to study new approaches, such as electronic commerce. Changes in business will affect the career opportunities found, too. Here are only two of the many careers found in business today.

Product Manager Product managers oversee every step of the process for developing a product. They participate in choosing the original design and in deciding how to sell the product. They make presentations explaining how the product works. Managers must often be able to do a number of things at one time.

Employment Interviewer
Employment interviewers interview job applicants and refer them for jobs. They work in human resources departments or for employment agencies. Interviewers must be able to communicate well with people and keep up to date on hiring laws. While it helps to be familiar with technology and different high-tech jobs, it is not a must for success in this field.

ACTIVITY

Find out if your school provides *job shadowing*. In job shadowing, you spend time with someone at the workplace. By watching and listening, you learn about the job.

Building Things

What Is Construction?

THINGS TO EXPLORE

- Define construction and give examples of different structures that people build.
- Explain how construction has changed with the development of new materials and different needs.

Techno Terms
constructed
construction
shelter

You've learned about the production of smaller products from cars to crayons. Now you will learn about the production of larger products, such as houses and bridges. This area of production is called **construction**.

The Construction Industry

The construction industry uses the resources of raw materials, money, land, and technology to produce the structures you're used to seeing. Construction is a large and complex industry that affects you in many ways.

How much do you know about building structures? Look around you at all the houses, office buildings, schools, and factories that people have **constructed** (built). Fig. 12-1. Most people think of construction as buildings. But construction also gives us highways, airports, bridges, dams, and tunnels, to name just a few. Sometimes structures, like the Washington Monument or the Lincoln Memorial, are important for their design or historical meaning. But most of the time, we design our buildings and other structures to meet a special need.

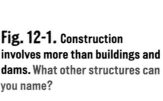

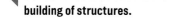
OPPOSITE Construction is the building of structures.

Fig. 12-1. Construction involves more than buildings and dams. What other structures can you name?

A Brief History of Construction

In the beginning, people needed a **shelter** (place out of the cold and rain). Early shelters were caves or structures made mostly of animal skins, twigs, branches, mud, or other natural materials that they found. Fig. 12-2. As people started to settle in one place, they needed more permanent dwellings. They needed shelters that wouldn't blow over in a strong wind and that could withstand different kinds of weather.

Though no one knows the name of the world's first structural engineer, early humans learned to build some amazing structures without machinery or iron tools. Have you seen pictures of the pyramids? The Great Pyramid at Gizeh, Egypt, was built about 2600 B.C. from huge limestone blocks cut with copper chisels and saws. The people who built it had no mortar or cement, so the entire pyramid is held together only by the weight of the stones. Later, the Romans built huge arenas, bridges, dams, aqueducts (waterways), and temples to serve their needs. You can still visit many of them today.

Fig. 12-2. Nomadic villagers in India build these teepee-like homes. Look up the word *nomad.* Why do you think nomads would prefer structures like these?

Fig. 12-3. Concrete is used in road construction.

By the nineteenth century, new materials changed the ways structures could be built. The Bessemer steel-making process, developed in the 1850s, was used to mass-produce high-quality steel. This steel made it possible to build structures that were stronger, more reliable, and longer lasting. Then new concretes that could be poured and would set underwater were developed. Concrete combined with steel produced *reinforced concrete* that had the best properties of both materials. Fig. 12-3.

Today, plastics, fiberglass, and improved metals enable architects and structural engineers to build even better and stronger structures. Better sources of power and new technologies also have helped the construction industry move ahead.

SECTION 1
TechCHECK

1. What is construction?
2. List five different examples of construction.
3. What developments have helped architects and structural engineers build better and stronger structures?
4. **Apply Your Knowledge.** Research different construction projects in your area. Find out why they are being built and what materials are being used to build them.

Making Plans for Construction

THINGS TO EXPLORE

- List information designers need to know before they can begin designing plans.
- Explain what a floor plan is.
- Identify different kinds of working drawings, such as site plans, floor plans, elevation drawings, and section views.
- Design and draw a floor plan.

Before any actual building begins, you have to plan your design. If you were an architect or a structural engineer for a project, you would meet with your *client* (the person you are building for) to discuss the client's needs. Some things you as the designer would need to know are

- What the building will be used for
- How much space is needed to build the structure
- How much money the client wants to spend on this project
- Where the *site* (building location) is
- Soil, water, and other conditions at the site
- Any special community building codes that must be followed

The Planning Process

After discussing your client's specific needs, you would start with some *preliminary* (first) sketches. From these you would make more detailed plans until you and your client were both satisfied. Then you would make accurate, detailed, scale drawings called **floor plans**.

INFOLINK

See Chapter 5 for more information about CAD.

You might use either CAD (computer-aided design) software or traditional drafting tools to make your drawings. Often, house plans and plans for other small buildings are drawn to a scale of 1/4 inch. In other words, each foot of the actual building is represented by 1/4 inch on the plans.

Working Drawings

Many kinds of working drawings are needed to build a house. You need a site plan, floor plans, elevation drawings, section views, and maybe other special information. Drawings made using **orthographic projection** show several views of the project in two dimensions. You can be looking down on top of the structure, looking directly at the front, or looking at the back or sides.

- **Site plan.** A site plan is drawn to show the property boundaries and the exact location of the structure. Fig. 12-4. Surveyors are sent to the site to accurately locate boundaries and to stake out the building site.

- **Floor plan.** This is a view looking down on the house without the roof. Fig. 12-5. You can see the location of the walls, all door and window openings, and the size of each one. You would have a separate floor plan for each floor of the building, including a second floor or a basement.

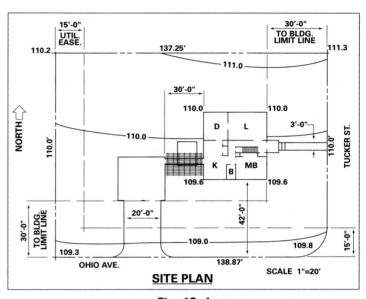

Fig. 12-4. Site plans are drawn using information from surveyor's instruments. Draw an approximate site plan for your school.

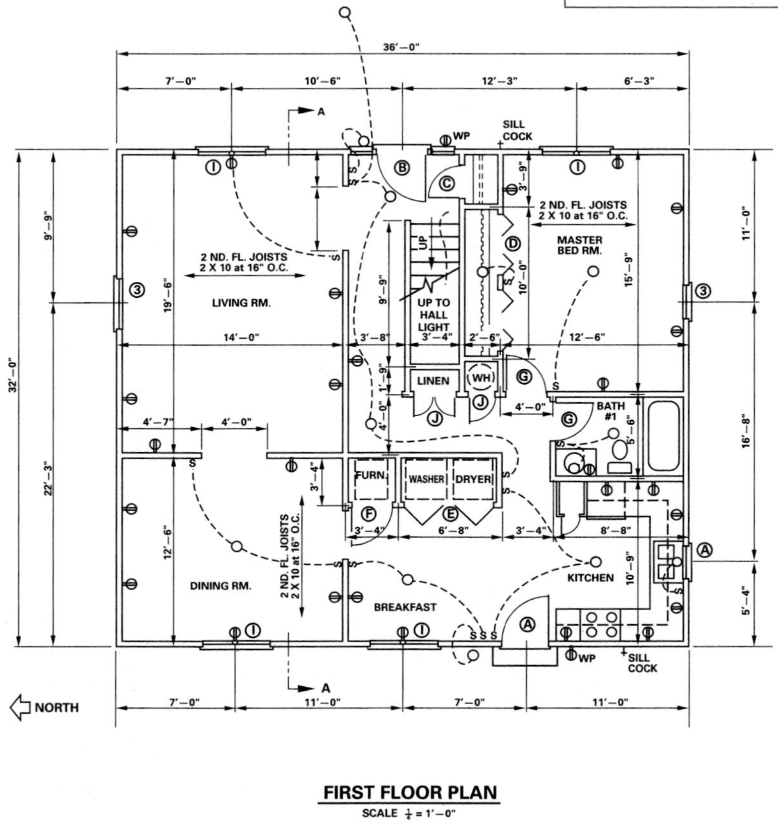

Fig. 12-5. This is a floor plan of a house. Draw a floor plan of your classroom.

FIRST FLOOR PLAN
SCALE ¼ = 1'—0"

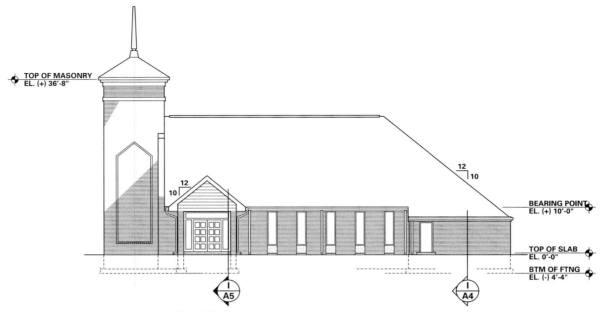

Fig. 12-6. This elevation shows the front of a church. What is the height of the tower masonry?

BUILDING BY THE RULES Designers and builders need to follow certain rules. For example, *zoning laws* regulate the type and size of building that can be constructed in a neighborhood. Almost all communities have *building codes*. These are laws that set the minimum requirements a building must meet. Most local building codes are based on guidelines published by national organizations. For example, the National Electrical Code (NEC) provides guidelines on wiring materials and methods.

Fig. 12-7. This section view shows the roof and outer walls of a house. Why do you think section views are necessary?

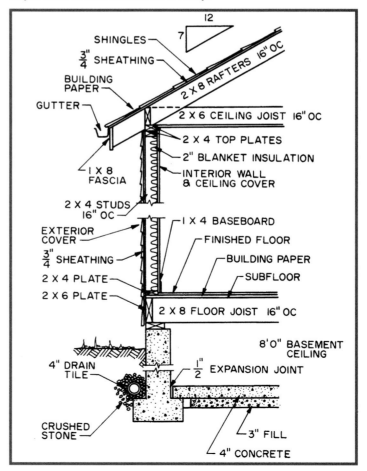

- **Elevation drawings.** These are views looking at a building from the front, the sides, or the back. Fig. 12-6.
- **Section views.** These are views looking into the building as if a part had been removed. Fig. 12-7.
- **Detail drawings.** These drawings are usually done to a large scale and have more information about a particular part of the project. Fig. 12-8.

A *pictorial drawing* shows a structure in three dimensions. It resembles a picture. Sometimes you might use it to show a client what a structure will look like when finished.

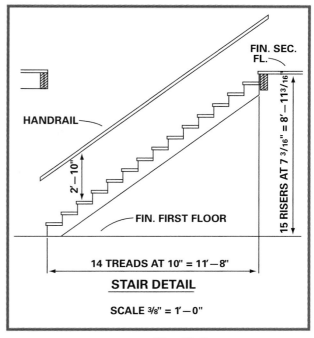

STAIR DETAIL

SCALE ⅜" = 1′—0"

Fig. 12-8. The detail drawing for a stair tells how many steps it must have. Why must each step be the same size?

Scale Models

You might decide to build a scale model. Models make it easier to show a client what you have designed. Models also make it easier and less expensive to test for such things as wind resistance and strength before the actual structure is built. Can you imagine building a tall skyscraper only to find out that the wind is so strong at the top that all the windows pop out? That has happened to buildings before!

SECTION 2

TechCHECK

1. What is a floor plan?
2. What are some things an architect needs to know in order to plan a structure for a client?
3. List five different kinds of working drawings you might need to build a house.
4. **Apply Your Knowledge.** Ask an architect to show you the types of working drawings used on a project. See if you can identify them, or bring samples to class.

ACTION ACTIVITY

Designing Your Dream House

Be sure to fill out your TechNotes and place them in your portfolio.

Real World Connection

House designs are often made by the owner of the house. There are many things that must be considered in the design of a house. The cost can be reduced and more livable space built in if you plan carefully. In this activity, you will design a dream house that you might build someday.

Design Brief

Design and draw the floor plan for a house that you would like to build some day. Your design must be practical and efficient. It must include the following:

- three bedrooms
- two bathrooms
- kitchen and dining area
- two-car garage
- utility room and laundry area
- living room or den
- closet space

Materials/Equipment

- graph paper
- drafting tools
- model-making materials (optional)
- CAD system (optional)

SAFETY FIRST

Follow the safety rules listed on pages 42-43 and the specific rules provided by your teacher for tools and machines.

Hints for House Design

- Imagine yourself walking through the house you've designed. Can you get to the rooms easily without bothering people in other rooms? Are the bedrooms located near noisy areas such as the kitchen or living room?
- Remember that safety is an important part of your design. There must be at least two outside doors in case of fire. The windows should be large enough for people to exit through in an emergency.
- Too many hallways are usually a sign of wasted space. The cost of a house is usually calculated by the square foot. You pay as much per square foot per hallway as you do for living space.
- If your design will include a second story or a basement, you will need a floor plan for each level. Be sure the stairs for each level are located carefully.
- Design the stairways so that furniture can be moved up or down easily.

Procedure

1. Make some preliminary sketches on graph paper. It is easiest to use 1/4" graph paper and a scale of 1/4" = 1'. Remember, your first idea may not be the best.

2. Architects and contractors use a set of symbols to draw floor plans. A wall, for example, is not drawn as a thin line. Walls are really 4 1/2" to 12" thick, depending on the construction materials used. The drawing should be made to show the actual thickness of walls. Some of the common symbols you might need to use in your design are shown in Fig. A.

3. Page 272 shows some general hints to help you design a house that will be livable and efficient. They are guidelines, not hard-and-fast rules.

4. When you have refined your preliminary sketch, it is time to make a finished drawing. If you use a computer, changing the plan will be very easy.

5. Complete your floor plan, and present your design to the class.

Evaluation

1. How many square feet (area) are in your design? Include all living spaces and hallways in your calculations.

2. Ask your teacher to find out the average cost per square foot for building a house in your area. Calculate the cost of building your house using this formula:

   cost = cost per square foot x area in square feet

3. How could the building cost of a house be reduced?

4. **Going Beyond.** Design a solar-heated house. Provide for large windows that face south. Make the inside area of the house from a material that can absorb heat during the day and radiate the heat at night.

5. **Going Beyond.** Build a scale model from your floor plan.

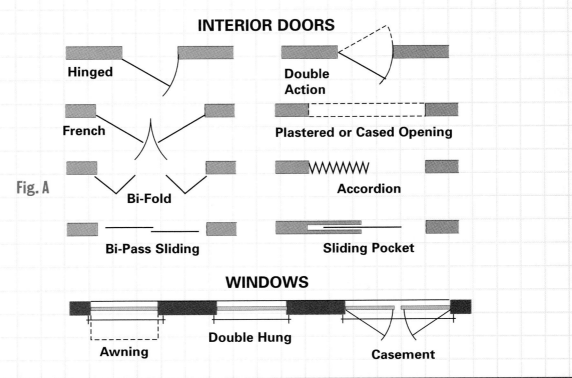

Fig. A

INTERIOR DOORS

Hinged

Double Action

French

Plastered or Cased Opening

Bi-Fold

Accordion

Bi-Pass Sliding

Sliding Pocket

WINDOWS

Awning

Double Hung

Casement

Structural Design

TechnoTerms

compression
dead load
dynamic load
live load
shear
static load
structural drawing
tension

THINGS TO EXPLORE

- Explain what structural drawings are and what information they provide.
- Identify the forces that work on structures and tell how they affect the structure.
- Explain the difference between static and dynamic loads.
- Tell what tension, compression, and shear forces are.
- Design, build, and test a model bridge.

Now that you've made the design plans, you need **structural drawings**. These drawings show structural parts. Kinds and sizes of materials that will be used, where parts go, and how the parts will be fastened should all be included. For example, if you were constructing a large building, you might need different kinds of glass for windows. The structural drawings would show where reflective glass should go or where you are supposed to put some wired glass for security reasons.

Forces Working on Structures

All structures, no matter what their shape or function, are in a game of tug-of-war. Nature's forces act on one side and the strength of the structure's design and materials act on the other.

Structural engineers have to plan structures that have enough strength to stand a sudden gust of wind, an extreme increase in temperature, an earthquake, the pull of gravity, and even the wearing effects of water. The San Francisco earthquake in 1989 caused skyscrapers to move, bridges to buckle, and buildings to collapse. Structural engineers were amazed that most of the area's structures made it through the quake. Fig. 12-9.

Fig. 12-9. A California earthquake destroyed this building. Research how an earthquake causes damage and share your findings with the class.

Loads Did you know that the forces working on structures are called *loads*? The structural engineer's first job is to figure out which loads will act on a structure and how strong they might be in an unusual situation.

Static loads are loads that are unchanging or slowly changing. Fig. 12-10. Static loads are broken down into two groups, dead loads and live loads. **Dead loads** include the entire weight of the structure itself—the beams, floors, walls, insulation materials, columns, and ceilings of a building, or the deck of a bridge, and so on.

Live loads are forces that a structure supports as it is used under normal weather conditions. Live loads include people, furniture, equipment, or stored materials. Cars and trucks that pass over a bridge or even the weight of snow, rain, or ice on a structure are all live loads.

Fig. 12-10. Structures must withstand both static and dynamic loads. Which do you think withstands more dynamic loads, a bridge or a dam? Explain your answer.

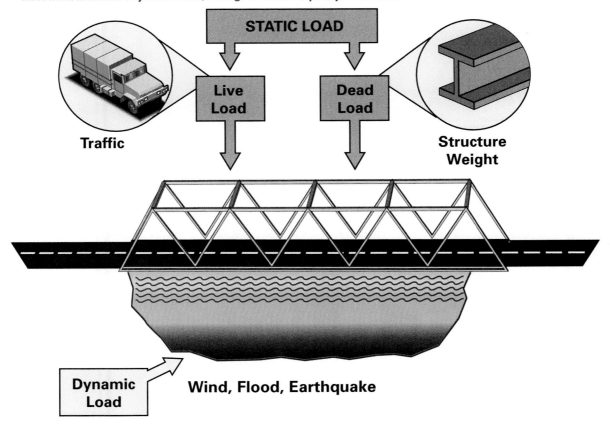

STATIC LOAD

Live Load

Dead Load

Traffic

Structure Weight

Dynamic Load

Wind, Flood, Earthquake

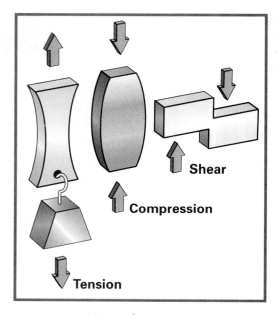

Fig. 12-11. Concrete is strong under compression but weak under tension. Steel bars are strong under tension but weak under compression. Why do you think both are used to make reinforced concrete that is used in roadways?

Shear

Compression

Tension

Dynamic loads are loads that change rapidly, like sonic booms, vibrations from heavy machinery, vibrations from people walking along a floor, a sudden gust of powerful wind, or an earthquake. Dynamic loads can be very dangerous because they happen so quickly and they produce forces greater than normal. Wind loads are extremely important, especially since we started building tall buildings. That's because the taller the structure, the more it is affected by wind. In very tall buildings, up to 10 percent of the structural weight goes into *wind bracing* (resistance to wind).

Compression, Tension, and Shear Other stresses affect structures too. Under **compression** an object is squeezed. Fig. 12-11. A standing column is always under compression. Concrete is the best material for withstanding compression. That's why it is used so much in construction.

SCIENCE CONNECTION

Thermal Expansion

Have you ever wondered why sidewalks are made in short sections instead of one long piece? The answer may surprise you. It has to do with a scientific principle called *thermal expansion.*

Tension is the opposite of compression. It is a pulling force that stretches materials. Tension is the main force at work in the steel cables of a suspension bridge or the inflated domes of some stadiums.

Shear is when opposing forces act on an object. The blades of scissors cut through paper by forcing it in opposite directions as the blades slide by each other.

When you're building structures, you need to remember how important the loads and other stresses are. It is important to plan for these before you start the actual building process.

SECTION 3
TechCHECK

1. What information should be included in a structural drawing?
2. How are static loads different from dynamic loads?
3. What do compression, tension, and shear forces do?
4. **Apply Your Knowledge.** Take a closer look at a bridge or highway overpass and find examples of tension and compression forces at work.

TechnoFact

THE NEW BUILDERS Robots and computers are changing construction. New architectural software and advanced graphics allow people to "walk through" a structure before construction starts. Construction robots are being used to apply plaster and to spray on fireproofing. Architects and engineers also want robots that can move around a construction site easily or use artificial intelligence to make decisions. They are not far away!

Thermal means "heat," and when something expands it gets bigger. The lines you see in sidewalks are called *expansion joints*. When the sun heats the concrete, it expands. The expansion joints are filled with a material that remains flexible, such as tar. Without expansion joints, the sidewalk would soon crack.

Expansion causes the roadways of bridges to get longer in hot weather and shorter on a cold winter day. The movement is so great that special metal finger joints are used to keep large cracks from forming.

ACTIVITY

Look for other types of expansion joints in the roadways of bridges or in concrete roads. Take pictures of examples you find.

ACTION ACTIVITY

What Holds Things Up?

Be sure to fill out your TechNotes and place them in your portfolio.

Real World Connection

The basic structure of a building is often covered with other building materials so that people can't see it. But it is the most important part of the building. Without it, the building would not stand up. However, the structure of a bridge is usually exposed so you can see exactly how it was built. In this activity, you will build two types of bridges and compare their strength.

Design Brief

Design and build a model of a suspension bridge and one of a truss bridge. Your bridges will be made to span a distance of 4 feet. Similar materials will be used for both.

Materials/Equipment

- plywood
- string
- hardboard
- wood glue
- scroll or band saw
- scissors
- power hand drill
- hot glue gun
- computer with CAD software (optional)

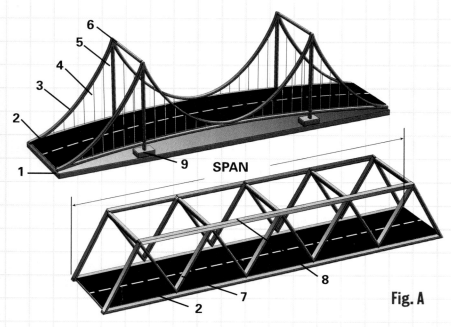

SPAN

Fig. A

KEY:
1. **Base: Plywood 3/4" x 6" x 48"**
2. **Roadbed: Hardboard 1/8" x 5" x 48"**
3. **String: 10 strands**
4. **String: 1 strand**
5. **Dowel Rod: 5/8" x 14"**
6. **Dowel Rod: 1/4" x 5 1/2"**
7. **Welding Rod: 1/16"**
8. **Wood: 3/8" x 3/8"**
9. **Wood: 2" x 4" x 6"**

Procedure

1. As a class, divide into two groups. One group will design and make a suspension bridge. The other group will design and build a truss bridge. Your designs might look like those in Fig. A. The bridge shown on top is a suspension bridge; on the bottom is a truss bridge.

2. Elect a job contractor from your group. The contractor will be responsible for organizing the building of your bridge. This is a big job. Your group may also elect an assistant contractor to help make the contractor's job easier.

3. Most large construction jobs have a definite date set for their completion. Sometimes, the contract between the builder and the contractor will have a penalty clause. This means that if the project is not finished on time, the contractor must pay a penalty.

4. Design your bridge, and make plans for how it will be built. You might use a computer with CAD software to help. When both groups have completed their design, the class should meet to discuss the completion date for both bridges. Your group should negotiate with your teacher to decide on the due date and a penalty clause.

5. Gather the materials that your group will need. Large construction jobs require a large storage area for construction materials. Your contractor will assign a specific job for everyone in the group to complete.

6. Complete the construction of your bridge. If your due date has passed, you might try to renegotiate the penalty clause so it won't hurt your grade as much.

7. Put the two bridges near each other so that their ends are supported on a block of wood about 3" off the floor. Start testing the bridges by placing weights (books or bricks, for example) in the middle of the roadbed. Be careful not to let the weights fall on you when the bridge fails.

Evaluation

1. How much weight did your bridge support?

2. What are the advantages and disadvantages of suspension bridges?

3. What are the advantages and disadvantages of truss bridges?

4. Where is the longest bridge that is closest to your school? What type of bridge is it?

5. **Going Beyond.** Design and build a drawbridge that will move out of the way of large ships.

6. **Going Beyond.** Use CAD software to design a bridge of the future.

SAFETY FIRST

• Follow the safety rules listed on pages 42-43 and the specific rules provided by your teacher for tools and machines.

• Remember to wear eye protection. Take your time, and work safely with machines and materials. Even though you are trying to work quickly, your first consideration is safety. This is true in this activity and on a real job, too.

Designing Communities

THINGS TO EXPLORE

- Explain how communities are affected by new construction.
- Tell why planning is an important part of community development.
- Design and build a model of a community.

TechnoTerms

community
development plan

What effect does modern construction have on you? Do you move around your **community** (area where you live) more easily because of a superhighway near you? How is shopping at the mall different from shopping in town? Fig. 12-12. Your answers to these questions might be different from someone else's answers. Some people might want to move away from an area that is being developed. But you might want to move into an area just because development has made it easy to get around.

Fig. 12-12. This roadway links these homes to a nearby shopping center. Draw a map of roadways linking your home to your school.

Development Requires Planning

Most construction is done to satisfy a need. The need might be for homes, shopping centers, parks, waste management, and roads. Most people care about their community. They want it to continue to meet their needs in the future. To make this happen, the community must have a **development plan** that shows the type of construction and where it will be located. The plan includes laws to control what kinds of construction can take place.

Planning boards, elected officials, and city planners plan for the future and guide community construction projects. They try to make the community a good place to live and work by planning what fits into it. How would you feel about a fast-food restaurant being built next door to your house? Fig. 12-13. Would you like the extra traffic?

Fig. 12-13. Many people like having restaurants close by. What would be the advantages and disadvantages?

Community Design Brings Changes

Designing a community is an enormous job. You have to make decisions that might not please all the people. This sometimes happens, for example, when a new highway is built through a city. Sometimes old buildings must be torn down to make room for the road. People must be moved to new neighborhoods away from friends and familiar surroundings. A quiet neighborhood might change to a busy place with new traffic.

Whenever possible, structural engineers and architects should meet with community people and planners to talk over these changes before any construction starts. That way, the people in the community feel they are part of any construction decisions.

Energy Conscious Communities

Several new kinds of communities are being built to save energy. The homes are energy efficient, and these communities have no roads or cars. People walk to the surrounding shops.

Earth-sheltered construction is an example of a design meant to help save energy. Part of the finished building is below ground, where the earth keeps heat from escaping during cold weather.

SECTION 4
TechCHECK

1. Why do we need to plan communities?
2. What are some changes that happen when new construction occurs?
3. What kinds of construction projects might a community need?
4. **Apply Your Knowledge.** Ask your city planner what new changes are coming for your neighborhood or town.

ACTION ACTIVITY

Designing a Community

Real World Connection

Proper planning and efficient use of land can make our lives in a community safer and more enjoyable. Community planners must consider many factors, such as the need for public buildings. Fig. A. In this activity, you will be a city planner. Your job will be to design an entire community that you would enjoy living in.

Design Brief

Design and build a model of a community. Your design should consider how people work and play as well as how they shop and move from one place to another. Your community should provide space for at least 20,000 people.

Materials/Equipment

- materials for making models, such as 1/8" x 1/8" balsa wood or Styrofoam plastic foam
- utility knife
- masking tape
- 4' x 4' plywood or particleboard
- 4' roll of butcher's paper
- 4-foot T-square or straightedge and triangles
- wood or hot-melt glue
- hot wire cutter
- hot glue gun
- scroll or band saw
- computer with CAD or city planning software

Fig. A

SAFETY FIRST

Follow the safety rules listed on pages 42-43 and the specific rules provided by your teacher for tools and machines.

Procedure

1. In this activity, you will be part of a planning team to design an entire town or city. Your group will need a leader to coordinate its activities. Be sure to name your city, and decide where it is located in your state or county.

2. Use a 4' x 4' square board as the base for your community. Choose an appropriate scale so the area included in your model will be large enough to represent a town. Consider the size of an average home in your scale. For example, if you chose the scale of 1 foot = 1 mile, a house would measure less than 1/8" x 1/8".

3. Make a rough sketch of how you would like your city to look. Following are some of the things you might want to include in your design:

- *Industrial*—light-industry area (small companies); large-industry area; electric power generation plant
- *Residential*—single-family, low-density housing (individual homes); multiple-family, high-density housing (apartments); public areas
- *Public*—open space, parks, bike and jogging paths; schools, libraries, vocational training centers, colleges or universities; transportation access (major highways, railroad stations, airports, boat docks); waste management.
- *Commercial*—office space, fire departments, police stations; shopping centers, service areas, retail stores

4. Cover a 4' x 4' square plywood or particleboard base with butcher's paper. Lay out the streets and highways. Include bridges or tunnels where they might be needed. Consider the location of industrial areas that need access to highways and residential (home) areas that need quiet. Where should shopping centers be located? How much park space do you think is needed? Where should park space be located? Do you think one or two large parks are better than many smaller parks?

5. With your teacher's help, cut balsa wood or Styrofoam models for the homes, apartment buildings, shopping centers, schools, factories, and so on. Glue the model buildings to the paper layout. Name and label streets and highways. Put any finishing touches on your model.

Evaluation

1. How many people live in the following areas of your city? What is the total population?
- high-density housing
- low-density housing

2. Would you like to live and work in your town? In which areas? Explain.

3. What would happen if a disaster such as a hurricane or a flood required an emergency evacuation of your town?

4. **Going Beyond.** Modify your town to hold two or three times the population for which it was designed. How would schools, police and fire departments, or roads have to change? Would high-rise apartment buildings solve the housing shortage, or would they create other problems? Why do you think there are zoning laws that control how structures can be built?

5. **Going Beyond.** What could be added to your city to make it a place that people would want to visit on a vacation?

12 REVIEW &

CHAPTER SUMMARY

SECTION 1

- Construction is the part of production that deals with building structures.
- Building structures has changed with the development of stronger, better materials such as steel and concrete.

SECTION 2

- Structures must be thoroughly planned before construction starts.
- Preliminary sketches are refined into floor plans, elevations, structural views, and section views.

SECTION 3

- Static loads either stay the same or change slowly; dynamic loads change quickly.
- Compression forces squeeze objects together and tension pulls objects apart. Shear tends to cut through objects.

SECTION 4

- Designing communities is done by city planners and zoning committees that try to meet the needs of the people.
- Some factors in designing communities include the location of industrial areas, residential areas, and commercial areas.

REVIEW QUESTIONS

1. What resources does the construction industry use to produce buildings and other structures?

2. What are elevation drawings?

3. Name the two types of static loads and give examples of each.

4. Why might someone like living by a freeway?

5. Why is planning a community a difficult job?

6. How could your city or town be changed to make evacuation faster and easier in an emergency?

CRITICAL THINKING

1. Contact the highway department, and ask how bridges are inspected.

2. Research how bridges must be maintained so they will be safe for many years.

3. Research one of these topics and make a model that demonstrates your findings:
   - geodesic dome
   - pneumatic structures
   - earthquake-resistant design
   - undersea structures
   - space stations or moon bases

4. Modify your community model to show what it will look like in the future. How could the city be changed to help prevent air and water pollution?

5. Ask a physician to discuss what tension and compression do to the human body. Share your findings with the class.

ACTIVITIES 12

CROSS-CURRICULAR EXTENSIONS

1. SCIENCE Design an underground house that uses the earth for insulation. What kinds of special materials or construction techniques would an underground house need?

2. MATHEMATICS Design housing that you think people would like to live in for a city with a large, crowded population.

3. COMMUNICATION Take pictures of storefronts in a business district of your town. Cut and glue the pictures together to show an entire block of storefronts. Place tracing paper over the pictures, and design a new front for each store that would make the business area more attractive. Show your design to a city planner or zoning committee.

EXPLORING CAREERS

Technology helps people build things faster, cheaper, and with fewer flaws. Here are two careers that involve building things.

Construction and Building Inspector

Construction and building inspectors inspect buildings, bridges, and highways while they are being built and after they are finished. They have a good eye for detail. They use their engineering skills to make sure that structures are built properly and safely and that they conform to the building code. A strong background in mathematics and good communication skills are needed.

Operating Engineer

Operating engineers operate power construction equipment, such as cranes, tractors, and derricks, used to move earth and erect structures. They must have good mechanical skills since they often repair and maintain the equipment. Operating engineers need to be in good physical condition. They should enjoy working outside and be able to follow instructions.

ACTIVITY

Suppose you are planning to build a home on a vacant lot. Make a list of job titles for all the workers you will need to build your home from start to finish.

Using Energy

Where Do We Get Energy?

THINGS TO EXPLORE

- Tell what energy is and give examples of potential and kinetic energy.
- Explain where energy comes from.
- Tell how we depend on energy.

TechnoTerms
energy
kinetic energy
potential energy
power
precipitation

Did you ever stop to think of the ways you use energy in a day? Every time you pick up a pencil or blink an eye, you are using energy provided by the food you eat. The bus or car you might ride to school in every day uses the chemical energy stored in gasoline. Fig. 13-1. When you flip on a light switch, you use electrical energy to produce light. What is energy, and where does it come from?

What Is Energy?

The definition of **energy** is the ability to do work. Some people confuse power and energy. They are related, but they are not the same. Simply having the energy to do work doesn't mean work gets done. **Power** is the amount of work done in a certain time period, or the *rate* of doing work.

Energy is never lost or destroyed, but it can be changed from one form to another. For example, a battery changes chemical energy to electrical energy. Fig. 13-2. A solar-powered car turns light energy into electrical energy. Technological processes often change one form of energy into another, more useful, form to do work for us.

Fig. 13-1. The oil used to make gasoline was produced over time from decayed plant and animal matter.

◀ OPPOSITE At this refinery, oil is changed to gasoline. The energy from gasoline is used to power cars, planes, and other vehicles.

Fig. 13-2. These racing cars are powered by electrical energy from batteries. Find out what other alternative fuels are being tried in automobiles. Report your findings to the class.

MATHEMATICS CONNECTION

Electricity Costs!

Your local electric company keeps track of the electricity you use at home with an electric meter. When you turn on any electric appliance, the dials of the meter start moving.

Kinds of Energy

There are two kinds of energy. **Potential energy** is energy at rest waiting to do work. A book about to fall off the edge of the table has potential energy. So does a compressed or coiled spring, like a spring on a pinball game.

When you release a spring the spring has kinetic energy. **Kinetic energy** is energy of motion. A moving bicycle also has kinetic energy. Can you think of some other examples of kinetic and potential energy?

Where Does Energy Come From?

Did you know that nearly all of the Earth's energy comes from the sun? The sun creates wind energy, water energy, energy for living things, and even fossil fuels.

Winds are a kind of kinetic energy caused by uneven heating of the Earth's atmosphere by the sun. The hot air rises and the colder, heavier air moves in under it. Sometimes the resulting winds are strong enough to power electric generators.

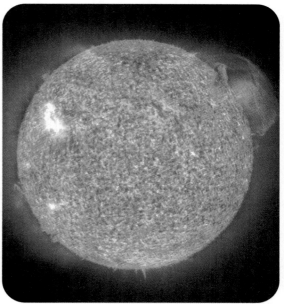

The sun is the ultimate source of Earth's energy. There's more energy packed into one hour of sunshine than we use in an entire year.

You can find out how much energy is used by an electrical device over a certain period. This is done by multiplying the power it requires by the amount of time it uses that power.

electrical energy used = power x time

The unit of electrical power is the kilowatt-hour (kWh). One kilowatt-hour equals 1000 watts of power used for one hour. The electric company charges you for each kilowatt-hour you use. For example, a hair dryer requires about 1000 watts per hour. If you use it 6 hours total in one month, you will have used 6000 watt hours or 6 kilowatt-hours. If your electricity costs you $.09 per kilowatt-hour, then it costs you about $.54 per month to use your hair dryer.

ACTIVITY

If the average power needed to run your color TV is 200 watts each hour and you use the TV 80 hours in one month, how many kWh did you use? If the electricity rate is $.09 for each kWh, how much would you pay to use your television?

Fig. 13-3. The energy you use to power a bicycle comes from the food you eat. Find out what unit is used to measure food energy. Report your findings to the class.

Most living things on Earth needs solar (sun) energy. Plant cells store solar energy during photosynthesis. Animals like us eat plants for energy needs. Fig. 13-3. Some decayed plant and animal materials become fossil fuels after millions of years.

The sun also keeps Earth's water cycle going. The sun heats water in a lake or ocean until it evaporates (changes to a gas). As the water vapor rises, it cools and forms clouds. As the vapor cools even more, it becomes liquid again and falls as **precipitation** (rain, snow, or sleet).

We Depend on Energy for Technology

Without energy our technology wouldn't be what it is today. Technology processes and systems depend on energy. Primitive people used muscle power as their main source of energy. They did only the things they themselves were strong enough to do. Then they discovered that animals could be tamed and used as energy sources, too. For thousands of years, most technologies depended on animal energy.

Then people learned how to use other sources of energy, such as the wind and falling water, to do work. Eventually, adding power to tools and machines made it possible to do work faster and more easily. Today, we depend on energy to be there when we want it. What would your life be like without advances in energy and power?

SECTION 1
TechCHECK

1. What is energy and where does it come from?
2. How is potential energy different from kinetic energy?
3. Tell how you depend on energy in your daily life.
4. **Apply Your Knowledge.** Choose a form of energy and tell how it can be changed to another form of energy to do a specific job for you.

Common Energy Sources

THINGS TO EXPLORE

- Identify common energy sources used in developed countries.
- Identify the major sources of energy used today.
- Build a working model of a turbine generator.

TechnoTerms
conventional
electric generator
nuclear fission
turbine

Today's **conventional** (most common) energy sources used in developed countries include the following:

- **Electricity.** Most electricity is produced by an **electric generator**, a machine that converts mechanical energy to electrical energy. You might have a portable generator on your bicycle that you power by pedaling.
- **Fossil fuels.** Coal, oil, and natural gas are fossil fuels. Fig. 13-4. These are the major sources of energy used today. They are produced deep in the Earth over time from decayed animals and plants. Burning fossil fuels produces a lot of heat energy that we can use. Fossil fuels are nonrenewable (cannot be replaced) in our lifetime.

INFOLINK

See Chapter 7 for more information about the use of fossil fuels.

- **Nuclear energy.** Nuclear energy is the energy found in atoms. In **nuclear fission**, atoms of materials such as uranium are split, releasing huge amounts of heat energy that is used to heat water. The steam from the hot water then spins a **turbine**. A turbine is a type of fan that can operate under high pressures and sometimes high temperatures. It is turned by the force of a gas or liquid striking its blades.

Fig. 13-4. Natural gas and oil come from wells deep in the ground. Find out where gas and oil deposits are found in the U.S. Report your findings to the class.

Fig. 13-5. Turbines inside a dam like this one in Arizona create hydroelectric power. Look up the prefix hydro. What does it mean?

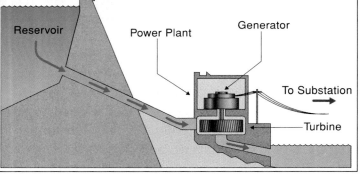

Reservoir Power Plant Generator To Substation Turbine

● **Hydroelectric power.** This power is made when water stored behind a dam passes through a turbine. Fig. 13-5. A generator working with the turbine produces electrical energy from mechanical energy.

One of our biggest challenges today is to find ways to use these conventional energy sources more efficiently. In the activity that follows, you will build a turbine generator that demonstrates how hydroelectricity and other forms of conventional electricity are produced.

SECTION 2
TechCHECK

1. Name four conventional energy sources used in developed countries.
2. What are the main sources of energy used?
3. What form of energy is changed in a generator to make electricity?
4. **Apply Your Knowledge.** With your teacher's help, contact your local power company. Ask a representative to speak to your class. Ask questions about the future of the power plant. How will it keep up with the growing need for electricity?

ACTION ACTIVITY

Making a Turbine

Be sure to fill out your TechNotes and place them in your portfolio.

Real World Connection

What would your life be like without electricity? Stereos, televisions, irons, toasters, and computers would be useless. Almost all of the electricity we use involves a turbine blade turning a generator. In this activity, you will make your own model turbine and put it to work creating electrical energy.

Design Brief

Build a working model of a turbine generator to demonstrate how conventional forms of electricity are produced. Your model will include a turbine blade that harnesses the energy of moving water or air. The turbine will be connected to a generator to change mechanical energy into electrical energy.

Materials/Equipment

- polycarbonate plastic (Lexan), 1/8" thick
- 1/8" steel welding rod
- epoxy glue
- 1 1/2- to 3-volt DC motor
- hookup wire
- 1/8"-ID (inside diameter) plastic shrink tubing
- wood, 1" x 2" x 6" (optional)
- rubber cement
- drill press and 1/8" drill bit
- scissors
- scroll saw
- measurement tools
- plastic strip heater
- compass
- 45° triangle

- pliers, bolt cutters
- digital multimeter, Fig. A
- heat gun
- compressed-air gun
- computer with CAD software (optional)

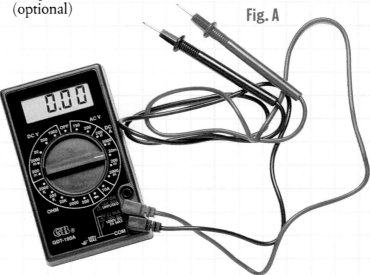

Fig. A

(Continued on next page)

Procedure

1. Work with two or three other students to make your turbine blade and generator. First, draw the plan for the turbine blade on a sheet of paper. Fig. B. Use a compass and a 45° triangle or a computer and CAD software to make your plan.

2. Cut out the finished paper design with scissors. With rubber cement, glue the design to a piece of 1/8" thick polycarbonate plastic. (Polycarbonate is a very strong thermoplastic.) The proper way to use rubber cement is to coat both surfaces lightly and let them dry. When the glue is dry, press the two surfaces together.

SAFETY FIRST
- Wear eye protection.
- Follow the general safety rules on pages 42-43 and specific rules for the machines you are using.
- Remember to ask your teacher for help.

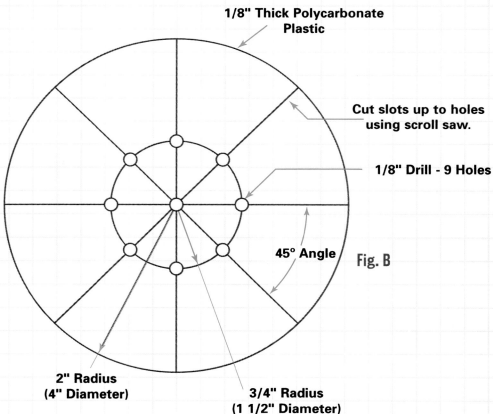

1/8" Thick Polycarbonate Plastic

Cut slots up to holes using scroll saw.

1/8" Drill - 9 Holes

45° Angle

Fig. B

2" Radius (4" Diameter)

3/4" Radius (1 1/2" Diameter)

3. Drill the eight 1/8"-diameter holes around the inside circle and another in the center. The holes will help prevent the plastic from cracking. Heat the plastic between the holes with a heat gun with your teacher's help. Use pliers to bend each fin of the turbine blade to a 45° angle.

SAFETY FIRST

The plastic must reach a temperature of 350° to 400° before it will bend. Do not touch the hot plastic or the heating element of the strip heater.

---

4. Using bolt cutters, cut an 8"-long piece of 1/8"-diameter steel welding rod. Be careful not to bend the welding rod because it will become the axle for your turbine.

5. Glue the turbine blade to the middle of the welding rod using epoxy glue. Follow the directions on the tube of glue carefully. Mix the two parts of epoxy together to start the chemical reaction that will make it harden. Be sure to read the label to see how long you have to wait for the glue to cure (harden). Ask your teacher for help.

6. After the epoxy has cured, attach the axle to the shaft of a 1 1/2- to 3- volt DC motor. This connection will be made flexible using a 1'-long piece of 1/8"-ID (inside diameter) plastic shrink tubing. The tubing can be shrunk to fit tightly around the shaft of the motor by applying heat. Ask your teacher for help in shrinking tubing.

7. The motor you are using has permanent magnets inside so it will also work as a generator. Instead of using a battery to run the motor, you will use the turbine to turn the generator to produce electricity. Connect the terminals of the motor to a digital multimeter. The multimeter should be set to read 0-3 volts DC.

8. In the first test of your turbine generator you will simulate a hydroelectric power plant. You will use the force of water running out of a water faucet to turn the turbine blade. Hold the motor/generator in a sink so it will be out of the water stream and low enough to prevent splashing. Watch the multimeter reading and see what voltage your turbine will produce. If the meter shows a negative number, reverse the positive and negative wires.

9. The test will simulate the use of a turbine being run by high-pressure steam. In a real power plant, steam would be generated by burning coal, oil, or natural gas to boil water. Nuclear power plants use radioactive fuel rods to boil water and make steam. High-pressure steam is very dangerous. Instead of using steam to spin your turbine, you will use compressed air from a compressed-air gun. Fig. C.

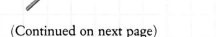

Fig. C

(Continued on next page)

10. Design and build a stand to hold the turbine blade. Your stand might be made of plastic or wood. Fig. D.

11. Connect the multimeter to the generator terminals as before. With your teacher's help, use a compressed-air gun to spin the turbine blade.

SAFETY FIRST

Compressed air can be dangerous. Do not point the gun at anyone or put your finger over the end of the gun. Keep your fingers and any long hair away from the spinning turbine blade. Do not exceed 20 psi.

12. Watch the multimeter reading to see how many volts your turbine generator can produce.

Evaluation

1. What was the voltage reading in the hydroelectric test? What was the reading using compressed air? Why was there a difference? Explain.

2. Where does the electricity you use come from? How is it produced?

3. How many generators like yours would it take to run a house outlet that drew 120 volts?

4. **Going Beyond.** Design and make a modification (change) of your turbine blade so that you could use wind energy to make electricity.

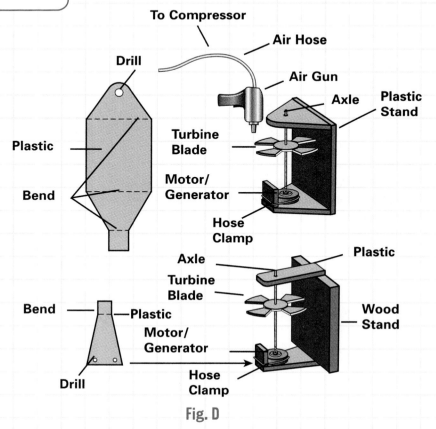

Fig. D

How Can We Save Energy?

THINGS TO EXPLORE

- Explain why we need to conserve energy.
- Identify ways technology can help save energy.
- Investigate a house or apartment for air leaks.

TechnoTerms
energy-intensive
insulated

Did you ever leave a light on in a room when you left? Have you seen people leave their cars running while they go into a store? Do you turn the heat up in your house but leave a window open for fresh air? These are some habits that we need to change in order to *conserve,* or save, our energy resources. Conservation can take many forms.

People are trying to find ways to conserve energy because many of our resources are nonrenewable. Even those like electricity, which we think will always be there, count on other energy sources for their production. In some places, power plants use coal or oil to make electricity. Both of these fossil fuels are running out. We must use our energy supplies more wisely and efficiently.

You live in an **energy-intensive** world. That means many of the things you do daily require energy. Every day more products that use energy are being developed to meet your needs. Sometimes these products make life easier, but they are not needed for your survival. Do you really need an electric can opener? How about an electric pencil sharpener? Some of the ways you can save energy are simple, but others are not so easy.

Making Things More Energy-Efficient

What can we do to save energy? We can make products that require less energy to operate. For example, one company has produced a new fluorescent light bulb that uses less energy than conventional bulbs and lasts longer. We can also make cars that burn less fuel. Using plastics for many parts instead of metals makes cars lighter. Not as much fuel is needed to move them.

TechnoFact

BRIGHT IDEA Can you imagine how reducing the size of the bulb used in a car's headlights by one inch could result in a new shape for cars of the future? The new light bulb will be far more durable and energy-efficient than the bulbs used now. It doesn't have a heated filament. It uses an electronically controlled spark to heat gases. The smaller bulb produces more light and less heat. Since the bulb itself is cooler, the assembly that holds it can be smaller, lighter, and made from plastics. That means cars can be more streamlined, or aerodynamic, and therefore have improved gas mileage.

In addition, we can change products so they waste less energy. Up to now, no machine can make total use of all the energy put into it. All products that use energy waste some. For example, about half of the energy in fossil fuels burned in engines is lost in the form of heat that doesn't do any useful work. This is true for all types of engines. Automobile manufacturers are trying to make car engines more efficient. Also, furnaces are now being produced that are twice as energy-efficient as the ones made five years ago.

Even home appliances can be more efficient. For example, water heaters can be better **insulated** so that heat stays in. Fig. 13-6. Maybe your dishwasher has a special energy-saving setting that doesn't use energy to dry your dishes. Many large buildings and schools have special switches that automatically lower room temperatures and turn lights off at night or on weekends when most people are not working. Your walls, ceilings, doors, and windows can be better insulated to prevent heat loss during cold weather. That saves a lot of energy.

Fig. 13-6. Wrapping a water heater with insulation saves energy. Stop at an appliance store and look at the EnergyGuide stickers that show how much energy a large appliance saves. Describe the stickers to the class.

Recycling

Have you been recycling cans, bottles, plastic, and newspaper? Recycling is another way to save energy because it takes less energy to recycle a used material than to process a raw material. Throwing away one aluminum can is like throwing away one-half gallon of gasoline!

SECTION 3
TechCHECK

1. What does *conserve* mean?
2. Why do we need to conserve energy?
3. How can technology help us save energy?
4. Apply Your Knowledge. List five ways you can conserve energy at school and at home. Share your list with your classmates.

ACTION ACTIVITY

Calling All House Detectives!

Be sure to fill out your **TechNotes** and place them in your portfolio.

Real World Connection

In cold climates, cool air leaking in from outside can add a great deal to the heating bill. Fig. A. In warm climates, hot air coming in adds to the cooling bill. The places where air enters can be found and fixed easily. In this activity you will investigate how finding air leaks in a house or apartment can save energy.

Design Brief

Draw a floor plan of your house or apartment. Investigate your home for air leaks and mark them on the floor plan. Then help correct any problems.

Materials/Equipment

* graph paper with 1/4" squares
* weather stripping (optional)
* thermometer
* flashlight
* screwdriver

SAFETY **FIRST**
Follow the safety rules listed on pages 42-43 and the specific rules provided by your teacher for tools and machines.

Fig. A

(Continued on next page)

Procedure

1. Sketch a floor plan of your house or apartment on graph paper. Use a scale of 1/4" = 1'. Use the architectural symbols from Chapter 12 to mark walls, windows, and doors.

> **INFOLINK**
>
> See Chapter 12 for more information about architectural symbols.

2. Check your home for air leaking inside through cracks and spaces around doors or windows. Wait until night and turn off the lights. Ask someone to go outside and shine a flashlight around windows and exterior (outside) doors. Then you watch from the inside to find any spots where the light shines through. If you live above the ground floor, ask your teacher what you should do.

3. Mark on your floor plan any areas where you saw light. These problems can be corrected using weatherstripping (Fig. B) or by installing storm windows, if appropriate. Ask your parents or caregivers if you can help correct these problems or if someone else can do it.

4. Measure the temperature inside and outside. Record the temperature difference.

5. Write a paragraph that explains your investigation and the steps you took to fix any problems. If possible, use a computer and word processing software.

Evaluation

1. Did you find any areas where air was leaking into your home? If you did, how did you correct the problem?

2. Find out what type of heat your house or apartment has. How could the heating bill be reduced?

3. What is the maximum temperature difference anyone in your house can remember between the inside and outside of your home?

4. **Going Beyond.** With your teacher's help, organize a group to test and weatherize the homes or apartments of poor or elderly people in your neighborhood. Ask local building suppliers if they would either donate the materials needed or sell them at a discount to the disadvantaged.

5. **Going Beyond.** Research how electronics will help conserve energy in the smart houses of the future.

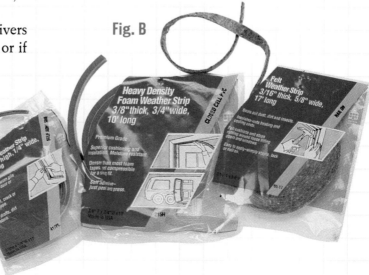

Fig. B

What Is Alternative Energy?

THINGS TO EXPLORE

- Identify alternative energy sources.
- Explain why alternative energy sources are important to us.
- Design, build, and test a photovoltaic battery charger.

TechnoTerms

alternative energy
biomass
geothermal energy
photovoltaic cell
solar cell

Besides trying to conserve our current energy sources, we need to look at **alternative energy** sources. They are also called *renewable resources* because we will not run out of them. They are constantly renewed through natural processes caused by the sun's energy. Alternative energy sources are important to you. Can you name some of them?

Solar energy, wind energy, biomass, tidal energy, and geothermal energy are examples of alternative energy sources. They provide energy with far less damage to the environment than nuclear or fossil fuel sources. They do not produce much waste or pollution. We need to find more ways to use them. They may replace or be added to nonrenewable energy supplies to do work for us.

Solar Energy

The sun provides the Earth with lots of energy, some of which can be used for heating purposes and to produce electricity.

Solar cells, or **photovoltaic cells**, make electricity directly from sunlight. Solar cells were developed to use on satellites in the 1950s, and they were very expensive then. Do you have a solar-powered calculator? Solar cells are now much cheaper to make, and calculators that contain them are inexpensive and powerful.

▲ Solar-powered calculators change light energy into electricity.

Fig. 13-7. This building uses the sun's energy for heating. Solar collectors are built into the roof. Look for solar collectors on homes in your neighborhood. How many can you find?

Another direct use of solar energy is using solar collectors to heat water. The hot water is then used to heat homes. Fig. 13-7. Some solar collectors produce temperatures high enough to be used in industry and for generating electricity. Active solar heating is much more effective in sunny climates.

Wind Energy

Wind is one of the most promising alternative energy sources. Many countries, especially those that get a lot of wind, are developing wind power technology.

The most important use of wind energy is to produce electricity. The wind turns a turbine shaft that is hooked to a generator. The turbine depends on a steady supply of wind averaging 10 miles per hour or more. Medium-sized wind-driven turbines have been the most efficient so far. In some places, batteries are used to store the energy for times when the wind isn't blowing. More than 20,000 wind turbines are now producing electricity around the world.

Wind energy can generate electricity at the same price as fossil fuels and nuclear power, but it is safer and doesn't cause pollution. "Wind farms," or collections of wind generators, in California have the power of two nuclear power plants but cost half as much as conventional power stations to operate. Fig. 13-8.

Fig. 13-8. Wind "farms" like this one in California use windmills to capture wind energy. What were early windmills used for?

Biomass

Biomass is living or dead plant or animal matter. Its main sources are wood, crops, animal wastes, and organic materials found in garbage. Almost half of the world's population depends on biomass to supply energy for cooking, heating, and light.

The energy in biomass can be released and used in many different ways. Garbage can be burned to produce lots of heat. Most poorer nations get much of their energy from burning wood or from animal waste when wood is scarce or expensive.

Biomass can be used to produce biofuels such as methane, methanol, and ethanol. Processes also exist that can change it into petroleum. Fig. 13-9.

Fig. 13-9. Ethanol is a type of alcohol produced from corn or other biomass. Check the pumps at a local gas station. Does the station sell gasoline mixed with ethanol? If so, what is the percentage of ethanol contained in the mixture?

Fig. 13-10. This geyser is at Yellowstone National Park in Wyoming. Geysers are hot springs that erupt into columns of steam and hot water. Find out if geothermal energy is used at Yellowstone and report your findings to the class.

Tidal Energy

Ocean tides have mechanical energy that can be changed to a form you can use. Turbines can produce electricity from the rising and falling tides. The water is first trapped and then released through the turbine. The energy available depends on the difference between the heights of high and low tides.

Special generators can also change the energy in waves into electric power. This source of energy has lots of potential. But high waves and strong winds pose problems that scientists must first solve.

Geothermal Energy

Geothermal energy is heat from beneath the Earth's crust. Fig. 13-10. When it is brought up to the surface as steam or hot water, it can be used directly to heat water. It can also be used to drive generators or steam turbines. The geothermal resources in the upper three miles of the Earth's crust are estimated to contain more energy than all the world's natural gas and crude-oil reserves. Until now we have used only a small percentage of this energy.

You can see that many energy resources are available to us. We will have to start using some alternative energy supplies because the nonrenewable ones are running out. Could you live with less energy? Most of us can, because we waste so much. We need to find more efficient ways to use and conserve our energy sources and, at the same time, to keep looking for new sources.

SECTION 4
TechCHECK

1. What is alternative energy?
2. List five types of alternative energy.
3. Why do we need to use alternative energy?
4. **Apply Your Knowledge.** Research other devices that are powered by solar cells besides calculators. Share your information with the class.

ACTION ACTIVITY

Using Sunlight to Charge a Battery

Be sure to fill out your **TechNotes** and place them in your portfolio.

Real World Connection

Wouldn't it be great if we could change sunlight directly into usable energy? Well, we can, thanks to a thin slice of silicon called a photovoltaic cell. Fig. A. In this activity, you will build a solar battery charger. This simple electrical circuit will change sunlight directly into electricity that you can use while the sun shines or store in a battery for use at night or on cloudy days.

Design Brief

Design, build, and test a photovoltaic battery charger. Your charger should produce approximately 1.5 volts DC. It must be able to charge a nicad battery.

Materials/Equipment

- silicon solar cells
- silicon diode
- hookup wire, rubber cement
- AA nicad battery, AA battery holder
- 1/16"-acrylic plastic (Plexiglas)
- aluminum foil
- electronic soldering pencil
- band or scroll saw
- drill press
- digital multimeter
- strip heater (optional)
- 1 1/2-volt DC motor, 1/16"-steel welding rod, gears, belts, pulleys, from machine dissection activity in Chapter 8 (optional)

Fig. A. Here, large groups of photovoltaic cells convert solar energy into electrical energy. "Solar farms" like this may become major sources of electricity in the future.

(Continued on next page)

Procedure

1. In this activity, you can work individually or in a group to make a solar battery charger. You will need to use the problem-solving steps you learned in Chapter 1 to solve the problems.

> **INFOLINK**
>
> See Chapter 1 for more information about problem solving.

2. A solar collector must be pointed at the sun to be able to produce electricity. You will need a way to adjust the collector so it will gather enough light to charge a battery. You may also need to help gather sunlight by using reflectors to direct the sunlight to the solar cells. Brainstorm ideas for how you will meet this need. Some possibilities are illustrated for you in Fig. B.

3. Refine your design and start construction. Think about the goals of your solar collector. It must
- be adjustable
- gather enough sunlight to operate at 1.5 volts or more
- be lightweight enough for a solar-powered car

> **SAFETY FIRST**
> Wear eye protection. Follow the general safety rules on pages 42-43 and specific rules for the machines you are using. Remember to ask your teacher for help.

Fig. B

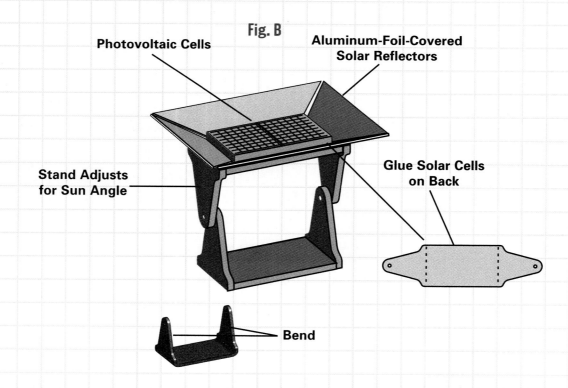

Photovoltaic Cells

Aluminum-Foil-Covered Solar Reflectors

Stand Adjusts for Sun Angle

Glue Solar Cells on Back

Bend

4. With the help of your teacher, solder the hookup wires to the solar cells according to the directions on the package. Handle the solar cells carefully. They are very thin and fragile. (They are also expensive.) Solder the hookup wires together to make a series circuit called a *solar array*. This arrangement will add the voltages of all the solar cells together. Solder the positive lead to a diode before it is connected to the battery holder. The diode will let electrons flow in only one direction. Fig. C.

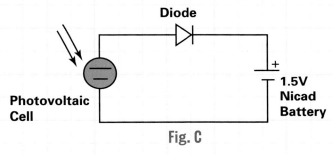

Fig. C

5. Carefully glue the solar cells to your collector using rubber cement. Do not press hard on the solar cells. They will crack easily if handled roughly.

6. Test your solar generator by pointing it toward the sun. Connect the leads of a digital multimeter to the positive (+) and negative (-) terminals of the battery holder. If you get a negative reading on the multimeter, you must reverse the battery holder. It is important that the polarity (+ or -) be correct for the charger to work.

SAFETY FIRST
Do not look directly into the reflections from the mirrors.

Evaluation

1. How many volts of electricity did your solar charger produce?

2. How did you use reflectors to trap more light? Measure the voltage output with and without the reflectors. What is the difference?

3. Try using your solor charger indoors. What is the difference in results?

4. **Going Beyond.** Use the solar collector as a part of a solar-powered car. You might decide to team up with another group to solve this problem. You will need to apply knowledge about mechanical systems, covered in Chapter 8, to connect a motor to the wheel or axle of your car. If your first attempt does not make the car move, can you connect two or more solar collectors together to get more power? If your class is successful in making more than one solar-powered car, set up a race to see which car goes the fastest or the farthest.

5. **Going Beyond.** Research the development of photovoltaic cells. What has happened to the efficiency and cost of producing solar cells? Where are they commonly used today?

6. **Going Beyond.** Research the solar-powered car made by General Motors called the *Sunraycer*. How is this car designed differently from an ordinary car? Do you think you will be able to buy a solar-powered car some day? Explain.

13 REVIEW &

CHAPTER SUMMARY

SECTION 1

- Energy is the ability to do work; power is the amount of work done in a certain period of time.
- There are two kinds of energy, potential energy (energy at rest) and kinetic energy (energy of motion).
- Most of the Earth's energy comes from the sun.

SECTION 2

- Conventional energy sources used in developed countries are electricity, fossil fuels, nuclear energy, and hydroelectric power.
- One of our biggest challenges today is to find ways to use these conventional energy sources more efficiently.

SECTION 3

- Technology is helping find ways to make machines and other products more efficient so they don't use or waste so much energy.

SECTION 4

- Alternative energy sources are being developed to replace nonrenewable energy supplies.
- Solar energy, wind energy, biomass, tidal energy, and geothermal energy are some alternative energy sources.

REVIEW QUESTIONS

1. What were solar cells originally developed for?
2. How are fossil fuels produced?
3. What is the difference between energy and power?
4. Does a coiled spring have potential or kinetic energy?
5. Name ways energy is wasted in your home or school.
6. What is biomass?

CRITICAL THINKING

1. Research the cost of electricity in your area. Find out how much the electricity used in your house costs every month. List ways you could reduce the electricity bill.
2. Design your own experiment that demonstrates conservation of energy. Ask your teacher for help in making a procedure for your idea.
3. Ask your teacher to show you plans for a super-insulated house. How are super-insulated houses built differently from normal houses? How long do you think it would take to pay for the additional cost of materials by saving energy?
4. Make a sketch of a hydroelectric power plant. Label the turbine, generator, dam, and reservoir.

ACTIVITIES 13

CROSS-CURRICULAR EXTENSIONS

1. **SCIENCE** With your teacher's help, contact a glass shop and ask them to save small pieces of mirror for a class project. Tape the edges of the mirrors to prevent cuts. Have each person in the class design and build an adjustable stand to hold the mirrors. Set all of the mirrors in the sun so that each one reflects sunlight toward a solar cell array. How does the voltage change when all the mirrors are adjusted properly?

2. **MATHEMATICS** Refer to Activity I. Calculate which would be more cost-effective, a large number of solar cells or a few solar cells and many mirrors.

EXPLORING CAREERS

The ability to harness and use energy allows us to heat our homes, drive cars, and create powerful lasers. Expanding energy technology is also creating career opportunities like those listed below.

Laser Technologist Lasers can be used to cut through steel, reshape a cornea in the eye, and read the bar codes on packages. Lasers are being used in medicine, manufacturing, and retail businesses. Laser technologists transport, set up, demonstrate, and maintain various types of laser equipment. They must have mechanical abilities, be able to work well with customers, and be willing to keep up with the changes in laser technology.

Nuclear Engineer Nuclear energy is used to generate electricity. It powers ships and spacecraft and is used in medicine. Some nuclear engineers manage power plant facilities. They conduct and monitor tests or oversee daily operations. Nuclear engineers need strong management and problem-solving skills and a strong background in mathematics and science.

ACTIVITY

Bring a product to class that has a bar code (Universal Product Code) on it. Find out what each of the numbers in the bar code represents.

Moving Things

What Is Transportation?

THINGS TO EXPLORE

- Define transportation and tell how it affects you.
- Explain how transportation technology is changing rapidly.
- Describe the different modes of transporting materials.

TechnoTerms
composite
mode
transportation

Take a quick look around your classroom. Everything you see had to be transported to this room. **Transportation** is the movement of people or goods from one place to another. Transportation is a big industry that affects you in many ways.

Do you own a bicycle or a scooter? Does anyone in your house depend on a car or bus to get to work? How does your luggage get from the airplane into the main airport terminal? How did the food you ate for breakfast get to the grocery store? Transportation is very important to your way of life. You depend on transportation such as trains, ships, automobiles, airplanes, or buses to get you where you want to go. Almost everything you use, from the food you eat to the chair you are sitting in, is there for you because of transportation systems.

Changes in Transportation Technology

Transportation technology is changing rapidly. Everything moves much faster today. At top speed, the first airplane built in 1903 could go only 30 miles per hour (mph). Now we have the *Concorde* jet, which can fly more than 1,200 miles per hour. Fig. 14-1 (page 312). In space, people travel faster than 17,000 miles per hour!

Most transportation systems today use a lot of computer technology. Computers also control lights, door locks, braking systems, and the amount of pollutants produced in many cars,

TechnoFact

BREAKING THE SOUND BARRIER Have you heard the term "Mach 1" before? Mach 1 is the speed of sound, or about 700 miles per hour. When an airplane approaches the speed of sound, the air it pushes ahead of it creates a shock wave. A loud noise is produced as the plane "breaks the sound barrier." Then the plane becomes very hard to handle. Chuck Yeager was the first person to fly at more than the speed of sound. The *Concorde* cruises at Mach 2, or twice the speed of sound. A hypersonic plane is being planned that will reach speeds of Mach 6 and up!

OPPOSITE Airplanes are our fastest form of everyday transportation.

Fig. 14-1. The Wright Brothers made the first long, controlled, engine-powered flight. The Concorde is a bit faster. Do a little research on the Concorde. When was it first used, and how many passengers does it carry?

trains, and airplanes. Even farms use mechanized transportation technologies. Computers help schedule trains, and one day satellite communication systems will help air traffic controllers monitor (keep track of) planes on long-distance flights so they don't run into each other.

Design engineers are also looking for ways to reduce the weight of vehicles, using aluminum, plastics, or lightweight steel. Plastics and **composites** (fiber-reinforced plastics) are used to make smaller aircraft that are light and strong. Using more plastics in cars and trucks will soon make many exterior parts recyclable. These new designs are making vehicles more fuel-efficient.

In the future, transportation systems that use alternative energies will be explored further to see if they are practical and economical. Fig. 14-2. Wind energy is now being used to supplement electricity on large ships. Photovoltaic cells, which change sunlight into electricity, are being used to power motors in cars and airplanes.

Can you imagine how transportation systems will change in the next 100 years?

Fig. 14-2. This Ford P2000 automobile uses both diesel and electric power. It is non-polluting, and its aluminum body is lightweight. Why do you think weight makes a difference? What other lightweight materials might be used to make car bodies?

Modes of Transportation

We use different **modes** (ways) of transportation. The mode you use depends on whether you want to move people or products on land, water, air, or in space.

Industry and business depend on transportation systems to move goods. We want the fastest, most economical ways to transport products so we don't waste products, time, or money.

As you explore the different types of land, air, and water transportation in this chapter, think about those you use a lot.

INFOLINK

See Chapter 17 for more information about space travel.

SECTION 1
TechCHECK

1. What is transportation?
2. How does transportation affect you in your daily life?
3. What are the different modes of transportation?
4. **Apply Your Knowledge.** Make a bulletin board display of different modes of transportation.

Land Transportation

TechnoTerms

AMTRAK
automated transit
   system
fifth wheel
piggybacked

THINGS TO EXPLORE

- Identify different kinds of land transportation.
- Explain what methods of land transportation are used for and give examples.
- Design and build a safe car to transport an egg.

When you think of land transportation, do you usually think of some kind of vehicle such as a car, a train, or a truck? Actually any transportation that moves on or beneath the Earth's surface is a form of land transportation. Pipelines, conveyors, moving sidewalks, escalators, and elevators all move products or people from one point to another.

Automobiles

The automobile has become a necessity for many Americans. We depend on the car to get us where we want to go.

Smart Cars Almost all the cars made in the world today have on-board computers with special sensors to help the different mechanical systems work right. Fig. 14-3. Some computers let you know how far the car can travel before it runs out of gasoline or the time of your arrival at a certain place. Computerized navigation systems in cars show local maps and indicate the best *route* (path) to take.

Car Safety *Passive car safety systems* don't require the passengers to do anything. For example, *air bags* operate automatically in case of a collision. Even the car itself can be designed to absorb or cushion the impact of a crash. Special areas, called crumple zones, are built into the front of the passenger compartment.

Fig. 14-3. Computer chips are installed in cars at the factory. Write a paragraph describing the tasks you think a "smart" car should do for itself. Which should the driver be responsible for?

Trucks

The trucking industry is big business in the United States. The big advantage to using truck transport is that goods can go directly from the producer or distributor of a product to where the product is going to be sold.

Most trucks are tractor-trailers. Fig. 14-4. They have a large tractor, which is the power plant, and one or more trailers, which hold the freight, or cargo. Tractor-trailers use powerful diesel engines that can go long distances before needing major repairs. Have you heard the term **fifth wheel**? The fifth wheel is really a large, disk-shaped hitch that hooks the trailer to the tractor.

Trailers today are made to carry special products, from refrigerated goods to melted chocolate! Sometimes they are **piggybacked** (carried on railroad flatcars) to a location. Then a truck picks them up and takes them to their final destination.

Buses

Many kinds of buses are used to move large numbers of people around. City buses, school buses, and motor coaches are the most common types. Fig. 14-5.

Fig. 14-4. Truckers drive long distances to deliver their cargoes. Many trucks now have computers on board that log the miles traveled as well as the time in operation. These trucks are also called tractor-semitrailers. Look up the prefix semi. How do you think its meaning is applied here?

Fig. 14-5. People in London, England, ride double-decker buses.

Fig. 14-6. Heavily populated countries are looking for better systems of mass transit. This high-speed maglev train is being tested in Japan. Why do you think mass transit systems are popular in heavily populated areas?

Trains

Trains have been used for many years to move people and products. In the United States, trains provided the fastest way to travel before the automobile was invented.

- **Modern passenger trains.** In Japan and France, passenger trains are still commonly used. Fig. 14-6. Trains in the United States have lost riders to buses, because buses are cheaper, and to airplanes, because they are faster. **AMTRAK** (*Am*erican *Tra*vel *Tra*ck) system provides all the long-distance rail service in the United States.

- **Mass transit rail systems.** Many types of mass transit rail systems can be located underground, above-ground, or at ground level. Subways, found in large cities, are underground rail systems. Subways often use tunnels, which are expensive to build. Monorails are transit systems that are sometimes elevated. They use a single rail.

Other mass transit rail systems are totally automated and do not have a driver. These **automated transit systems** (ATS) are often used at airports, remote parking areas, and shopping centers.

INFOLINK

See Chapter 20 for information about maglev trains.

- **Freight trains.** In the United States, railroads are used mainly to move freight. The big advantage of trains is that they can move large loads over long distances economically and efficiently. Most locomotives today are powered by diesel-electric power or gasoline turbine engines like those used in airplanes. For efficiency, computer systems keep track of every freight car in the rail network.

Pipelines

Some transportation systems, such as pipelines, do not use vehicles to move things. Certain kinds of materials, such as natural gas or oil, can be moved by pipeline very economically. Most pipelines are buried under ground and the product moves through them in only one direction.

Conveyors

Conveyor belts can move people or products. One of the most popular is the "people mover," or moving sidewalk, found in major airports. Have you ever walked on one of these? Fig. 14-7.

Fig. 14-7. Moving sidewalks in airports help weary travelers cover long distances quickly. Where else do you think moving sidewalks would be helpful?

SECTION 2
TechCHECK

1. List six forms of land transportation.
2. What is the main advantage to using truck transportation?
3. For what purpose are railroads used most often in the United States?
4. **Apply Your Knowledge.** Make a photo-collage from magazine pictures of all the modes of land transportation you can find. Share it with your class.

ACTION ACTIVITY

Designing for Car Safety

Be sure to fill out your TechNotes and place them in your portfolio.

Real World Connection

Automotive engineers have been working for years to make cars safer. But current safety features can't prevent thousands of people from being injured each year in car crashes. There is a great need for improved cars designed with safety in mind. For this activity you will design and build a safe model car.

Design Brief

Design and build a model of a car with safety in mind. Use a raw egg to represent the driver. The egg "driver" must be able to "see" out the front windshield. The egg can have a foam seat, seat belt, and shoulder harness but cannot be covered with foam. The balsa-wood frame can be designed with crumple zones and should be able to protect against both front and side impact. Design your car so that it is easy for the driver to get in and out. Then test your design on a crash test track.

Materials/Equipment

- car test track
- balsa wood
- hot glue gun
- rubber bands
- plastic wrap
- spray paint
- masking tape
- plastic wheels
- dowel rod for axles
- bungee cord
- raw eggs
- plastic sandwich bags
- camcorder (optional)
- computer with simulation software (optional)
- Styrofoam plastic foam

SAFETY FIRST

- Follow the safety rules listed on pages 42-43 and the specific rules provided by your teacher for tools and machines.
- Ask your teacher for permission to use power tools. Use them with caution and only with teacher supervision.
- Wear safety glasses when using power tools and when testing your car. Place the raw egg in a plastic sandwich bag to prevent messy accidents.

Procedure

Part 1 · Building Your Car

1. Ask your teacher if you should work in small groups or individually. Ask your teacher to help build the track.
2. Design a model car that would fit into a 12" long x 8" wide x 6" high box. Fig. A. Make a sketch of your ideas.
3. Build the frame of your car out of balsawood strips. Use a hot glue gun to assemble the frame.
4. Use Styrofoam plastic foam to make a seat for the driver. Foam can also be used to make bumpers.
5. Use rubber bands to make the seat belt and shoulder harness.

6. Think of new ways to protect your driver.
7. Cover your car by stretching plastic wrap around the frame. Wrap the plastic so that the seams are on the bottom of the car.
8. Use masking tape to protect the windshield, side, and rear windows from paint.
9. Follow teacher directions to spray paint your car. Be sure to use a spray booth.
10. Wait for the paint to dry. Peel off the masking tape to expose the "glass" areas.
11. Install the wheels and axles.
12. Cut a hole for the door so that your "driver" will fit in.

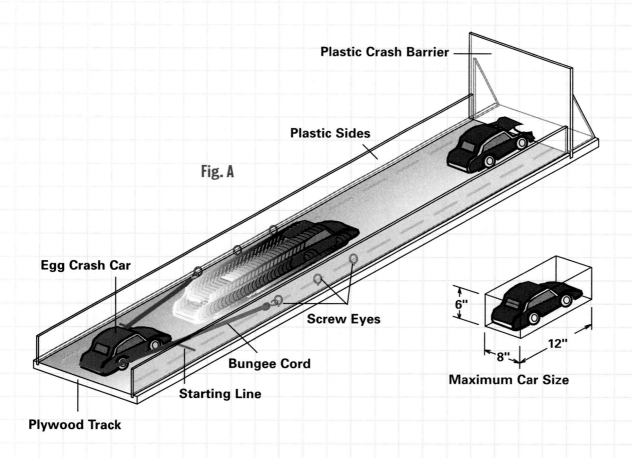

Fig. A

Plastic Crash Barrier

Plastic Sides

Egg Crash Car

Screw Eyes

Bungee Cord

Starting Line

Plywood Track

6"

8"   12"

Maximum Car Size

(Continued on next page)

ACTION ACTIVITY

Part 2 · Testing Your Car

1. Start with the bungee cord hooked to the screw eyes set for the shortest distance (weakest pulling force). Wrap an egg in a plastic sandwich bag and place it in your car. Place your car on the test track.

2. Pull the cord and your car back to the starting line.

3. If possible, record the crash using a camcorder on the side and behind the crash barrier. Release the car and watch the crash.

4. Examine the egg for cracks. If your egg survived the first test, continue with the rest of the tests until it cracks.

5. Put the bungee cords on the second set of screw eyes. This will make the car hit the barrier harder.

6. Use a car that has already been eliminated to crash into the side of your car to test for side impact safety.

7. Put the bungee cords on the third set of screw eyes. This will make the car hit the barrier harder still.

8. Review the videotape to evaluate the safe design of your car.

9. From now on, remember to buckle up!

Evaluation

1. List three different car safety devices.

2. What is a crumple zone?

3. What is a passive safety system?

4. **Going Beyond.** Carefully analyze the crash by playing your video a frame at a time or use a computer to digitize the video.

5. **Going Beyond.** Design and test a highway safety device, such as a crash barrier that helps to absorb impact forces. Use empty 35mm film canisters or other items to create the barrier. Re-test your car.

6. **Going Beyond.** Think of other test methods that would simulate real road conditions.

7. **Going Beyond.** Find out how auto manufacturers do crash testing. Fig. B. What kinds of information do the crash tests provide?

Fig. B

Water Transportation

THINGS TO EXPLORE

- Describe historical changes in water transportation.
- Identify today's water transportation vessels and tell what each one does.
- Explain why computers are important to water transportation.
- Build a hovercraft you can ride.

TechnoTerms
air cushion vehicle (ACV)
barge
hovercraft
hydrofoil

Using water for transportation is nothing new. The first boats were powered by muscle only. For short trips close to shore, people moved boats with poles, paddles, and oars.

As people learned how to use wind power to move vessels (boats), they could travel longer distances. After the steam engine was invented, ships no longer had to depend on wind, currents, or muscle power. Steam-powered ships could sail any time, whether the wind was blowing or not. Also, the ships could be larger, and that meant larger cargoes could be carried.

Modern Water Transportation

Today's modern ships have lighter-weight steel hulls and more efficient, powerful engines. Fig. 14-8. Some commercial ships have a double bottom, so the ship can safely carry liquid cargo or fuel for the ship's engines. Modern ships are built with prefabricated sections that are then welded together.

Fig. 14-8. At one time all ships used wind power. Now sailboats are only for recreation. What material was used to make early sailing ships?

Fig. 14-9. Hydrofoils skim over the water thanks to the lift created by the hydrofoil "leg." The shape of a hydrofoil is similar to that of an airplane wing.

Computers play a big role in water transportation today. They can determine a vessel's position and plot a course for the ship. The computer can store a complete set of charts that take up less space than paper charts and are easier to update.

In general, water transportation is slow, but it is also less expensive than air or land transportation.

- **Passenger ships.** The use of passenger ships for ordinary travel has declined. Small ocean liners or cruise ships are popular for vacations. Ferries are used to move people short distances over water.
- **Hydrofoils.** A **hydrofoil** is a passenger ship that moves above the surface of the water. Fig. 14-9. Hydrofoils can go very fast because there isn't much friction or resistance from contact with the water.
- **Hovercraft.** Also known as **air cushion vehicles** (ACVs), **hovercraft** ride on a cushion of air. High-speed fans driven by gas turbines push air under the boat. The air is trapped around the outside edges of the vehicle, so it is actually lifted above the surface of the water. You get a very fast, smooth ride. Hovercraft can travel over ice, snow, dry land, or marshes, too.
- **Submarines.** Submarines are ships that travel either on or beneath the surface

Fig. 14-10. Submersibles can study the deepest parts of the ocean. Some are robots; others carry scientists. What do you think can be gained by studying the ocean floor?

of the water. By changing the amount of air in their tanks, submarines can float at any level. The newest submarines are powered by nuclear energy. Smaller robotic submersibles are being used to explore deep-water areas. They are controlled by computers on ships at the surface. Fig. 14-10.

- **Cargo ships.** Most ocean-going ships in use today are cargo ships. Fig. 14-11. Specialized cargo ships carry products from one port to another. Cargo ships that carry liquids, such as oil, are called tankers.
- **Barges.** Barges have flat bottoms and blunt ends. They haul cargo on inland waterways such as canals, rivers, and lakes.
- **Tugboats.** Tugboats pull barges or ships into and out of harbors. They need powerful engines in order to move large ships such as ocean liners.

Fig. 14-11. A crane lowers cargo into the hold of a ship. Some method is usually used to hold the cargo in place. What do you think could happen if the cargo in a ship were to suddenly shift?

SECTION 3
TechCHECK

1. What invention made it possible for ships to sail any time without waiting for the wind?
2. List seven types of water transportation vessels.
3. In what ways are computers used in today's ships?
4. Apply Your Knowledge. Research hydrofoils and find out how they can move so fast.

ACTION ACTIVITY

Building a Hovercraft You Can Ride

Be sure to fill out your TechNotes and place them in your portfolio.

Real World Connection

Hovercraft are vehicles supported on a cushion of air. They can move over land or water at high speeds. Hovercraft are used to transport people quickly across open water in some parts of the world. In this activity, you will make a hovercraft that you can actually ride. Fig. A.

Design Brief

Design, build, and test a hovercraft that will support the weight of a person. It must meet the following requirements:

- Your design must be safe. Electrical extension cords must be the proper size for the vacuum motor. To prevent possible electrical shock, your hovercraft will not be tested over water or on a wet surface
- The area of the base of the hovercraft must be large enough to support the weight of a person.
- A switch must be provided so the rider can stop the vacuum motor and stop the hovercraft.
- The bottom of the hovercraft must be smooth and free of nails or screw points that could cut the plastic or scratch the floor.
- A chair or seat must be mounted to the hovercraft for people to sit on.

Materials/Equipment

- 3/8" plywood
- polyethylene (Visqueen) plastic (6 mil)
- duct tape
- wood screws (flathead)
- abrasive paper (60 grit sandpaper)
- plastic coffee can lid
- saber saw
- stapler
- utility knife
- vacuum cleaner

SAFETY FIRST
Do not use the hovercraft outside or near water.

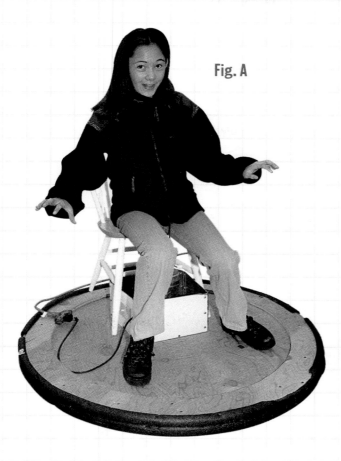

Fig. A

Procedure

1. In this activity, your class will be divided into two groups. Your teacher will help each group with its hovercraft design. Fig. B.

2. Make a list of the tasks needed to complete your hovercraft. Assign the tasks to individuals or small groups on your team. Some tasks might include

- Mark the plywood for cutting. Cut the plywood shape and sand the edges to remove sharp splinters.
- Drill a hole large enough for the vacuum motor exhaust port or hose.
- Measure and cut the polyethylene sheet to the desired size using a utility knife. Mark and cut the holes to release the air. Be careful not to scratch the workbench or floor while cutting.
- Staple the polyethylene sheet to the plywood. Screw or staple a plastic coffee can lid to the center of the plastic. This reduces the friction of the plastic against the floor.
- Design and build a mount for the vacuum motor.

3. Assemble the hovercraft using the parts built by each group. Place duct tape around the edges to help prevent rips in the plastic.

4. Attach the emergency stop switch with your teacher's help.

5. Attach the vacuum motor to the base with wood screws. Be sure the screws do not go through the plywood and cut the plastic. Seal around the motor and plywood with duct tape.

(Continued on next page)

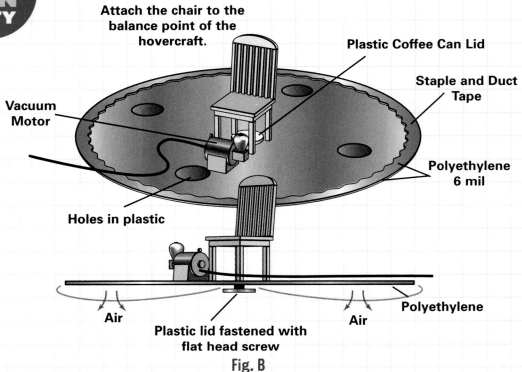

Attach the chair to the balance point of the hovercraft.

Plastic Coffee Can Lid

Staple and Duct Tape

Vacuum Motor

Polyethylene 6 mil

Holes in plastic

Polyethylene

Air

Air

Plastic lid fastened with flat head screw

Fig. B

6. With your teacher's help, test your hovercraft without anyone riding it. Clear a path for the hovercraft test run. Be sure to keep the extension cord out of the way. Test the emergency stop switch so the driver will have brakes to stop the hovercraft. If everything is working, have the first test driver climb aboard.

7. No one should attempt to ride the hovercraft while standing. Only one person should ride at a time. Start the vacuum motor and give the rider a gentle push. The test driver should practice balancing on the hovercraft so the weight is distributed evenly. Take turns test driving your hovercraft, and suggest ways to make it better and safer.

Evaluation

1. What are the advantages and disadvantages of hovercraft compared with cars?

2. Do you think you will have a choice of buying a car or a hovercraft in the future?

3. How could the hovercraft be made so it would not need to be plugged into an extension cord?

4. **Going Beyond.** Research other designs of hovercraft. See if you can find out which hovercraft is the fastest and which is capable of lifting and transporting the most weight.

5. **Going Beyond.** Design and test a safe method of propelling your hovercraft forward and backward.

6. **Going Beyond.** Design and test a way to steer your hovercraft.

Air Transportation

THINGS TO EXPLORE

- Describe how air transportation has changed since the early days of flight.
- Explain the difference between lighter-than-air vehicles and heavier-than-air vehicles.
- Test an airfoil design using a wind tunnel.

TechnoTerms

aerodynamics
airfoil
blimp
dirigible
heavier-than-air
vehicle
lighter-than-air
vehicle

Probably the most imaginative transportation ideas have come in the field of flight. Even great inventors like Leonardo da Vinci of Italy thought people should be able to fly by flapping some kind of wing device. What really started air transportation was dreams, not need.

Lighter-than-Air Vehicles

You can divide aircraft into lighter-than-air vehicles and heavier-than-air vehicles. **Lighter-than-air vehicles** float in air. **Heavier-than-air vehicles** must supply power to fly.

The first successful aircraft of any kind was the hot air balloon designed by Frenchmen Joseph and Etienne Montgolfier in 1783. They did not know what really made their balloon go up. They thought maybe some unknown, mysterious gas was released from burning wood! Fig. 14-12.

This principle of lighter-than-air flight was used in many more designs as people experimented with hydrogen-filled and hot-air-filled balloons. The first human flight was also made in 1783, when two people remained at an altitude of 3,000 feet for 25 minutes.

Fig. 14-12. Many people like to take rides in hot-air balloons. What causes a hot-air balloon to rise? Explain.

In the early 1900s, lighter-than-air vehicles called **dirigibles** carried passengers and freight around the world. The *Hindenburg* was the largest at more than 800 feet in length. It could carry 100 passengers. But dirigibles were filled with hydrogen, which burns very quickly when it is ignited. The *Hindenburg*, like many other dirigibles, came to a tragic end when the hydrogen exploded and the dirigible burned.

Today's **blimps**, such as the Goodyear blimp, are filled with helium gas, which is safer than hydrogen. Most are used for advertising or photography and sometimes freight.

Heavier-than-Air Vehicles

After people got one foot off the ground, there was no stopping them. The Wright Brothers' flight in 1903 was the first long, controlled, engine-powered flight. They had experimented with gliders and had even built a wind tunnel to test different *aerodynamic* (streamlined) shapes. They also experimented with how to control a plane in flight. This meant studying the forces that work on an airplane and how to adjust for them.

Aerodynamics is the branch of science having to do with forces created by air. Fig. 14-13. An airplane has to overcome both gravity and drag (air resistance). Wings that provide *lift* can

SCIENCE CONNECTION

Bernoulli's Principle: What Keeps Airplanes Up?

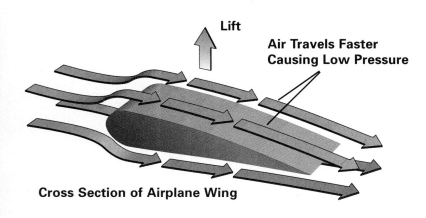

Lift

Air Travels Faster Causing Low Pressure

Cross Section of Airplane Wing

Over 200 years ago, a Swiss mathematician named Daniel Bernoulli discovered a scientific principle that is one factor in how airplanes fly.

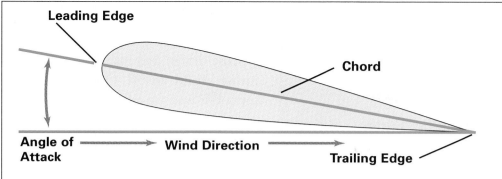

Angle of attack: The angle between the chord and the direction of the wind.
Chord: A straight line from the leading edge to the tail.
Drag: The resistance of an object to the flow of air, often determined by the object's shape.
Leading edge: The point on the wing that is farthest forward. The spot where wind hits the airfoil first.
Lift: The force created by the airfoil. It must be greater than the force of gravity so that an aircraft can fly.
Trailing edge: The back edge of the airfoil.

Fig. 14-13.
Another name for an airplane wing is airfoil.

overcome gravity. But once in the air, the plane is held back by drag. Drag is overcome by the thrust of the engines and a streamlined shape. The less drag on an airplane, the less power is needed to propel it.

Bernoulli's Principle states that as the speed of a fluid increases, its pressure decreases. You're probably asking what that has to do with an airplane's flying. Well, a fluid can be a liquid or a gas. Air is a fluid.

As air flows over an **airfoil** (wing), the wing's shape and *angle of attack* cause the air to speed up above the wing's surface. As the air speeds up, its pressure goes down, creating a low-pressure area above the wing. This low-pressure area creates *lift*, drawing the aircraft upward. This allows the aircraft to overcome gravity and to fly.

Bernoulli would be amazed to learn how technology has put his discovery to use! His scientific principle has helped the Wright Brothers and every flier since.

ACTIVITY

Place a ball above a leaf blower or hair dryer (on cool) that is blowing air. To demonstrate Bernoulli's Principle, see if you can balance the ball in the same position while slowly rotating the blower.

Jet Aircraft

Many technological advances in airplane design and manufacture came during World War II. One of the most important was the jet engine.

Today, stronger, lightweight materials make new designs possible. There are many different kinds of planes, and each is built for a special job. Some airliners can take off and land on short runways. Jumbo jets carry hundreds of people over long distances but need a long runway.

Fig. 14-14. This helicopter takes scientists for a view of an active volcano. Why do you think a helicopter would be ideal in this situation?

Helicopters

Helicopters are able to land people and supplies in places where other types of transportation can't go. Fig. 14-14. They can fly straight up while taking off or straight down while landing. Helicopters can also hover (fly in one place) in the air and change directions very quickly.

Helicopters are used for traffic control in large cities, where they monitor traffic jams or accidents. They are also used in the construction industry to move lumber or prefabricated sections.

SECTION 4
TechCHECK

1. Name two lighter-than-air vehicles.
2. What do heavier-than-air vehicles need in order to fly?
3. Describe two early air vehicles.
4. Apply Your Knowledge. Make a lighter-than-air vehicle, such as a hot-air balloon, and see how many ways you can find to make it lift.

Testing Airfoil Design in a Wind Tunnel

Real World Connection

When the Wright brothers were designing their airplane, they tested wings in a wind tunnel. Today, huge wind tunnels are used by NASA to test aircraft at supersonic (faster than the speed of sound) speeds. Models are often used, and many design problems can be solved by wind-tunnel testing before production starts.

For this activity you will design, build, and test an airfoil. Your test will show how airfoils create lift to make aircraft fly.

Be sure to fill out your TechNotes and place them in your portfolio.

Design Brief

Design, build and test an airfoil in a wind tunnel.

Materials/Equipment

- Styrofoam plastic foam
- 1/16"-welding rod
- paper
- scissors
- pins
- hot wire cutter
- wind tunnel (Fig. A)

SAFETY FIRST
Follow the safety rules listed on pages 42-43 and the specific rules provided by your teacher for tools and machines.

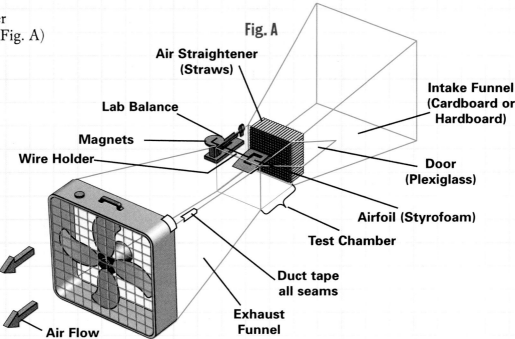

Fig. A

Air Straightener (Straws)

Lab Balance

Magnets

Wire Holder

Intake Funnel (Cardboard or Hardboard)

Door (Plexiglass)

Airfoil (Styrofoam)

Test Chamber

Duct tape all seams

Exhaust Funnel

Air Flow

(Continued on next page)

Procedure

1. Design an airfoil pattern on a piece of stiff paper. Cut out your design with scissors. Pin the pattern to a piece of Styrofoam plastic foam.

2. Use a hot wire cutter to cut out your airfoil. Follow the paper pattern carefully to make a smooth cut.

3. Locate the leading edge and trailing edge of your airfoil. (See Fig. 14-13.) Draw a straight line between the leading and trailing edges to show the chord.

4. Mount your airfoil on a wire holder made from a 1/16" welding rod. Place the wire holder through a hole in the test chamber of your wind tunnel. Attach the airfoil to the proper place. Fig. B. Adjust the airfoil so that the angle of attack is zero.

5. Make a data table to record all the measurements you are going to make. Your data table might look like the one in Fig. C.

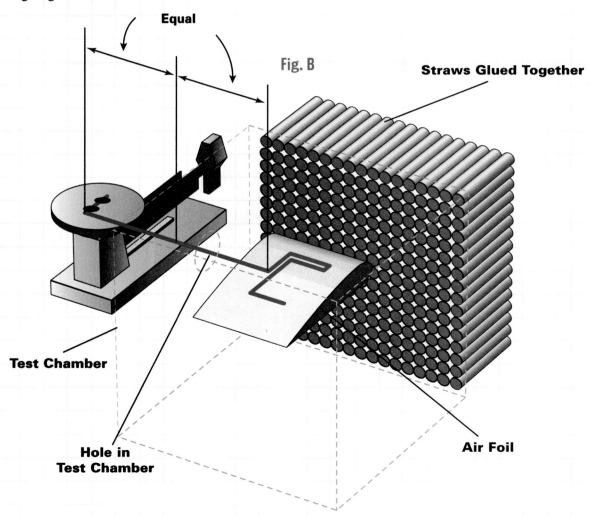

Equal

Fig. B

Straws Glued Together

Test Chamber

Hole in Test Chamber

Air Foil

6. You are now ready to test your wing for lift. Turn the wind tunnel fan on to a low speed. See if the wing goes up in the test chamber. Record the results.

7. Continue to test your airfoil at medium and high speeds. Record the results.

Evaluation

1. Did your airfoil fly?

2. Did you see any relationship between the lift and the wind speed? Explain.

3. Is there a relationship between the angle of attack and the amount of lift? Explain.

4. **Going Beyond.** Make a graph that illustrates the effect of wind speed on the amount of lift. Make another graph that illustrates the effect of angle of attack on the amount of lift.

5. **Going Beyond.** Design and test a method to see any turbulence around the test shape. You might try a vaporizer or humidifier to make a safe "smoke" source.

6. **Going Beyond.** Design and test other wing shapes. Make a data table for each shape. Can you predict the performance of an airfoil by looking at its shape? Explain.

Angle of Attack			0°			5°		10°	
Wind Tunnel Test Data	A-Balance setting (fan off)	B-Balance setting (fan on)	C-Difference (B-A)	D-Balance setting (fan on)	E-Difference (D-A)	F-Balance setting (fan on)	G-Difference (F-A)		
Fan Speed									
Low									
Medium									
High									

Fig. C

14 REVIEW &

CHAPTER SUMMARY

SECTION 1

- Transportation is the movement of people or goods from one place to another.
- There are different ways, or modes, of transporting, or moving materials on land, water, air, and in space.
- Transportation systems that use alternative energies, such as wind energy or solar energy, are being explored.

SECTION 2

- Land transportation includes cars, buses, trucks, railroads, pipelines, conveyors, moving sidewalks, escalators, and elevators.
- Many types of mass transit rail systems are used to move many people at one time.

SECTION 3

- Before the steam engine was invented, ships had to depend on wind, currents, or muscle power to move them.
- Water transportation includes passenger ships, cargo ships, hydrofoils, hovercrafts, submarines, barges, and tugboats.

SECTION 4

- Hot air balloons, dirigibles, and blimps are examples of lighter-than-air vehicles.
- Airplanes and helicopters are heavier-than-air vehicles that need power to fly.

REVIEW QUESTIONS

1. Does where you live make a difference in the transportation systems you use most? Explain.

2. Figure out how far the Trans-Alaska pipeline moves crude oil starting at Prudoe Bay and ending at Valdez.

3. What kind of materials are carried in barges?

4. What is the meaning of leading edge, trailing edge, chord, and angle of attack?

5. Why do car designers try to make cars lighter?

CRITICAL THINKING

1. How does Bernoulli's Principle affect car design?

2. Research the design of a jet engine. Make a chart or bulletin board display to show how it works.

3. Make a video showing how planes fly that elementary school students would enjoy.

4. With the help of your teacher, arrange a visit to an airport and investigate aircraft maintenance and air traffic control.

5. Write a major oil company and ask what that company is doing to prevent oil spills.

ACTIVITIES 14

CROSS-CURRICULAR EXTENSIONS

1. MATHEMATICS Find the relationship between the area of the hovercraft and its carrying capacity.

2. SCIENCE Design a method for measuring the drag (air resistance) on an object such as a model airplane, car, or rocket. Try your idea in the wind tunnel. Can you make a general statement that would apply to all shapes and the amount of drag they encounter?

3. COMMUNICATION Ask one of the following resource people to speak to your class: airline pilot, railroad engineer, air traffic controller, automobile race driver, or test driver.

EXPLORING CAREERS

Technology has changed the way people order, ship, and track the delivery of products. This fast-paced field includes new careers, as well as many that have been around for some time.

Air Traffic Controller Air traffic controllers control the movement of air traffic to make sure that planes keep a safe distance apart. Some controllers are responsible for traffic around a single airport and others handle flights between airports. Air traffic controllers must be able to work well under pressure and be able to give clear instructions.

Diesel Mechanic Diesel mechanics and service technicians repair and maintain the diesel engines that power heavy vehicles such as trucks, locomotives, and bulldozers. They perform routine checks to prevent problems from occurring. Diesel mechanics use a variety of tools, including handheld computers, power tools, and hand tools. They must be able to interpret manuals and stay up to date on new engine components.

ACTIVITY

You must transport a truckload of oranges from Los Angeles, California, to Burlington, Vermont. Plan the route and method you would use. Explain your choices.

Finding & Using Information

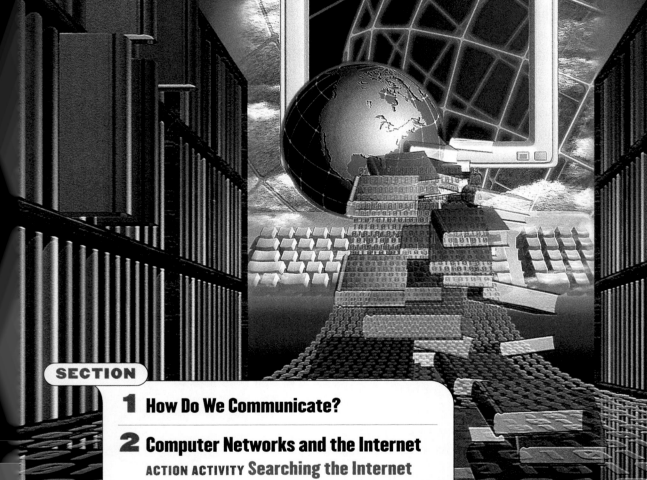

How Do We Communicate?

THINGS TO EXPLORE

- Define communication and give examples of how we communicate.
- Explain how a communication system works.
- Identify technologies used in communication.

TechnoTerms
communication
electronic noise
interference

Can you imagine your world without television, telephones, books, signs, newspapers, tape recorders, or computers? It would be pretty hard to go through a day without using any form of communication. What is communication? **Communication** is the process of exchanging information either by sending it or by receiving it. Fig. 15-1.

Sometimes your message is sent to another person; an animal, such as your dog; or a machine, such as a robot. Sometimes, even machines communicate with other machines, such as one computer to another. But before you are really communicating, the message has to be sent, received, and understood.

OPPOSITE Books, computers, and other resources are like banks that store thousands of years worth of human knowledge.

Fig. 15-1. Television is a powerful communication tool. What kinds of information does television bring us?

In this chapter, you will explore different ways to communicate, from using the Internet, to technical writing and drawing, to using communication satellite technology. As you read the chapter, think about the ways you and your family use technology to communicate. How does technology help you stay in touch with friends and relatives?

Parts of a Communication System

A communication system is like any other system. It has input, a process, and output. The *sender* creates the input, or message. The process is how the message is sent. The output is the form of the message that the *receiver* gets. Feedback from the receiver lets the sender know if the message was understood. Fig. 15-2.

INFOLINK

See Chapter 8 for more information on systems.

There is always a chance for interference anywhere in a communication system. **Interference** is anything that gets in the way of the message being understood. For example, sometimes it is hard for people to communicate if they do not understand the same language. Other forms of interference might be **electronic noise** such as static. A scratch on a record or a CD can also cause noise interference.

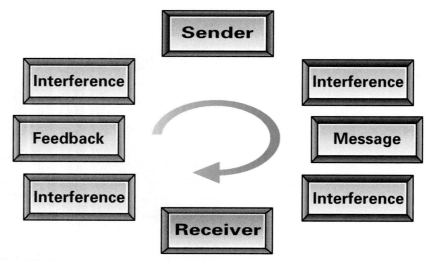

Fig. 15-2. The communication process makes a complete circle, from sender to receiver and back to sender. How does a teacher receive feedback from students?

Technology Brings Change to Communication

People have been communicating with each other since ancient times when they first used signs and symbols. Today, communication is changing very quickly. That's because we are developing new technologies that help us communicate in many exciting ways. You can communicate with digital photography, animation, and video. You can also use devices such as a *laser* (*l*ight *a*mplification by *s*timulated *e*mission of *r*adiation) in a laser disc player or CD player.

INFOLINK

See Chapter 5 for information about digital photography, animation, and video.

Did you ever think you would see a laser being used to get information at a grocery store? You have seen bar codes—a series of black lines and white spaces—on products. When you buy an item, a laser inside an optical scanner at the checkout counter reads the bar code. The scanner laser senses the lines and spaces and changes the information into on or off pulses of electricity. Each product has its own code. The code is sent to a computer, which then gives the price for that item. Bar codes help store owners in many ways. The computer can use the bar codes to keep track of price changes, how many of that item are in stock, and to let the store know when it's time to order more.

TechnoFact

LASERS ARE EVERYWHERE
Lasers have been around for a long time. The first laser was made by Theodore Maiman in 1960 using a ruby rod with mirrors at both ends. Today's lasers use materials such as helium, neon, argon, carbon dioxide, or organic dyes to produce different wavelengths of light. Lasers have become common in many applications: supermarket checkouts, laser disc players, CD players, optical data storage for computers, computer printers, and laser surgery.

SECTION 1
TechCHECK

1. What is communication?
2. Explain how a communication system works.
3. List three technologies that help us communicate today.
4. **Apply Your Knowledge.** Research how a laser works. Report your findings to the class.

Computer Networks and the Internet

THINGS TO EXPLORE

* Describe what a computer network is.
* Tell what the Internet is and how it works.
* Explain how a search engine works.
* Search the Internet for specific information.

Another very important way to communicate uses computers. When you connect computers together you create a **network**. When you connect many computers together, like in a school or office, it is called a *local area network* or *LAN*.

LANs often use a main computer to provide information to other computers on the network. The main computer is called the *server* and the other computers are called *clients*.

What Is the Internet?

If you take thousands of LANs and millions of home computers and connect them together, you have the world's biggest network, called the **Internet**, or "the Net."

The Internet actually started as a military communication system to be used in case of a nuclear war. It then expanded into a method for scientists to share information. Today, millions of people use the Internet to order products and send messages electronically using **e-mail** (electronic mail). Fig. 15-3. All of the words and many of the illustrations in this book were sent to the publisher using e-mail. Have you ever sent an e-mail message?

Fig. 15-3. E-mail is also used by electronic "stores" to send you a record of your purchase. Have you ever purchased a product online? If so, describe your experience.

This tells you the protocol, or set of rules, that computers use to understand the information; http stands for hypertext transfer protocol.

The information after the double slashes identifies the computer that stores the information you are looking for; "www" indicates the World Wide Web.

http://www.test.org/files/file.html

Single slashes mean that you are going to a certain directory or file at the site.

This is the name of the file that will appear on your screen.

The **World Wide Web** (**www**) is part of the Internet developed to present Internet information in a format that is easy to use. Because of this ease of use, the Web has become very popular.

Websites Many businesses, companies, schools, and even individuals have a location on the Web called a *website*. Websites are made up of web "pages" that are created on computers. A website can be found using its address, or **URL** (**uniform resource locator**). Fig. 15-4. URLs also make it easy to link pages together.

Making a web page is easy. Ask your teacher if you could make a web page for your technology education class.

Searching Made Easy People use software called browsers to access the Internet. Netscape and Internet Explorer are examples of browsers. Once you are connected, you can find information using **search engines**. Search engines take a key word or phrase and find matching files for you to view.

TechnoFact

HYPERSTUFF
"Hyper" words are used to describe the way the Web works. Here are a few. **HTML** (hypertext markup language) is the language used to write web pages. Hypertext is a system of writing and displaying words so they can be linked in many ways. Web pages are transferred over the Net using http (hypertext transfer protocol). Pictures, sounds, video clips, and animations seen on the Net are called hypermedia.

SECTION 2
TechCHECK

1. What is the Internet?
2. How do you search for information on the Internet?
3. What is a network?
4. **Apply Your Knowledge.** With your teacher's help, find the latest information on the next space flight planned by **NASA**. Print the results of your search and post the information on the bulletin board or print it in the school newspaper.

ACTION ACTIVITY

Searching the Internet

Be sure to fill out your TechNotes and place them in your portfolio.

Real World Connection

Have you ever searched for a book in the library only to find that it was already checked out? Have you ever talked with your parents or friends about the latest news and wondered how they heard about something that you knew nothing about? The Internet has quickly become the first source of information for millions of people around the world. In this activity, you will learn to search the Internet for specific information. Fig. A.

Design Brief

Use a search engine on the Internet to find information on a technology-related topic. Refine your search so it will be limited to only the specific information you are looking for. Put the information you find into a report or computer presentation.

Materials/Equipment

- computer with network or modem connection to the Internet
- word-processing software
- printer
- presentation software (optional)

SAFETY FIRST

- Follow the safety rules listed on pages 42-43.
- Check with your teacher to see if you and your parents need to sign an agreement for appropriate use of the Internet in school. Like the real world, the Internet is full of exciting things, but some of them are not appropriate for everyone. Make sure you have teacher permission and supervision while you use the Internet.

Procedure

1. Choose a topic related to technology that interests you. Here are some ideas:
   - Mars exploration
   - automobile crash testing
   - mountain bike design
   - wind surfing
   - microgravity
   - surfing conditions
   - snowboard designs
2. Boot your computer and start a browser such as Netscape or Internet Explorer.
3. Click on "Search" and type a key word or phrase in the appropriate box. Start the search and look at the number of matching files.
4. View the matching files and choose the ones that most closely match your needs.
5. Read the information and evaluate it. Remember to compare information from different sources in your evaluation.
6. Copy and paste images from your search results into a word processor for a printed report or into presentation software for a computer presentation.

Evaluation

1. List the names of three search engines.
2. Why is it important to know how to search for information on the Internet?
3. What do you think might happen if the Internet stopped working?
4. How do your parents or teachers use the Internet?
5. **Going Beyond.** Check with your teacher about creating a set of Internet "bookmarks" (URLs saved to the browser) related to your topic that others can use for further information.
6. **Going Beyond.** Research the weather for your area on the Internet. Print a weather map and forecast each day and post it in the school office for others to see.

Fig. A

Technical Writing and Drawing

THINGS TO EXPLORE

- Explain what technical writers and technical illustrators do.
- Identify the three types of pictorial drawings.
- Describe how computers help the process of writing and drawing.
- Create a procedure using technical writing and drawing skills.

TechnoTerms

edit
isometric drawing
oblique drawing
perspective drawing
technical illustrator
technical writer

Have you ever tried to put together a model of a rocket or use a computer software program only to get frustrated trying to follow the written directions? It isn't easy to write instructions or manuals that everyone can follow. Adding graphics or pictures is one way to make directions clearer.

Technical Writing

Sometimes when you are writing directions, they seem crystal-clear to you, but when other people read them they get confused. Even writing directions for making a peanut butter and jelly sandwich is not as easy as you may think. If someone had never seen a sandwich before, your directions would have to include how to put the two pieces of bread together!

Technical writers are trained to write technical manuals and instructions. Fig. 15-5. They have people test their instructions to make sure they haven't left out a step that they assumed everyone else already knew. In order to make sure your own messages are clear, you should have someone else **edit**, or check and correct, your work.

Fig. 15-5. The main goal of technical writers is to make material easily understood. Find an example of directions that come with a product. Are they clear? How could they be improved?

Fig. 15-6. The drawing done by a technical illustrator can make instructions easier to understand. Find an example of instructions in a how-to book in which a drawing plays an important part. Show the drawing to the class and explain its value.

Computers have made the editing process much easier than it used to be. If there are things that need changing in your instruction manual, a word-processing program will let you change them easily as well as check spelling and grammar.

Technical Illustration

Anytime you can use pictures or drawings to back up the words in a manual, the information will make will make more sense to more people. Technical drawings or illustrations help people understand the sizes and shapes of objects. **Technical illustrators** are people skilled in making this type of illustration. Fig. 15-6. Technical illustrators show objects in either pictorial or orthographic drawings.

Kinds of Pictorial Drawings A pictorial drawing shows a three-dimensional view of an object. There are three kinds of pictorial drawings. Fig. 15-7A.

- **Isometric drawings.** These drawings show an object as if you were looking at it from an edge. The object is angled and tilted slightly toward you. Lines that show the width and depth of the object are drawn at 30° angles from the horizontal.
- **Oblique drawings.** These drawings show one surface as if you were looking straight at it. Two other surfaces are shown at an angle.
- **Perspective drawings.** These are the most realistic drawings. Parts of an object that are farthest away look smaller.

> **INFOLINK**
>
> See Chapter 12 for more information about pictorial drawings.

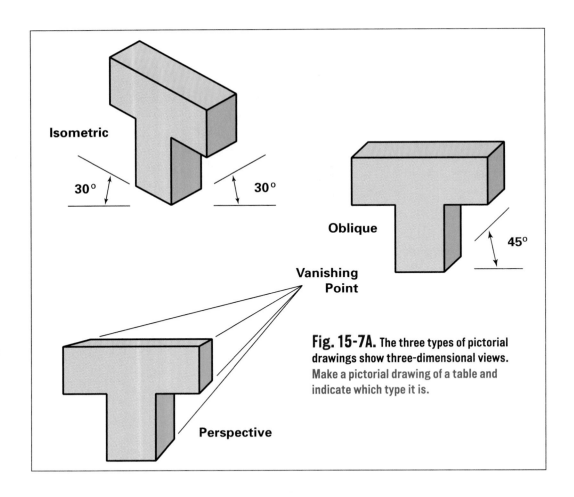

Isometric

30° 30°

Oblique

45°

Vanishing Point

Perspective

Fig. 15-7A. The three types of pictorial drawings show three-dimensional views. Make a pictorial drawing of a table and indicate which type it is.

How Ads Get Your Attention

What catches your eye first when you look at a billboard, a magazine ad, or a picture in a book?

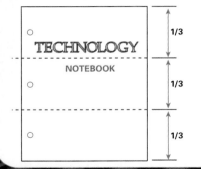

TECHNOLOGY

NOTEBOOK

1/3

1/3

1/3

You look at so many different things in a day that it is really a challenge for graphic artists to catch your attention. If you're going to communicate well with pictures and symbols, you need to do some of the things graphic artists do.

• **The line of golden proportion.** You can find the line of golden proportion by dividing the height of your drawing paper into three equal sections. The line one-third down from the top is the line of golden proportion. Half of any text or graphic should be above this line, and half should be below this line. If you arrange your pictures or writing this way, it

Orthographic drawings are another way to show an object. Here three views (usually the top, front, and side) of an object are shown as if you were looking straight at each one. Orthographic drawings are useful because they show the exact shape of each view. They also show how the views relate to each other. For example, in Fig. 15-7B, note that the top view is directly above the front view.

INFOLINK

See Chapter 12 for more information about orthographic drawings.

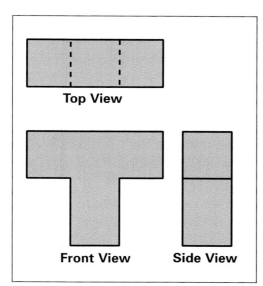

Top View

Front View   Side View

Fig. 15-7B. Orthographic drawings show views of an object. The dashes in the top view are "hidden" lines. They show the location of this object's base, even though you could not see the base when looking at the top.

makes them much more interesting. This method is also called the "rule of thirds."

- **Rhythm.** Rhythm guides the viewer's eyes toward a certain area. Sometimes you might number items to guide the eye. Other times you might use arrows to guide the eye to a specific spot in the picture. In a photograph, shadows can guide the eye to different areas.

- **Unity.** Unity is present in an ad when everything works together. For example, if you use too many different type styles or sizes in one ad, it looks confusing. If you are adding graphics, you need to think about how they fit your message.

Fig. 15-8. The photo on the left is the original. On the right is the altered version. **What has been changed?**

How Computers Help Computer-aided design (CAD) has improved technical drawing. CAD allows technical designers to create very precise, or accurate, drawings that are easier to edit than hand drawings. Then the drawings are printed. Making drawings that are realistic is also important in communicating an idea. Graphic artists can electronically create exciting posters or magazine layouts all on the computer screen.

Graphic artists used to have to work long hours to put photographs, artwork, and text together. Desktop publishing lets you put text and graphics together easily. Changes can also be made easily. A special effect can be added to a regular photo. That photo can then be added to another picture using a *digital image processor*. Fig. 15-8. For example, you could combine a picture of yourself with a picture of the beach even if you've never been there! In animation, graphic artists need to create a character or scene only once, save it on the computer, and call it up for changes in each scene.

(INFOLINK)

See Chapter 5 for more information about desktop publishing.

SECTION 3
TechCHECK

1. What do technical writers and illustrators do?
2. Why is it difficult to write clear instructions?
3. Name the three types of pictorial drawings.
4. How do computers help technical writers and illustrators?
5. **Apply Your Knowledge.** Check the manual that comes with your home or school VCR or television. Is it easy to use? How could you change it to make it better?

ACTION ACTIVITY

Writing and Drawing Technical Information

Be sure to fill out your TechNotes and place them in your portfolio.

Real World Connection

You know how frustrating it is when you don't understand directions or the directions don't make sense. Now you have a chance to write a procedure clearly and completely so anyone can understand it. In this activity, you will be working both as a technical illustrator and a technical writer.

Design Brief

Write a procedure for drawing an object located in your classroom. Your written instructions will be given to another technology student in another class. That person will follow your procedure, compare his or her drawing with your original drawing, and evaluate your instructions.

Materials/Equipment

- paper, pencil
- drafting equipment
- computer with CAD and word-processing software (optional)

SAFETY FIRST
Follow the safety rules listed on pages 42-43 and the specific rules provided by your teacher for tools and machines.

Procedure

1. In this activity, you will work individually. Choose an object that is visible in your classroom. The object should be recognizable from a front view.

2. Use drafting tools or CAD software to make a front view of the object. Keep your drawing simple. You may need to leave out some details and make only an outline. You should draw in just enough detail so the object can be recognized easily.

3. Ask someone else in the class to look at your drawing. Can it be identified easily? If not, you need to refine your drawing or put in more detail so there won't be any confusion.

4. Next, write directions for another student (in another class) to make a drawing exactly like yours and to identify the object.

(Continued on next page)

5. Write the directions on paper or use word-processing software. Create them carefully so both drawings will be identical. Keep in mind that sometimes a direction can have more than one meaning. Then it is called *ambiguous*. As you follow these sample directions, check for ambiguous directions. Fig. A.

Are These Directions Clear?

- Tape the paper to a drawing board so the long edge is parallel to the T-square.
- Starting from the upper left corner, measure 5 1/2" to the right and 1" down. Mark this position as point A.
- Draw two 75°-angle lines, 1/2" long, down to the left and right of point A.
- Join the end of the lines in step 2 with a horizontal line, forming a triangle.
- From the lower corners of the triangle, draw two parallel vertical lines 4 3/4" long.
- Join the lines with a short horizontal line to complete the outline of the object.

Fig. A

6. Compare your drawing with another student's work. Could yours be improved?

7. If you used CAD software to make your drawing, your directions will be slightly different. In this case, you should include such things as scale and screen position. The hard copy (printout) of both drawings should be identical if your directions were clear and accurate.

Evaluation

1. Ask your teacher to give your directions to a student in another class. Compare the results of the drawing made from your directions with your original. Are they the same?

2. Were there any ambiguous directions in your list? Explain.

3. How do you think directions could be made simpler for very complex tasks?

4. **Going Beyond.** Make a pictorial drawing of your object rather than the front view. Was it easier or harder to write the directions for a pictorial drawing? Explain.

5. **Going Beyond.** Use walkie-talkies to give directions to another person instead of writing them or typing them. Was this method easier or harder? Explain.

6. **Going Beyond.** Use e-mail to send your instructions to a student in another school. Also have another student in another school send instructions to you. Exchange the resulting drawings by e-mail or fax machine.

Electronic Communication

THINGS TO EXPLORE

- Identify electronic communication systems you use.
- Explain how electronic communication systems work.
- Tell what GPS is and how it is used.
- Use a GPS to set up a satellite dish.

TechnoTerms

communication
satellite
electronic communi-
cation system
fax (facsimile)
machine
fiber-optic system
geosynchronous
global positioning
system (GPS)
modem

D id you talk long distance today to anyone? Or did you watch a news program live from another country by satellite? Communication technology has changed rapidly because of improvements in electronics. Today you use many **electronic communication systems**, such as radios, televisions, telephones, modems, satellites, and fax machines. These systems all use electronic or electromagnetic signals to carry messages through cables or through the air. In this chapter we will explore fiber optics, telephones, modems, fax machines, satellite communications, and global positioning systems (GPS). Chapter 16 will cover radio and television.

INFOLINK

See Chapter 16 for information about radio and TV.

Fiber Optics

Fiber-optic systems use light to carry information. What's so special about fiber optics? For one thing, fiber-optic cables are smaller, weigh less, and cost less than the many copper wires used on long telephone routes.

A *fiber-optic cable* is made out of many thin strands of glass fibers. Fig. 15-9. These fibers are sometimes thinner than a strand of your hair! This cable can carry light for long distances without loss of power. A strong light source such as a laser or a *light-emitting diode* (LED) is the transmitter. The fiber-optic cable carries the information that is received by a device that converts light energy into electricity. The cable can carry over 1000 circuits at a time.

Fig. 15-9. Fiber-optic cables can carry many more messages than copper wire. Look up the meaning of the word optic. Share your findings with the class.

Fig. 15-10. How telephones have changed! What do you think the crank on the side of the old-fashioned phone is used for?

Telephones

What would you do without a telephone? Alexander Graham Bell never knew that his invention would end up being one of the most used electronic communication tools today. Fig. 15-10.

The telephone's mouthpiece changes sound into electrical signals. The ear piece changes the electrical signals back into sound. Since many telephone conversations take place at the same time, switches and lots of wires have to connect one town to another. Today, conversations are being carried over fiber-optic cables. This is because one pair of fiber-optic cables can carry over 1,000 conversations at one time!

Cordless telephones contain a small radio transmitter that sends a radio signal to the base of the telephone. The base is connected by wire to the telephone lines. *Cellular phones* also send radio signals. But cellular phones can transmit over an entire city, where cordless phones work only over short distances. Telephones are also connected to computers. In some places, you don't have to look up a number in a telephone book anymore! The computer does it for you.

TechnoFact

FASTER, FASTER! Fiber optics really speeds up how fast computers can communicate with each other using modems. Using fiber optics can increase the speed by a multiple of 41,667!

Modems

You learned about modems in Chapter 4. It is important to understand that a **modem** changes computer signals into audio tones so they can travel over telephone lines. At the other end of the line, another modem turns the audio tones back into computer signals.

INFOLINK

See Chapter 4 for more information about modems.

Fax (Facsimile) Machines

Did you know that since the 1920s, weather maps have been sent by fax? **Fax (facsimile) machines** quickly send graphics or pictures electronically. An optical scanner moves across a page and changes information into electrical impulses. These impulses are sent over telephone lines to another location and then changed back into a picture. Corporations and businesses use fax communication to order materials and to send letters.

Satellite Communication

You now have instant access to any part of the world by means of satellites. **Communication satellites** are relay stations for television and radio. Have you ever heard of the Clarke Belt? It is an area 22,300 miles over the Earth's equator where many communication satellites are orbiting. Their orbits are **geosynchronous**, which means they stay in the same place above the Earth at all times. Fig. 15-11. They travel at the same speed as the Earth turns, so it appears as if they are fixed, or *geostationary*.

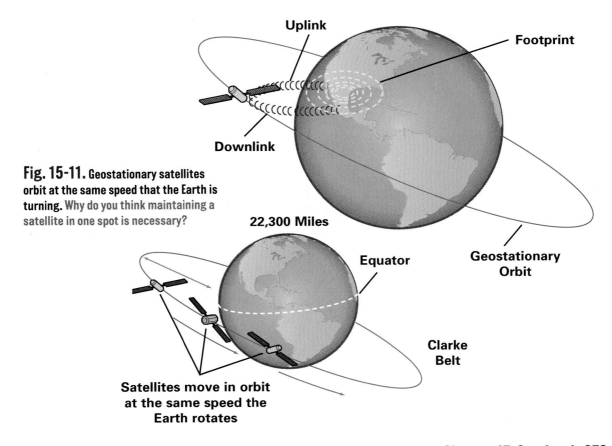

Fig. 15-11. Geostationary satellites orbit at the same speed that the Earth is turning. Why do you think maintaining a satellite in one spot is necessary?

Uplink

Footprint

Downlink

22,300 Miles

Equator

Geostationary Orbit

Clarke Belt

Satellites move in orbit at the same speed the Earth rotates

Satellite signals are *microwave signals*. Antennas on board each satellite focus the microwaves onto a special place on the Earth's surface. The microwaves travel in a straight line because the satellite remains in the same spot above the Earth. Signals going from the Earth to the satellite also follow a straight path.

A transmitting station sends a transmission on one frequency, called the *uplink,* to the satellite. Inside the satellite, a receiver takes in the signal and changes it to another frequency called the *downlink*. The area over which signals can be picked up is called the *footprint*. Satellite footprints show where each satellite's power levels are the best. If you have a satellite receiver, you can pick up transmissions. Some commercial uplink sites scramble their transmissions so people cannot receive them for free.

The Global Positioning System (GPS)

Imagine an electronic device that could tell you your location anywhere on Earth in latitude, longitude, and even altitude. What if that same device could map your travel path and guide you to your desired location or bring you back to where you started?

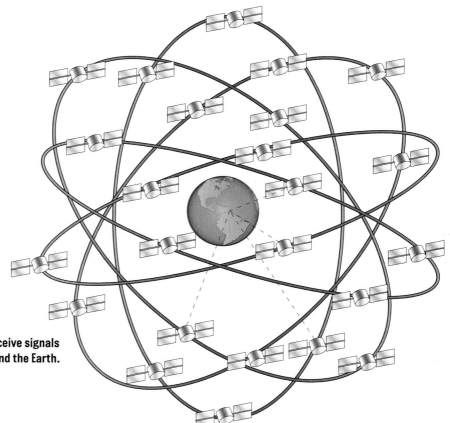

Fig. 15-12. GPS receivers receive signals from 24 satellites in orbit around the Earth.

Now imagine such a device being small enough to fit into your pocket or even be a part of your wristwatch. That incredible technology exists today, and it's called the **global positioning system (GPS)**. Fig. 15-12.

The handheld electronic devices are actually called *GPS receivers*. They receive signals sent from 24 GPS satellites in orbit around the Earth. Each of the satellites has a very accurate atomic clock on board. GPS receivers use this accurate time signal to calculate positions and travel paths.

The longitude and latitude of a location are its *coordinates*. GPS receivers can store dozens of specific coordinates for many locations. Each location is called a *waypoint*. Waypoints are points along a path or route that help the GPS receiver keep track of your exact movement.

You don't have to be sitting still to use GPS signals. Airplanes use GPS to find their location and locate airports. GPS is even available in some cars to help find the best route and prevent drivers from getting lost. Many people can enjoy the accurate navigation ability of GPS while hiking, fishing, or boating. Law enforcement officials use GPS for search and rescue operations. Geologists and forestry workers use GPS to locate and map specific sites.

The future of communication depends a lot on what you as the consumer need and expect from it. Who knows? Maybe you'll be able to turn your own room at home into a live TV studio using the telephone lines and a video camera. Won't it be fun to send a "live" greeting to a friend somewhere far away?

TechnoFact

PLAY IT SAFE In just a few years, GPS receivers have increased in capabilities and decreased in cost. As GPS technology gets even smaller and cheaper, more people may use it. However, the directions that come with GPS receivers warn users not to rely on them for all their navigation needs. If you are thinking of a trip that requires navigation skill, take a compass and a map in addition to a GPS device. The batteries might go dead just when you need them!

SECTION 4
TechCHECK

1. What electronic devices do you use to communicate?
2. How do electronic communication devices work?
3. What is **GPS** and how is it used?
4. **Apply Your Knowledge.** Use a GPS device to track your path as you walk around the perimeter of your school.

ACTION ACTIVITY

Where in the World Are You?

Be sure to fill out your TechNotes and place them in your portfolio.

Real World Connection

GPS satellites and direct-TV satellites are in a geostationary orbit above the Earth. If you combine the two technologies, you can learn about how both GPS and direct-TV broadcasting work.

Design Brief

Use a GPS device to find your location, Use the coordinates of your location to aim a direct-TV satellite dish and receive a TV signal. Fig. A.

Materials/Equipment

- GPS receiver
- compass
- level
- direct-TV satellite dish (18" or smaller)
- satellite receiver
- television or monitor

SAFETY FIRST

- **Check with your teacher to see where you should locate the satellite dish. You will be setting up the dish for only a short time.**
- **There is no need to go up on a roof or mount the dish permanently.**
- **Do not attempt to do this activity in the rain or if there are storms in the area.**
- **If necessary, use only approved extension cords.**
- **Follow the safety rules listed on pages 42-43 and the specific rules provided by your teacher for tools and machines.**

Fig. A

Procedure

1. Ask your teacher where you should locate the satellite dish. Work with a partner.
2. Set up the mounting post of the satellite dish so it is plumb (straight up and down). Use a level to check it.
3. Use a GPS receiver to determine the latitude and longitude of your location. Fig. B. Write the coordinates on your TechNotes.
4. Turn on the satellite receiver and go to the setup menu. Choose the option of using latitude and longitude.
5. Enter your coordinates. The receiver will display the elevation and azimuth for your location. The azimuth is the direction measured in degrees from north. The elevation is the angle of the satellite dish measured in degrees.
6. Adjust the angle of the dish to the proper *elevation*. Use the compass to point the dish in the direction given in the *azimuth*.
7. Set the satellite receiver using its signal strength meter. Make small adjustments in the dish until a strong signal is received.
8. Set the satellite receiver to the free preview channel and show your teacher that you have been successful.

Evaluation

1. List three uses for a GPS receiver.
2. How is it possible to adjust a direct-TV satellite and not need to change its position to continue receiving a signal?
3. What is azimuth and how is it measured?
4. **Going Beyond.** With your teacher's help, take the GPS device on a field trip. Check its accuracy on the return trip.
5. **Going Beyond.** Research where direct-satellite broadcasting originates.

Fig. B

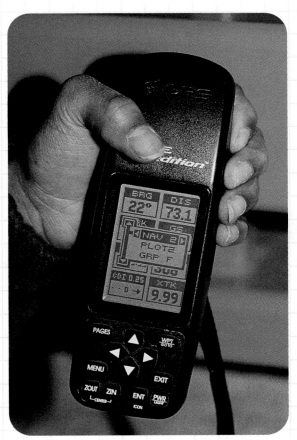

15 REVIEW &

CHAPTER SUMMARY

SECTION 1
- Communication is the process of exchanging information.
- A communication system has input, output, process, feedback, and interference.

SECTION 2
- Networks are groups of computers linked together.
- The Internet consists of millions of computers linked together.

SECTION 3
- Technical writers write instructions or manuals.
- Technical illustrators show objects in either pictorial drawings or orthographic drawings.
- Computers allow technical designers to create very precise drawings that are easy to edit.

SECTION 4
- Electronic communication systems such as radio, television, telephone, fiber optics, modems, satellites, and fax machines are used today.
- Satellites in geosynchronous orbit stay in the same place above the Earth at all times.
- GPS technology can be used to find your location anywhere on Earth.

REVIEW QUESTIONS

1. How does today's instant communication, such as a live television news broadcast, make our world seem smaller?
2. What is an Internet browser?
3. What two electronic communication devices do you use most at home?
4. Why is it difficult to write a technical manual?
5. How might people use a GPS receiver?

CRITICAL THINKING

1. What are some positive and negative effects on society of today's communication technology?
2. Research ways to make a poster using a computer and a digital camera. Does your school have the technology to do this yet? How could you use this process to make photographs for your school newspaper?
3. Send a set of written directions for drawing an object or making a product to students in a foreign country. Get a set of directions from a foreign student. Use a translation dictionary to understand the instructions. Exchange the drawings through the mail. Were there more errors in understanding the instructions when you had to translate the language? Why?

CHAPTER 15

ACTIVITIES

CROSS-CURRICULAR EXTENSIONS

1. COMMUNICATION With your teacher's help, design a website for your school. Ask other teachers to have their students make web pages for each subject.

2. SCIENCE Research how to make a hologram and try to make one.

3. MATHEMATICS Research how microwaves and other electromagnetic radiation are measured.

EXPLORING CAREERS

When your parents were in school they had to go to the library to look up information. Today we are expanding our ability to create, find, and use information. Careers like those listed below involve the search for information.

Central Intelligence Agency Engineer The CIA employs engineers to develop technology for collecting and analyzing information. This involves both in-house development and working with private companies to acquire technology that has already been developed. CIA engineers must be interested in the latest technology. They also must have good communication skills.

Medical Research Technician Techniques used in medical research today were found only in science fiction 20 years ago. Medical research technicians can be involved in many different tasks and may work on projects as part of a team. Most technicians conduct several experiments at the same time. Medical research technicians must keep accurate records.

ACTIVITY

Suppose you have just been offered a new job. Write down all the questions that you think you should ask before deciding whether or not to accept the position.

CHAPTER 16
Producing TV/Radio Programs

How Radio and TV Work

THINGS TO EXPLORE

- Describe the early days of radio and television.
- Explain how radio signals are sent and received.
- Explain how a television changes signals into audio and video information.
- Give examples of how television and radio impact your life.

TechnoTerm
audio
frequency
hertz
video

You have enjoyed TV and radio programs, but how often have you really thought about all the work that goes into making them? How different would your life be without TV or radio? Let's take a quick look at the history and technology behind TV and radio.

Radio

In 1901, Guglielmo Marconi became the first person to transmit and receive wireless signals across the Atlantic Ocean. Right away, people could see how radio communication was going to help them. Fig. 16-1.

Do you ever turn on the radio to listen to music or hear the latest news or the weather report? Have you ever picked up a station hundreds of miles away while riding in a car?

◀ **OPPOSITE** Much goes on behind the scenes before your favorite TV programs can be broadcast.

Fig. 16-1. When these operators see something important on their radar screens, they radio the information to other locations in the submarine. Why do you think radios are used? Why don't people simply deliver the information in person?

In radio broadcasting, information is sent through the air by means of transmitters to a receiving antenna someplace else. The messages travel at a certain **frequency**, which is measured in cycles per second. One cycle per second is called one **hertz**. Each radio station transmitter uses a different frequency, and there are many frequencies to choose from. Transmission happens very fast. Radio waves travel at about 186,000 miles per second, no matter what the frequency!

Television

Americans were first introduced to TV in a futuristic exhibit at the 1939 World's Fair. The first successful TV transmission was made by John L. Baird. Inventors Philo Farnsworth and Vladimir Zworykin came up with a combination of technologies that led to the TV we know today.

Did you know we watch an average of three hours of television every day? Television is used to inform, entertain, and sell. Many shows bring subjects like history and science alive. While you're watching, you also become one of millions of possible consumers for the products advertised during the shows. You can even shop at home by calling a toll-free phone number and using a credit card to order hundreds of products.

COMMUNICATION CONNECTION

Waves That Carry Information

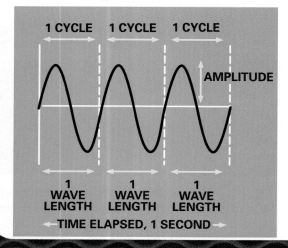

Energy can travel in waves that are both electrical and magnetic. The waves can differ in frequency and length, but they all travel at the same speed. This range of waves is called the electromagnetic spectrum.

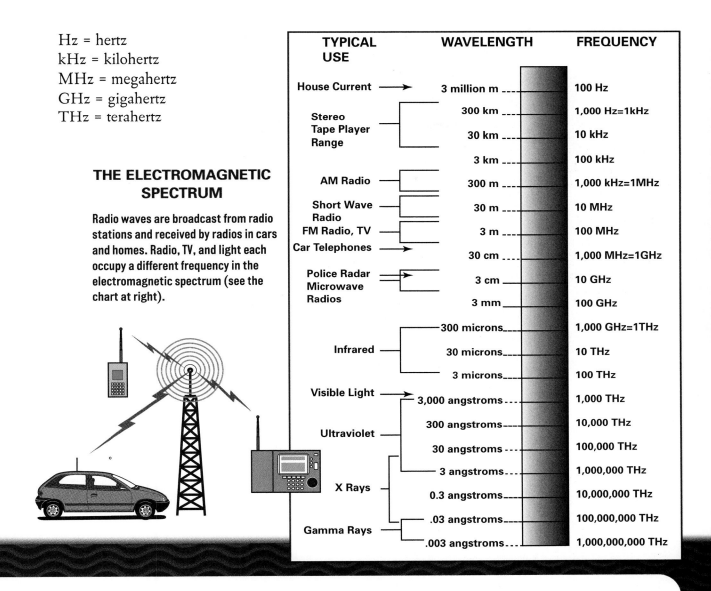

TYPICAL USE	WAVELENGTH	FREQUENCY
House Current ⟶	3 million m	100 Hz
Stereo Tape Player Range	300 km	1,000 Hz=1kHz
	30 km	10 kHz
	3 km	100 kHz
AM Radio ⟶	300 m	1,000 kHz=1MHz
Short Wave Radio	30 m	10 MHz
FM Radio, TV	3 m	100 MHz
Car Telephones ⟶	30 cm	1,000 MHz=1GHz
Police Radar Microwave Radios ⟶	3 cm	10 GHz
	3 mm	100 GHz
Infrared	300 microns	1,000 GHz=1THz
	30 microns	10 THz
	3 microns	100 THz
Visible Light ⟶	3,000 angstroms	1,000 THz
Ultraviolet	300 angstroms	10,000 THz
	30 angstroms	100,000 THz
X Rays	3 angstroms	1,000,000 THz
	0.3 angstroms	10,000,000 THz
Gamma Rays	.03 angstroms	100,000,000 THz
	.003 angstroms	1,000,000,000 THz

Hz = hertz
kHz = kilohertz
MHz = megahertz
GHz = gigahertz
THz = terahertz

THE ELECTROMAGNETIC SPECTRUM

Radio waves are broadcast from radio stations and received by radios in cars and homes. Radio, TV, and light each occupy a different frequency in the electromagnetic spectrum (see the chart at right).

All waves have amplitude and frequency. Amplitude is the amount of energy the wave carries. Frequency is the number of waves that pass a given point in one second. Shorter waves have a higher frequency than longer waves.

Frequency is measured in cycles per second, or hertz. The waves shown in the drawing have a frequency of 3 hertz.

Electromagnetic waves can be used to carry radio, TV, and other signals, as shown in the diagram above.

Fig. 16-2. An antenna receives a TV signal and transmits it through wires to the receiver inside the house.

Television communication changes **audio** (things you can hear) and **video** (things you can see) information into electrical signals. The signals are sent out and then changed back into pictures and sound by your TV receiver. Fig. 16-2. The signals might be sent to your home by cable, broadcast directly from a transmitter, or bounced off a satellite.

A monitor is like a TV without the ability to change channels. Many TV's today are really a TV and a monitor together.

Combination cameras and recorders called camcorders are often used to record video and audio onto tape. At home, you can watch the videotape or record over it on a video cassette recorder (VCR).

SECTION 1
TechCHECK

1. When was the first wireless radio message sent and received?
2. How are radio and TV signals sent and received?
3. How are electrical signals for television changed back into sounds and pictures?
4. **Apply Your Knowledge.** Make a poster showing how television or radio works.

Pre-Production

THINGS TO EXPLORE

- Explain what pre-production is and list some of the planning steps.
- Tell why it is important to have good talent and a good script.
- Write a script and rehearse a show.

TechnoTerm
ad lib
pre-production
public service
   announcements
   (PSA)
talent

D id you know that many hours of planning and rehearsing are required for every minute of TV that we watch and radio that we hear? This work is called **pre-production**. Pre-production has a direct effect on the quality of the finished show. It might include

- script writing
- drawing storyboards
- making cue cards or programming a teleprompter
- rehearsal

The Talent

The people at the microphone or in front of the camera are generally called the **talent**. What would happen if the talent had no idea of what to say or do? Some creative performers can easily work without a script. They **ad lib**, or make up, what they say and do. However, most of what we watch on TV and listen to on the radio has been planned in great detail in a script. The plan is then practiced in a test run called a rehearsal.

Dialogue is sometimes written with large letters on big sheets of paper called *cue cards*. The talent can read the words at a distance in case they forget their lines. A *teleprompter* is a computer used to display the dialogue on a sheet of glass in front of a camera. Fig. 16-3. The talent can read the dialogue, but it doesn't show on camera.

Fig. 16-3. This student is practicing with a teleprompter. The dialogue is entered into the computer and printed out on the screen for the talent to read. Why do you think TV directors use hand signals while a program is being taped?

Writing for TV and Radio

Good talent can help make a show interesting. However, watching TV or listening to the radio would be pretty dull if the shows didn't have a story line or a theme. Careful script writing can keep viewers' or listeners' attention.

A script includes written words, called *dialogue*, and other details necessary to stage the program. Fig. 16-4. Radio scripts might include dialogue for the news, commercials, and music segments. Special announcements that warn people of the dangers of smoking or remind them to wear seatbelts are called **public service announcements (PSAs)**. TV scripts include special instructions for camera operators, lighting technicians, special effects, and set design. The set is the stage area that can be designed to look like anything from a living room to a space station.

Fig. 16-5 shows a storyboard. A *storyboard* is a simple cartoon drawing showing the main "scenes" in a video.

In the activity for this section, you will write a script and rehearse it as part of pre-production. All of the pre-production steps are important to make sure your production is a good one.

SAMPLE SCRIPT			
CAMERA	SUBJECT	DIALOGUE	SPECIAL EFFECTS
CLOSE-UP	YOLANDA SITTING AT ANNOUNCER'S DESK	YOLANDA: NOW LET'S GO TO OUR WEATHER REPORT.	VFX: DISSOLVE TO TIM AT WEATHER DESK

Fig. 16-4. VFX is short for *video effects*. What camera movements and special effects would you use in a short video on how to feed a dog or other pet?

A New Way to Capture the Moment

Real Action-in Color

Never Better, Never Easier

Fig. 16-5. Storyboards do not usually show much detail. How do you think having a storyboard would help a TV director?

SECTION 2
TechCHECK

1. What are the possible steps in pre-production?
2. What is talent?
3. Why are good talent and a script important to a production?
4. **Apply Your Knowledge.** Ask your local radio station broadcasters if they follow scripts or ad lib most of the time.

ACTION ACTIVITY

Writing a Script and Rehearsing a Program

Be sure to fill out your TechNotes and place them in your portfolio.

Real World Connection

Have you ever watched an educational program on TV or heard one on the radio? TV and radio can teach us many things. In this activity, you will be the script writer, talent, and part of a TV or radio pre-production team to produce your own program. Fig. A. The program will be completed in later activities. Your finished program will be shared with other classes to teach them something you have learned about technology.

Design Brief

Write a script and rehearse a video or radio program (or both) that is informative as well as entertaining. Your program should teach the audience about some aspect of technology and run from five to ten minutes. Your program should include a commercial for a product or a public service announcement.

Materials/Equipment

- props, studio lights, costumes (as needed)
- camcorder or audio recorder
- stereo, tape deck, CD player, mixer, microphone (optional)

SAFETY FIRST

- **Keep away from hot studio lights. Watch where you walk to avoid tripping on mic (microphone) cords or video cables.**
- **Follow the safety rules listed on pages 42-43 and the specific rules provided by your teacher for tools and machines.**

Procedure

1. Work in a group of three to five people to plan your program. In a later activity, your group will team up with another group for the actual production. While your group is performing in front of the camera, the other group will be operating the equipment. When your group is finished, you will switch places.

2. Brainstorm ideas that you could use for your program. Remember it must be both instructional and entertaining. Here are some possible topics and themes.
   - The fast growth of technology
   - How technology has affected our world
   - Making a pneumatic robot arm
   - A technology quiz or game show
   - A news program or soap opera
   - A how-to program

3. Write a script for your idea. Everyone in your group should have a part. Ask your teacher to approve any special props, music, or costumes that you think you might use in your production.

4. Be sure to include a commercial for a product or a public service announcement. Many commercials run for 1 minute. Some are 30 seconds or up to 2 minutes long. Design your commercial to fit one of these times. The commercial message can also be educational.

5. Use stereo equipment to record your audio or video commercial.

6. Rehearse your program and refine your presentation for the next activities. Time the length of your production.

Evaluation

1. List items that might be included in a script.

2. Why do you think commercials should be short?

3. Name some occupations related to TV and radio production.

4. **Going Beyond.** Team up with a language arts and drama class to write and produce your production.

5. **Going Beyond.** Design and build a set to simulate a scene from a movie you have seen. Use the set in your video production.

6. **Going Beyond.** Record sound effects while recording a story on audiotape. Try to make realistic sounds related to your story, such as doors slamming, people walking, or horses galloping.

Fig. A

Radio Broadcasting

THINGS TO EXPLORE

- Tell what AM and FM radio frequencies are.
- List the parts of a radio program.
- Describe common jobs in radio broadcasting.
- Broadcast a radio program.

TechnoTerm
amplitude modula-
tion (AM)
dead air
frequency modula-
tion (FM)
silence sensors

The radio stations we listen to differ in how the radio signals are sent and received. AM stands for **amplitude modulation**. FM means **frequency modulation**. Fig. 16-6. Both types of stations usually specialize in one style, or format, of programming such as classical, jazz, rock, or country music. No matter what type of format is used, workers such as announcers, combo operators, engineers, or radio personalities, are needed. Fig. 16-7.

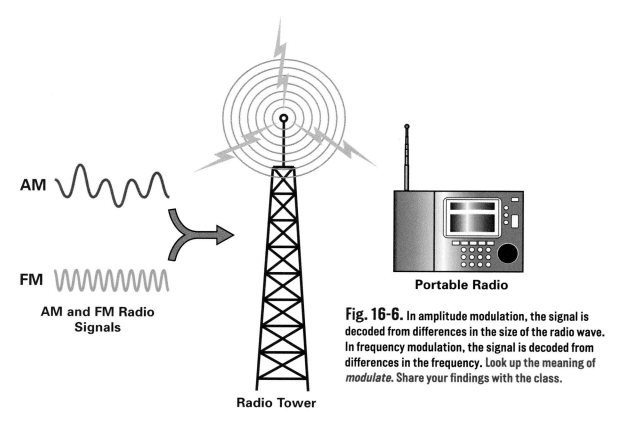

AM

FM

AM and FM Radio Signals

Radio Tower

Portable Radio

Fig. 16-6. In amplitude modulation, the signal is decoded from differences in the size of the radio wave. In frequency modulation, the signal is decoded from differences in the frequency. Look up the meaning of *modulate*. Share your findings with the class.

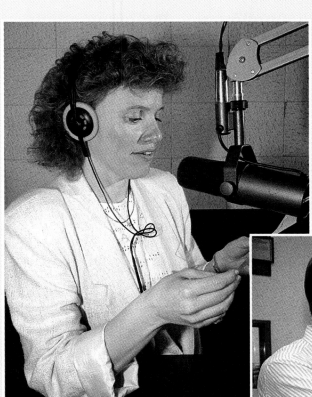

Fig. 16-7. An announcer (top left) reads the news. A combo operator (bottom left) is an announcer as well as the operator of equipment. Engineers (middle) operate control room equipment. Radio entertainers (bottom right) usually have colorful personalities.

One job that audio engineers do is to operate the audio mixer.

A radio program is made up of parts that usually come at a special time. Some of these include station identification, introduction, music, weather, sports, and news. Listeners become used to the schedule and know when to tune in.

Radio stations must stay on a strict time schedule because some of the programming, like the news, comes from a "live," rather than a pre-recorded, source. Live news is often sent to many radio stations at the same time using a satellite signal. Professional announcers try to avoid cutting off a song in the middle to play the news.

Have your ever been listening to a radio program and noticed silence for a short time? This is called **dead air**. Dead air is usually caused by equipment problems or by someone not flipping a switch at the right moment. Radio professionals try to avoid it. If listeners don't hear anything for even a short time, they often change the station. **Silence sensors** are electronic devices that detect silence and automatically trigger a different audio source, such as a recording of music.

SECTION 3
TechCHECK

1. What are two types of signals used for radio? What is the difference between them?
2. List parts of a radio program that usually happen at a certain time.
3. What are some occupations related to radio production ?
4. **Apply Your Knowledge.** Use an audio recorder to record a simulated radio broadcast of a school sports activity. Play the tape for the class.

ACTION ACTIVITY

Broadcasting a Radio Program

Be sure to fill out your TechNotes and place them in your portfolio.

Real World Connection

Today, we often listen to the radio for music, news, weather, and sports. If you chose to write an audio script earlier, in this activity you will use a school radio station to record and transmit your audio program.

Design Brief

Record or transmit a five- to ten-minute radio broadcast. Include a commercial or public service announcement. Identify yourself and give credit to the production staff.

Materials/Equipment

- CD player
- audio tape recorder
- mixer
- microphones
- stereo amplifier/speakers
- headphones
- FM transmitter (optional)

SAFETY FIRST

Follow the instructions that came with the transmitter. Follow the safety rules listed on pages 42-43 and the specific rules provided by your teacher for tools and machines.

(Continued on next page)

Procedure

1. Work in groups of two to four. Do not play inappropriate songs or use offensive language. Ask you teacher to pre-approve your script and the lyrics (words) to songs. Read through your approved script and time how long it takes.

2. Your teacher will demonstrate how to use the audio equipment. Fig. A. Practice recording and using the mixer and microphones (mics). Use the audio mixer to combine your voice and music.

3. Set the stereo speakers so the entire class can hear your broadcast. Turn off the speakers in the room in which you are broadcasting or the mics will pick up the sound from the speakers and make a loud hum called *feedback*.

4. Record your broadcast on audio tape so you can listen to yourself later.

Evaluation

1. What is feedback?

2. How can you prevent feedback?

3. What do you think would happen in a real broadcast if there was dead air for a long time?

4. What does an audio mixer do?

5. **Going Beyond.** Visit a radio station to see how professionals broadcast.

6. **Going Beyond.** Ask if it would be possible to set up a real radio station at your school.

7. **Going Beyond.** Research digital radio on the Internet. With help from your teacher, listen to Internet radio from other countries. Research the future of digital radio.

Fig. A

Video Production

THINGS TO EXPLORE

- Describe in-camera, linear, and non-linear editing.
- Describe the difference between first- and second-generation video copies.
- Produce a TV program.

TechnoTerm
character generator
in-camera editing
linear editing
non-linear editing

Y ou probably will be surprised at how much planning and rehearsing it takes to make a professional-looking TV show. Even with the best script writing and many rehearsals, all TV programs require changes. The process of making changes is called editing.

Kinds of Video Editing

The simplest way to produce a video program is to plan each shot in sequence and tape them in order. This process is called **in-camera editing**, but it is not always the best method to use. As your productions become more complex, in-camera editing will no longer be possible. The quality of your final program may depend on how well it is edited.

During **linear editing**, segments from two or more tapes are combined into a finished program. The original (first generation) tape must be fast-forwarded or rewound until the right segment is found. Next, the segment is copied onto another tape (second generation). Each time the material is copied, some of its quality is lost. By the time a third generation tape is made, its quality is often unacceptable.

Today's digital video camcorders store information in digital form. Editing is done by plugging the digital camcorder directly into a computer. Fig. 16-8. "Clips" (segments) are digitized and stored on the computer's hard drive. A simple point and click method is then used to arrange the clips in the proper sequence.

Fig. 16-8. The camcorder on the table can be plugged directly into the computer, and the tape is edited on-screen. Tape a 30-second commercial you see on TV and analyze it. How many "cuts" (jumps) from scene to scene does it contain?

Because the clips are digitized, special effects are easy to add as well. Since you can use the computer to jump to whatever part of the video you want to without waiting for the tape to forward or rewind, digital video editing is called **non-linear editing**.

Editing video is like editing a written story. In writing, it is often necessary to move sentences or paragraphs. Using word-processing software makes the process of editing your writing easier. Using non-linear video editing software does the same thing for video, making desktop video production possible.

Video Titles and Text

The beginning title or the credits at the end of a program can add a professional look to your production. The text you see on TV is created by a special computer program called a **character generator**. Technicians type in the required words before air time. The text can appear unmoving on the screen, it can crawl sideways, or it can scroll up or down.

Putting It All Together

The actual production of a TV program is an exciting thing to watch. In addition to the talent getting their lines correct, all of the equipment must work just right. Live broadcasts are especially interesting because the show must go on no matter what. For example, take a close look at the tiny lapel microphones that newscasters wear clipped to their clothing. Notice that they often wear two mics together. This is a safety measure in case one mic stops working while they are on the air. This way they can keep going and no one even knows there has been a problem.

SECTION 4
TechCHECK

1. How do first generation and second generation copies of a video differ?
2. Why is in-camera editing not always workable?
3. What is the difference between linear and non-linear editing?
4. **Apply Your Knowledge.** Tape an activity and practice in-camera editing as you tape.

ACTION ACTIVITY

Producing a Video

Real World Connection

Putting the whole show together takes lots of time and planning. In this activity, you will actually record the video your group planned in the first activity.

Be sure to fill out your TechNotes and place them in your portfolio.

Design Brief

Produce a five- to ten-minute video on a technology-related topic. Include at least one commercial or public service announcement. Fig A. Make the video look and sound as much like a real television program as possible. Show your completed video to the entire class.

Materials/Equipment

- props, costumes (as needed)
- studio lights (as needed)
- camcorder, audio recorder
- audio and video tapes
- stereo, tape deck, CD player, mixer, microphone (optional)

SAFETY FIRST

- Keep away from hot studio lights.
- Watch where you walk to avoid tripping on cords or video cables.
- Follow the safety rules listed on pages 42-43 and the specific rules provided by your teacher for tools and machines.

Fig. A

(Continued on next page)

ACTION ACTIVITY

Procedure

1. Your teacher will demonstrate the proper use of the video equipment (Fig. B) and the connectors needed (Fig. C).

2. Gather or make any props that you will need. Some things you might include are
   - graphic title for the introduction
   - graphic for audio commercial
   - credits for the actors, actresses, production staff, and so on
   - costumes or backgrounds to make your production look real

3. Rewind the tape you are using and set the counter on the camcorder or VCR to zero. This way you will be able to find your place on the tape.

4. Be careful not to record over another production. Write down the starting and finishing times, or index numbers, so that your production can be found easily to show to the entire class.

5. Make cue cards to help the talent remember their lines.

6. Tape your program. Fig. D. Keep a close eye on the picture and listen to the sound being recorded using headphones.

7. Show your finished program to the entire class.

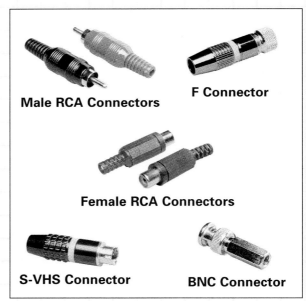

Male RCA Connectors

F Connector

Female RCA Connectors

S-VHS Connector

BNC Connector

Fig. C

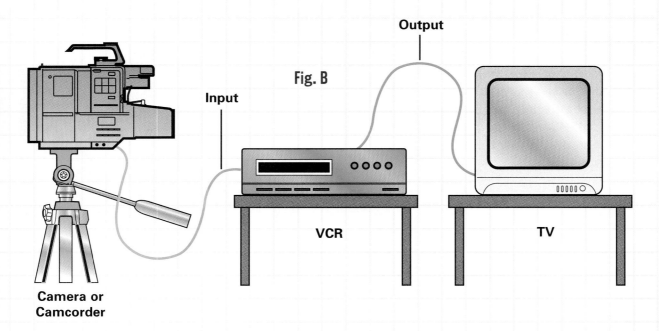

Output

Fig. B

Input

VCR

TV

Camera or Camcorder

Evaluation

1. Why do newscasters often wear two mics at the same time?
2. Why must everything work correctly the first time in live programs?
3. Why is it important to rewind tapes and zero the counter?
4. **Going Beyond.** Make a news video about other activities in the technology room.
5. **Going Beyond.** Ask your teacher if you could show your video to other classes. Find out if you could set up a studio in your technology room. See if you could produce a school news program that the entire school could watch every day.
6. **Going Beyond.** Check with your teacher about going on a field trip to a TV production studio in your area. Write a report or prepare a slide show that compares that studio with the studio in your technology classroom.

Fig. D

REVIEW &

CHAPTER SUMMARY

SECTION 1

- In 1901, Guglielmo Marconi became the first person to transmit and receive a wireless message across the Atlantic.

- Radio stations have an assigned frequency measured in hertz.

- Philo Farnsworth and Vladimir Zworykin invented technologies that led to the TV we know today.

SECTION 2

- The work that goes into planning a TV or radio program is called pre-production and includes script writing, storyboards, and rehearsal.

- Careful script writing can help avoid production problems and keep the viewers' or listeners' attention.

SECTION 3

- The radio stations we listen to are either AM or FM.

- Some of the parts of a radio program might include station identification, introduction, music, weather, sports, and news.

SECTION 4

- The quality of a television program depends on how well it is edited.

- The text we see on TV is created by a special computer program called a character generator.

REVIEW QUESTIONS

1. Who was the first person to transmit and receive a wireless message?

2. Why is it important to rehearse a script?

3. What is the simplest way to edit a video program?

4. What happens to the quality of a videotape when it is copied?

5. What is a character generator?

CRITICAL THINKING

1. What did you think was the hardest part of writing a script? Explain.

2. Do you think that you look and sound the same on television as you do in real life? Explain.

3. Many professional actors and actresses get very nervous just before performing. Why do you think some people get "stage fright"?

4. Research how long it takes to produce one television show. Call your local station to find a list of the jobs related to television production.

ACTIVITIES

CROSS-CURRICULAR EXTENSIONS

1. **COMMUNICATION** Produce a promotional video about your school that could be shown to new students. Ask the foreign language teachers to help make the video in other languages. Team up with a language arts and drama class to write and produce your production.

2. **SCIENCE** Research high-definition television. How is it different from the TV you usually watch? What do you think television will be like 50 years from now?

3. **MATHEMATICS** Create a display of the electromagnetic spectrum. Include the wave lengths of the different radiation waves in exponential powers of 10.

EXPLORING CAREERS

Cable, fiber optics, and satellites allow people access to TV broadcasts all over the world, not just in their local areas. TV is also becoming more interactive. With larger international audiences and expanded services, the career opportunities in this field will grow as well.

Digital Film Editor Film editing has been changed by technology. Editors today sit behind a computer holding a mouse in one hand and a script in the other. Editors combine their artistic abilities with technology to produce quality films. They must have patience and diligence to work on a film and see it through from start to finish.

Broadcast Technician Broadcast technicians work behind the scenes. They set up, operate, and maintain the electrical and electronic equipment used in television and radio broadcasts. They may work in the studio or on location to ensure smooth broadcasts. A background in computers and electronics and the ability to work well with others are necessary.

ACTIVITY

As part of a team, develop the format for a new radio or television program for an audience you think is currently not being reached. Share your program ideas with the class.

CHAPTER **17**

Traveling in Space

THINGS TO EXPLORE

- Identify major milestones in space history.
- Explain how the development of the Space Shuttle changed the space program.

Techno Terms

NASA
Skylab
Sputnik

D o you ever look up at the sky at night and watch the stars and planets? Fig. 17-1. Do you wonder what it's really like out there? For a long time, space travel was just a dream. Now it's the real thing. We've already explored much of the Earth, so it seems natural to go beyond Earth. For this reason, space is sometimes called the "final frontier." Fig. 17-2 (pages 384-385). The challenges are great. How do we send scientific instruments, machines, and people into space and bring them back safely?

Pioneers of Rocketry

Stories of rocket-like devices come from as far back as 400 B.C. The Chinese in the thirteenth century used a form of gunpowder to launch a simple form of a solid-propellant rocket they called "fire-arrows."

◀ **OPPOSITE** In this NASA photo, the Space Shuttle has docked at the *Hubble Space Telescope* to make repairs.

Fig. 17-1. The Lagoon Nebula is a bright, misty region found in the southern constellation of Sagittarius. This photo was taken by the *Hubble Space Telescope*. Find a photo of this nebula taken by an earthbound telescope and compare the two. Was the *Hubble* worth the investment?

Adventures in Space

Yuri Gagarin, a Russian cosmonaut, became the first human to attempt space flight. His ship, the *Vostok 1*, entered orbit on April 12, 1961.

Alan Shepard, Jr. became the first American in space on May 5, 1961. His flight in the *Freedom 7* lasted 15 minutes, 28 seconds, and he traveled at a speed of 5,134 mph.

Virgil "Gus" Grissom became the second American to escape the Earth on July 21, 1961. Although the flight was successful, the capsule of the *Liberty Bell 7* sank in 15,000 feet of water right after splashdown. But Grissom escaped safely, and the capsule itself was recovered from the ocean in 1999.

Buzz Aldrin, Neil Armstrong, and Michael Collins

began a mission on July 16, 1969 which would put people on the moon. "One small step for man...one giant leap for mankind," were Neil Armstrong's words as he stepped from the *Apollo 11* module onto the lunar surface. The ship returned to Earth on July 24. Without an atmosphere or weather to disturb them, the footprints left on the moon by Buzz Aldrin and Neil Armstrong will stay there for thousands of years.

Neil Armstrong   Michael Collins   Buzz Aldrin

Fig. 17-3. Tourists in Russia view a Sputnik satellite in a museum. Sputnik 1 weighed 184 pounds and was only 23 inches in diameter. Find out the size of modern communication satellites. How do they compare to Sputnik?

Modern rocketry really began with the following three people. They are called the first true pioneers of space travel. They took the ideas and dreams of people like Jules Verne, the science fiction writer, and tried to make them work.

- In 1898, Konstantin Tsiolkovsky, a Russian schoolmaster, suggested using liquid fuels for rockets. Even though he never launched a rocket, he explained the principles by which rockets could fly in space.
- An American, Robert H. Goddard, launched the first liquid-propellant rocket in 1926. The flight, rising only 41 feet, lasted 2.5 seconds, but it was a start.
- Hermann J. Oberth, a German, also worked with long-range rockets between 1917 and 1955. His writings got scientists and engineers excited about building rockets.

Early Space Travel

New developments in technology helped people move from dreaming about space travel to making it happen. The launch of *Sputnik* by the Soviet Union in 1957 was an exciting moment for the entire world. Fig. 17-3. With *Sputnik*, we had entered the Space Age.

Fig. 17-4. The rocket launcher used for this Apollo mission could be used only once. This made missions more costly. Watch a video of the film Apollo 13. Report what you learned about space missions to the class.

Skylab and the Space Shuttles

Outer space has become an important area for research and development for the United States. In 1973, **Skylab**, the United States' first space station, was launched. Three different crews worked and lived in *Skylab* as it orbited 270 miles above Earth.

Until 1981, **NASA** (National Aeronautics and Space Administration) used rocket launchers, such as those for *Mercury*, *Gemini*, *Apollo*, and *Skylab*, only once. Fig. 17-4. Then space shuttle technology was developed. A shuttle takes off like a rocket but lands like a glider on a runway. The *Columbia* became the first reusable space vehicle. Even the solid rocket boosters were rebuilt after every mission. Russia's *Buran* worked the same way.

Then, in 1986, the shuttle *Challenger*, carrying a crew of seven, exploded shortly after it was launched. Fig. 17-5. That disaster reminded us that there are still many problems to solve in making space travel safe.

There are many *milestones* (important events) in space history. Robot spacecraft have traveled to the planets. Satellites launched by rockets and the shuttles have enabled us to investigate our planet, forecast weather, and communicate instantly with other people around the world. Which events that have taken place during your lifetime have affected space travel most?

Fig. 17-5. This photo shows *Challenger* before the accident. A teacher was on board and died along with the astronauts. Do you think civilians should continue to go on shuttle missions?

SECTION 1
TechCHECK

1. What is a milestone?
2. Name the three pioneers of modern rocketry.
3. What makes the Space Shuttle different from other spacecraft?
4. **Apply Your Knowledge.** Do you think life exists beyond the Earth? Write a paragraph explaining your answer.

Putting Things into Space

THINGS TO EXPLORE

- Using the correct terms, explain how rockets work.
- Apply Newton's Laws of Motion to model rocketry.
- Safely use pneumatic force to launch model rockets.

TechnoTerms
escape velocity
gravity
Newton's Laws of
Motion
thrust

What does it take to get objects into space? Putting objects like satellites and space shuttles into orbit takes lots of energy to overcome the Earth's gravity. Fig. 17-6.

Escaping Earth's Gravity

In general, **gravity** is the force of attraction between objects. Space transportation systems are used to provide power and speed to overcome Earth's gravity. In order to get into Earth orbit, you must travel more than 17,500 miles per hour.

Some people think the Space Shuttle can go to the moon. That's not true. The moon is about 240,000 miles away. But the shuttle's powerful engines can put it only in low Earth orbit, or about 300 miles out. The farther out an object is going, the more energy is needed to get it there. If a space vehicle is to move farther away from the Earth, it needs an extra push, or **thrust**, from its engines. It must reach **escape velocity** (25,000 miles per hour).

Fig. 17-6. Gravity pulls the shuttle downward as a 50,000 pound weight. The fuel required for a shuttle launch is burned by the two booster rockets on either side. Would there be any advantage to many small launches versus one big launch in delivering payloads to space? Explain.

How Rockets Work

Rockets work differently than jet planes. Rocket engines are designed to work outside Earth's atmosphere. They actually work better in space, where there's a near vacuum, than they do in air.

A rocket engine must carry all of its own fuel and oxygen. The fuel may be either a solid or liquids carried as two separate chemicals. When the chemicals mix, they burn. The hot gases produced then push out of the rocket to produce thrust. Fig. 17-7. **Newton's Laws of Motion** explain how things move on Earth. They also apply when we're trying to get off the Earth and into space. (See pages 398-399.)

Designing, building, and testing model rockets will give you a chance to explore how Newton's laws work. Some model rockets have solid-fuel engines. In the first activity for this section, you will use pneumatic force (compressed air) to launch your rocket. You will be able to find how high your rocket goes by using an altitude gauge and mathematics.

Fig. 17-7. This photo shows a close-up view of the booster rockets used to launch *Endeavor*. Look up the meaning of the word *endeavor*. Why do you think it was chosen as the name for a shuttle?

TechnoFact

NO "OFF" SWITCH
Did you know that once a solid-fuel rocket is started, it can't be turned off? A liquid-fuel rocket works differently. It can be turned on and off.

SECTION 2
TechCHECK

1. What is thrust?
2. What is gravity?
3. What causes the thrust that pushes a rocket away from Earth?
4. Apply Your Knowledge. Using the **NASA** Internet site, research the rockets used to put satellites into space. Print pictures of various launch vehicles and make a poster describing them.

ACTION ACTIVITY

Building an Air-Powered Rocket

Real World Connection

You probably have seen old movies of early attempts to fly. People learned about rocket and airplane flight mostly by trial and error. They tried their idea and if it didn't work, they learned a valuable lesson. They knew what not to try again.

Many design concepts apply to small rockets as well as large ones. The design concepts that are most important for a rocket to fly correctly are center of gravity and center of pressure. In this activity, you will work as an aerospace engineer to design a launch system and a rocket. Your rocket will be made of paper and use compressed air to fly.

Design Brief

Design, build, and test a model rocket and launch system that can safely launch a paper model rocket. You will use plumbing fittings, valves, and copper tubing to make a safe launch pad. Compressed air will provide the thrust. To determine the height your rocket reaches, you will use an altitude gauge. (See Activity 2 on page 394.)

Materials/Equipment

- 3/4" copper tubing (1 piece 3' long; 1 piece 2' long)
- 3/4" copper tubing-to-pipe adapter
- 1/4" acrylic sheets (4 pieces, 5 3/4" x 14")
- portable drill; 3/16", 5/16", and 7/8" drill bits
- 3/4" ball valve
- 1/4" air hose
- 1" U-bolts (2)
- 2" x 6" pine, 5 1/2" long
- abrasive paper, 180 grit

- Teflon pipe-sealing tape
- 3/4" to 1/4" pipe-reducing bushing
- 3/4" copper elbow
- 3/4" plywood (36" x 6")
- 1/2" dowel rod (36" long)
- wood screws, string
- solder flux, solder, propane torch
- paper, white glue, brush, file folders
- tubing cutter

SAFETY FIRST

- Follow the safety rules listed on pages 42-43 and the specific rules provided by your teacher for tools and machines.
- Use only paper to make nose cones.
- Follow your teacher's instructions while using a propane torch for soldering. Wear safety glasses. Keep flammable materials away from the torch and open flame. Follow the directions for using solder flux.
- Wear a clear face shield and keep at least ten feet away from the pad during launch. Warn those watching not to try to catch the rocket as it comes down.

Fig. A

Procedure

Part 1 · Building the Launch Pad

Work in small groups of 2 to 4 students. One group can make the copper tubing assembly (Fig. A), another the launch pad, and so on.

1. *Copper tubing assembly.* With your teacher's help, cut the copper tubing to the lengths listed on page 390.
2. Clean the ends of the tubing with the abrasive paper and apply solder flux.
3. Solder the two tubes to the elbow. Solder the tube-to-pipe adapter to the end of the shorter tube. Ask your teacher for help.
4. Smooth any rough edges on the long end of the tube with abrasive paper.
5. *Launch pad base.* Cut the plywood base to the size listed. Sand the edges.
6. Measure and mark the center of the 36" dowel rod.

7. Drill a 3/16" screw hole in the center of the dowel. Use a wood screw to attach the dowel to the center of the base as shown in Fig. B. The dowel rod will steady your launch pad.
8. Drill 4 5/16" holes for the U-bolts to hold the short tube as shown.
9. *Safety shield.* Cut 4 pieces of acrylic plastic for the safety shield.
10. Cut the 2" x 6" pine to length and drill a 7/8" hole in its center to fit over the long copper tube.
11. Drill 3/16" screw holes in the plastic and assemble with wood screws as shown in Fig. B.
12. *Air hose and valve.* Assemble the bushing, air hose, and valve as shown in Fig. B.
13. Use Teflon pipe-sealing tape on all threaded connections to prevent leaks.
14. Connect the valve to a long air hose so that the launch can take place outside.

(Continued on next page)

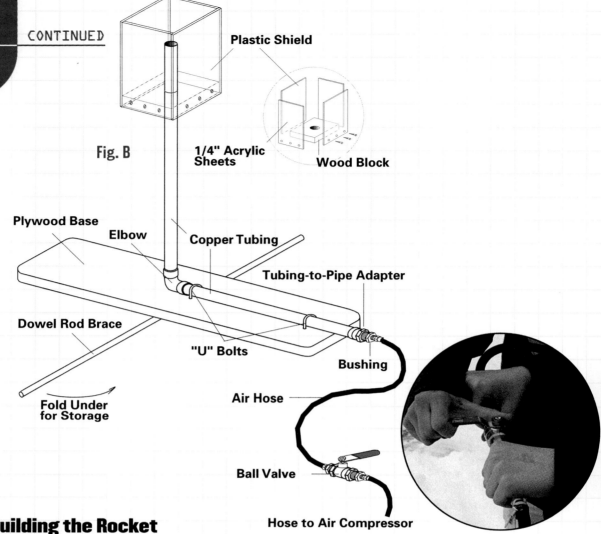

Fig. B

Plastic Shield

1/4" Acrylic Sheets

Wood Block

Plywood Base

Elbow

Copper Tubing

Tubing-to-Pipe Adapter

Dowel Rod Brace

"U" Bolts

Bushing

Fold Under for Storage

Air Hose

Ball Valve

Hose to Air Compressor

Part 2 · Building the Rocket

1. Roll a piece of paper around a piece of copper tubing to form the rocket body. Tape the paper and remove the tube. Cover with a coat of glue. See Fig. C.

2. Cut another piece of paper as shown to form the nose cone. Cover the paper with white glue and roll it into a cone shape that will fit your rocket body.

3. After the glue dries, attach the nose cone to the body and coat the rocket with white glue. Let the glue dry.

4. Cut 3 or 4 fins out of file folders as shown. Glue the fins to the rocket body. Let the glue dry and brush another coat over the entire rocket.

5. Test your rocket for stability. Fig. D.

SAFETY FIRST
When you launch the rocket, stay at least 10 feet away from the launch area.

Part 3 · Launching Your Rocket

1. Wear a clear face shield and warn others to stay clear of the launch area.

2. Launch the rocket.

3. Have a student use the altitude gauge to determine the maximum height, or *apogee*, of the rocket's flight. Follow the instructions for calculating the altitude. Show your work on your TechNotes.

Evaluation

1. What is the apogee of a rocket?
2. What do real rockets have in common with model rockets?
3. Use a video camera to tape a launch. Play the tape for the class.
4. **Going Beyond.** Design, build, and launch a rocket that has a parachute recovery system to slow its descent.
5. **Going Beyond.** Try to make longer rockets using two or three sheets of paper instead of one.

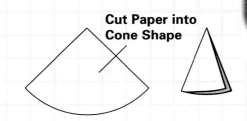

Cut Paper into Cone Shape

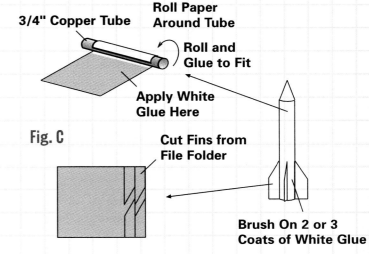

3/4" Copper Tube

Roll Paper Around Tube

Roll and Glue to Fit

Apply White Glue Here

Fig. C

Cut Fins from File Folder

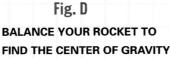

Brush On 2 or 3 Coats of White Glue

Fig. D

BALANCE YOUR ROCKET TO FIND THE CENTER OF GRAVITY

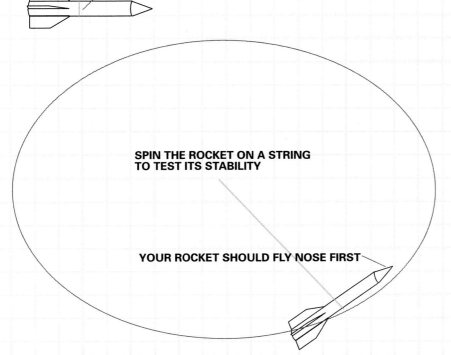

SPIN THE ROCKET ON A STRING TO TEST ITS STABILITY

YOUR ROCKET SHOULD FLY NOSE FIRST

ACTION ACTIVITY

Making an Altitude Gauge

Be sure to fill out your TechNotes and place them in your portfolio.

Real World Connection

Mathematics is used to solve many engineering problems. Trigonometry is used to solve many of those dealing with measurement.

Design Brief

Design, build, and use an altitude gauge to determine the maximum height your rocket reaches. The gauge will be used to measure an angle so you can use trigonometry to calculate the altitude.

Materials/Equipment

- calculator
- trigonometry table (see Appendix)
- protractor
- string
- cardboard tube
- weight
- thread

SAFETY FIRST

Follow the safety rules listed on pages 42-43 and the specific rules provided by your teacher for tools and machines.

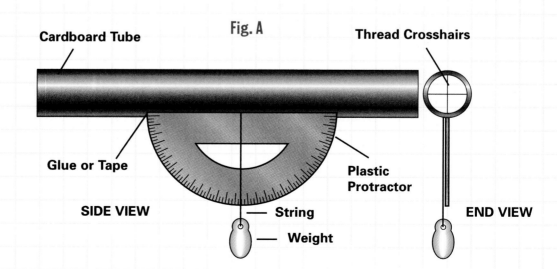

Fig. A

Cardboard Tube

Thread Crosshairs

Glue or Tape

SIDE VIEW

Plastic Protractor

String

Weight

END VIEW

Procedure

1. Roll a piece of paper into a tube, or use a cardboard tube from foil or plastic wrap. Tape two pieces of thread to one end to make cross hairs for accurate sighting. Use tape or glue to attach a protractor to the side of the tube. Attach a string and weight as shown in Fig. A.

2. Measure a distance of 100 feet from the launch pad. Someone in your group should stand with the altitude gauge at this location and be ready to sight the rocket's highest point.

3. As the rocket is launched, hold the altitude gauge to sight the maximum altitude. Measure the angle created using the protractor on the altitude gauge.

4. The height of your rocket's flight will be measured using trigonometry ratios for right triangles. This sounds hard, but most of the hard work has actually been done for you. First, subtract the angle you measured on your altitude gauge from 90°. This will give you the angle from the ground to the highest point your rocket reached.

   angle from ground = 90° − angle on altitude gauge

5. To calculate the height, you will use a trigonometry function called *tangent* (TAN). The formula is easy:

   TAN angle __° = opposite side (altitude) ÷ adjacent side (100 feet)

6. The tangent of the angle you measured can be found in the trigonometry table in the Appendix. To make it even easier, this formula can be changed to look like this:

   TAN angle __ x 100 feet = __ altitude (feet)

7. Record your angle and the maximum height of your rocket's flight.

Evaluation

1. Why do you need to subtract the angle you measured on the altitude gauge from 90°?

2. How could the design of your rocket have been changed to make it go higher?

3. **Going Beyond.** Use your altitude gauge and the trigonometric formula to measure the height of other objects, such as flagpoles, trees, or your school. Fig. B.

Fig. B

Unstaffed Space Flights

THINGS TO EXPLORE

- Explain why unstaffed space flights are important.
- Describe sensor probes used to find information in space.
- Design, build, and test a Mars rover vehicle.

TechnoTerms

Hubble Space Telescope
Mars Pathfinder
sensor probe
Voyager

*U*nstaffed space missions are those having no people on board. Thanks to new technology, exploration doesn't always require people to travel in space. Fig. 17-8. This means the missions can be smaller, faster, and cheaper—NASA's way of exploring space on a budget.

There are many spacecraft out there sending us important information about space. They can even take close-up pictures of your town and school from thousands of miles above the Earth!

Sensor Probes

Information about the moon, planets, and even comets can be gathered by sensor probes. A **sensor probe** is a device that can measure such things as temperature and radiation. A probe also can analyze the chemicals or elements, such as hydrogen and helium, found in space. The data are sent back to Earth where scientists can evaluate what they mean.

Fig. 17-8. This composite illustration shows a remote-sensing satellite put into orbit by the European Space Agency. Its purpose is to study climate change on Earth. Find out which countries contribute to the European Space Agency.

Satellites and Other Spacecraft

Satellites are objects that orbit a body in space. For example, the moon is a satellite of Earth. Artificial satellites and other spacecraft can be sent to explore and map the surface of planets. They are equipped with sensor probes and other instruments, such as spectrographs and cameras, that can detect light and other forms of radiation. Fig. 17-9. These spacecraft have sent back fantastic images of moons, atmospheres, and planet surfaces.

The two unstaffed *Voyager* spacecraft, launched in 1977, visited Jupiter, Saturn, Uranus, and Neptune. Fig. 17-10. Now nearing the edge of our solar system, they will continue to operate until about 2020. *Voyager 1* is the most distant human-made object in space. Science information from *Voyager* is returned to Earth through the Deep Space Network (DSN) antennas located in California, Australia, and Spain.

Fig. 17-9. When white light passes through a prism, it is separated into different colors and wavelengths. The occurrence of different wavelengths gives scientists information about the source of the light. For example, sensor readings tell scientists that there is water ice on the north pole of Mars. Look at light that has passed through a prism. What colors do you see?

Fig. 17-10. Spacecraft carry many sensors. What do you think is the purpose of the dish on the *Voyager* spacecraft?

Fig. 17-11. This photo of the planet Pluto and its moon Charon was taken by the *Hubble* in 1994 when those worlds were 2.6 billion miles away from Earth—equivalent to seeing a baseball at a distance of 40 miles! Find out who Pluto and Charon were in Greek mythology.

TechnoFact

SNAPSHOTS OF THE UNIVERSE While the *Hubble* is sending us pictures from deep space, we can continue to look at the sky through large telescopes on Earth's surface. The largest is the Keck telescope, located on top of Hawaii's Mauna Kea volcano. Through it you can see two-thirds of the way to what may be the most distant objects in the universe.

The *Hubble Space Telescope* and the Chandra X-Ray Observatory are satellites in orbit around Earth. They allow us to see objects in deep space much better than we can from Earth, where the atmosphere distorts the view. The satellites are able to send back detailed pictures. Fig. 17-11.

Robots

Sometimes, robots can be used for exploration. The **Mars Pathfinder** mission used the *Sojourner* rover, a remote-controlled robot vehicle developed by NASA, to explore Mars' surface. Fig. 17-12. *Pathfinder* landed with the help of parachutes and inflated cushions. The *Sojourner* vehicle was only about the size of a microwave oven.

SCIENCE CONNECTION

Newton's Laws of Motion

Sir Isaac Newton (1642-1727) described physical motion with three scientific laws. The laws describe the motion of all objects, whether on Earth or in space. These are his laws stated in their simplest form:

• First Law of Motion: *Objects at rest will stay at rest, and objects in motion will keep moving in a straight line unless acted on by an unbalanced force.* A rocket on the launch pad is balanced and at rest. The surface of the pad pushes the rocket up, while gravity tries to pull it down. As the engine starts, the rocket thrust causes things to become unbalanced. The rocket travels upward. In space, a spacecraft will travel in a straight line if all the forces stay balanced. If it gets close to a larger body such as a planet, the planet's gravity will pull on the spacecraft and cause its path to curve.

Fig. 17-12. *Sojourner* roamed over the surface of Mars and sent data back to scientists on Earth. This photo was taken by a camera mounted in the *Pathfinder* module. Look up the definition of sojourner. Was it an appropriate name for the vehicle?

SECTION 3

TechCHECK

1. Why do we have unstaffed space missions?
2. What is a sensor probe used for?
3. Where are the *Voyager* spacecraft?
4. **Apply Your Knowledge.** Research photos of the Martian surface provided by the *Pathfinder* mission.

- **Second Law of Motion:** *Force is equal to mass times acceleration (F = ma).* To make something as large as a rocket move quickly, you need a lot of force. The amount of force or thrust produced by a rocket engine is determined by how much (mass) rocket fuel is burned and how quickly the gas escapes from the rocket.

- **Third Law of Motion:** *For every action, there is an equal and opposite reaction.* A rocket can lift off the launch pad only when gas is pushed out of its engine. The rocket pushes on the gas, and the gas pushes on the rocket. The action is the gas coming out of the engine. The reaction is the movement of the rocket in the opposite direction.

In space, even tiny thrusts will cause the rocket to change direction because there is so little gravity or other forces.

ACTIVITY

What happens when two football or hockey players collide? Explain which of Newton's Laws are working.

ACTION ACTIVITY

Building a Simulated Mars Rover Vehicle

Be sure to fill out your **TechNotes** and place them in your portfolio.

Real World Connection

NASA's *Pathfinder* mission used a robot rover that moved around on the surface of Mars. Refer to Fig. 17-12. Robotic exploration of space can add to our knowledge without putting people in danger. In this activity, you will build and control a simulated Mars rover that will send a video signal for you to analyze.

Design Brief

Design, build, and use a simulated Mars rover to explore a model of a Martian landscape. The rover will consist of a remote-controlled model truck with a wireless video camera transmitter and receiver. Fig. A. The rover will send a video signal to your tech classroom. You will analyze the video to determine the surface characteristics of your model planet.

Materials/Equipment

- remote-controlled (R/C) 4-wheel-drive model truck with radio controller
- 12-volt DC video camera
- 12-volt DC video transmitter
- 12-volt DC rechargeable power-tool battery and plastic cap
- video receiver and TV monitor
- rocks, sand, plaster, foam model-making materials
- velcro fastener tape
- plastic cable ties
- variety of screws and nuts

Fig. A

SAFETY FIRST

Follow the general safety rules on pages 42-43 and the specific rules provided by your teacher for tools and machines. Double-check all electrical connections to prevent short circuits when a wire goes directly from negative to positive. Use caution while soldering electrical connections.

Procedure

Divide into groups for the following tasks.

Part 1 · Building the Mars Rover

1. Remove the plastic body from the R/C model truck.
2. Design a system to hold the video camera securely to the front of the vehicle. Use velcro fastener tape or cable ties.
3. Connect the video transmitter to the vehicle with Velcro fastener tape.
4. Attach the power-tool battery using plastic cable ties.
5. Fasten the wires and connectors as shown or follow the wiring diagram that came with your video camera. Be careful not to get positive and negative wires mixed up.
6. Punch a hole in both sides of the plastic battery cap. Tighten two screws to be used as electrical contacts as shown in Fig. B.
7. Assemble the wires and test the system using a receiver to show the video signal on a TV monitor.

Part 2 · Creating the Mars Landscape

1. Ask your teacher where to build your Mars landscape. If weather permits, work outside.
2. Try to make your landscape look like Mars. Study Fig. A or research photographs in your school's library or on the Internet. Use rocks, sand, plaster-covered foam, or other model-making materials to make it look real.
3. Design backdrops to look like the Martian horizon.

Part 3 · Exploring Mars

1. Take turns controlling your Mars rover with a radio transmitter from a different room or a different area of the tech lab.
2. Set up your mission control area to view the wireless video signal coming from your rover.
3. Study and analyze the images received.

Evaluation

1. What was the name given to NASA's Mars rover?
2. Why is it safer to send robots to explore planets?
3. **Going Beyond.** Connect the video signal from your Mars rover to a computer. Send the images from your rover to another school using the Internet.

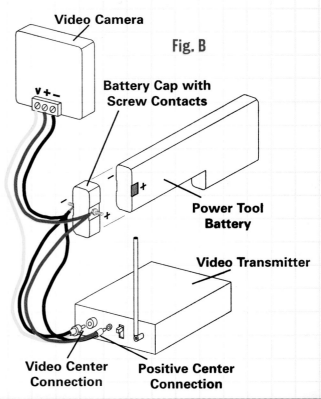

Video Camera

Fig. B

Battery Cap with Screw Contacts

Power Tool Battery

Video Transmitter

Video Center Connection

Positive Center Connection

17 REVIEW &

CHAPTER SUMMARY

SECTION 1

- Milestones in space history include the launch of *Sputnik*, humans walking on the moon for the first time, and the Space Shuttle missions.
- Early space pioneers such as Konstantin Tsiolkovsky, Robert Goddard, and Hermann Oberth developed modern rocketry.
- The development of the Space Shuttle made reusable launchers possible.

SECTION 2

- Escape velocity is the speed needed to get spacecraft away from the Earth.
- Newton's Laws of Motion explain how rockets work on Earth and why they also work in space.
- Launchers give the spacecraft extra power, or thrust.

SECTION 3

- Unstaffed space missions are used to find information about planets and other space objects.
- Satellites and other spacecraft use sensor probes to test for such things as temperature and radiation.
- Robots, such as the Mars *Sojourner* vehicle, can be remotely controlled from Earth.

REVIEW QUESTIONS

1. What does escape velocity have to do with gravity?
2. Who was the first person in space?
3. What must rocket engines carry with them in addition to fuel?
4. What kinds of instruments do satellites carry?
5. State Newton's Laws of Motion.

CRITICAL THINKING

1. Design and conduct three experiments that demonstrate Newton's Laws of Motion.
2. Explain the main difference between rockets and jet engines.
3. Discuss the advantages and disadvantages of unstaffed space flight.
4. Make a sketch of a Mars rover that could scoop a sample of Mars soil.

ACTIVITIES 17

CROSS-CURRICULAR EXTENSIONS

1. SCIENCE Design a method to safely measure the thrust of a model rocket.

2. SOCIAL STUDIES Research space achievements. Make a chart of what you think are major advancements in space technology.

3. MATHEMATICS Use the height or altitude gauge to measure how high the tallest part of your school is.

4. COMMUNICATION Write a science fiction short story describing a trip to Mars.

EXPLORING CAREERS

Travel between Earth and the moon may become routine in your lifetime. Many believe that vacations on the moon and trips to resorts orbiting Earth are also real possibilities. Work opportunities may exist in the future that we haven't even imagined today.

Aerospace Engineer Aerospace engineers design, develop, and test spacecraft and aircraft. They also develop new products and technologies for use in space exploration, and they may supervise their manufacture. Aerospace engineers have strong interests in mathematics and science and must be willing to continue their education throughout their careers.

Astronomer Astronomers research the basic nature of the universe and apply that knowledge to problems in navigation and space flight. They may be involved in developing and testing instruments used in flight. Astronomers are imaginative, have curious minds, and have a strong interest in physics.

ACTIVITY

Make a list of three things that you would take with you on a trip into outer space. Explain why you think you will need the items.

CHAPTER 18
Living and Working in Space

The Benefits of Space Travel

THINGS TO EXPLORE

- Explain what a space spinoff is and give examples.
- Evaluate the benefits of space travel.

TechnoTerms
alloy
space spinoff

In the last chapter, you learned about space travel. What are the benefits of space travel for you right now? They probably include new products and processes, as well as increased knowledge.

New Products and Processes

Space spinoffs are new technologies and products developed during a space project that can be used for additional purposes. Fig. 18-1. Battery-powered tools such as drills and screwdrivers came from the Apollo moon program. A new fiber called PBI, used to make spacesuits safer from fire, is now used in airline seat cushions. Even a special mapping system using computers, a satellite, and scanners to detect forest fires is a spinoff from NASA and the Jet Propulsion Laboratory.

Fig. 18-1. The technology for this digital thermometer was developed for the space program. The temperature sensor contains a thin coating of gold, which provides rapid, accurate results.

◄ OPPOSITE Astronauts float as they work inside a space capsule.

Space is also an ideal place for making certain products that are difficult or impossible to manufacture on Earth. Certain drugs for diabetes and blood diseases are easier to make in space. On Earth, gravity gets in the way but not in space.

New alloys are easier to make in space, too. An **alloy** is a mixture of two or more metals. On Earth, a lighter metal floats on top of a denser one. But the process of making an alloy works best in space because the metals flow together. Once the new alloys are made, they can be used anywhere—even back on Earth. Fig. 18-2. Once the problems of living in space have been solved, people can use the special space environment to make more things we can use in space and on Earth.

Increased Knowledge

Exploring and living in space add to our knowledge of Earth, our own solar system, and the larger universe. How would you like to take a vacation on the moon or Mars? Are you curious to see how things work in space and how people live? What kinds of sports do you think you could play on the moon? The possibilities for the future are very exciting. Fig. 18-3.

Fig. 18-2. This shows a computer chip (magnified) without its protective covering. Semiconductor materials used to create chips may be improved through techniques developed in space. What other materials might be improved in this way?

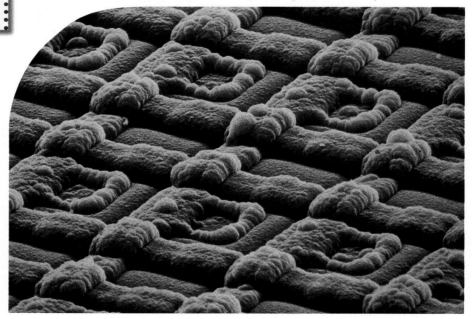

Fig. 18-3. Scientists hope to put astronauts on Mars in the near future. Do some research to learn when they hope to complete the mission.

Is the Space Program Worth It?

Planning for and building things in space is expensive and takes time. Many people feel the space program is worth the price. They can see the benefits to medicine and other industries here on Earth. They see space programs as being valuable in helping us to learn more about Earth from a different viewpoint. Other people do not want their tax dollars spent on space travel when there are many problems on Earth to solve. What do you think the future of space travel should be?

SECTION 1
TechCHECK

1. What is a space spinoff? Give three examples.
2. What is an alloy?
3. What are the advantages and disadvantages of the space program?
4. **Apply Your Knowledge.** Research NASA's website to find the current NASA projects. Choose a project and do one of the following:
   - Write a paragraph describing it.
   - Write a script about it for a school radio or TV newscast.
   - Write a story about it for a school newspaper.

Space Physics

THINGS TO EXPLORE

- Explain how the effects of gravity change when you move away from the Earth.
- Experiment with different materials in a vacuum.
- Create a simulated spacesuit glove.

TechnoTerms

free-fall
microgravity
vacuum

Physics is the science dealing with physical properties and processes. If things worked the same in space as they do on Earth, space physics would be simple. That's just not the case. To understand space physics, you need to understand the effects of gravity and atmosphere.

Gravity in Space

Earth's gravity pulls everything toward the center of the Earth. Things have weight because of gravity. Just how much you weigh depends on how strongly Earth's gravity is pulling on you.

Microgravity In space, people feel the effects of **microgravity**. Some people call microgravity *zero gravity* or *zero-g*. That does not mean there is no gravity. It just means the effects of gravity are reduced compared to what they are on Earth. Fig. 18-4.

Earth's gravity extends for a very long distance out into space because the Earth is so big. The moon is about 240,000 miles from Earth. What do you think holds the moon in orbit around the Earth? The Space Shuttle orbits only a few hundred miles above the Earth. If Earth's gravity holds the moon in orbit, then it affects the Space Shuttle, too.

Fig. 18-4. Astronauts on the moon had to learn how to walk all over again! The pull of gravity there is only one-sixth the pull of gravity on Earth. So you can have the same mass (amount of matter) but weigh only one-sixth as much.

Weightlessness All Earth-orbiting spacecraft are actually falling slowly toward the Earth. Why don't they fall straight down? The answer is in the spacecraft's speed and the shape of the Earth. The spacecraft is traveling forward, and the Earth curves away from it. Fig. 18-5. The astronauts appear to be weightless because they are in **free-fall**. Fig. 18-6. They still have weight, but they can't rest against the floor of the shuttle because the floor is falling at the same speed as they are.

Newton's Laws of Motion explain how things move on Earth. They also work for the space environment. Sometimes the results are different though! You can do tumbling stunts in space that would be impossible to do on Earth—thanks to microgravity. However, it makes living and working in space much harder.

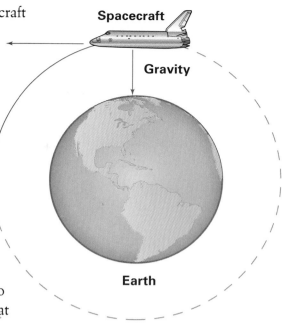

Fig. 18-5. This drawing shows how the speed of the spacecraft, gravity, and the shape of the Earth contribute to weightlessness.

INFOLINK

See Chapter 17 for more information about Newton's Laws of Motion.

Fig. 18-6. Weightless conditions can sometimes cause motion sickness. Have you ever ridden in an elevator that made you feel weightless for an instant? Describe that feeling.

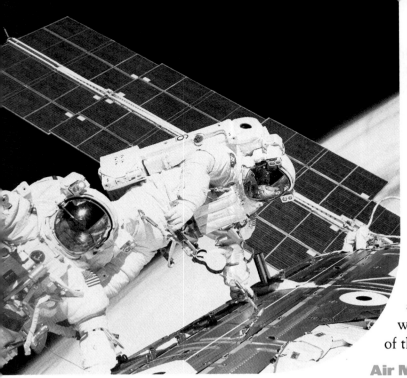

Fig. 18-7. Space suits protect astronauts as they work in space, but the air pressure inside makes it hard to move around. Can you think of design improvements that would make space suits easier to work in? Explain.

Atmosphere in Space

Gravity pulls on the air in our atmosphere and holds it near the surface of the Earth. If you could weigh the amount of air in the atmosphere at sea level, it would weigh 14.7 pounds for every square inch of the Earth's surface.

Air Molecules The higher you go above sea level the less air there is in the atmosphere. Scientists have measured the number of air molecules at different distances from the Earth. At sea level, there are a million, million, million air molecules in one cubic centimeter. Farther away from Earth the number gets smaller and air pressure is less. At 20 miles

MATHEMATICS CONNECTION

Light as a Feather—or a Brick!

If a person weighs 100 pounds on Earth, what would that person weigh on other planets?

You can use the chart shown here to calculate weights for different planets and even the moon. Simply multiply the weight on Earth by the gravity factor given in the chart.

Example: A 100-pound person on Earth would weigh 16.6 pounds on the moon (100 x 0.166 = 16.6).

above the Earth, the air becomes too thin for jet engines. At 50 to 600 miles away from Earth, there are one million molecules per cubic centimeter. When you reach 1,200 miles in altitude, there is only one air molecule per cubic centimeter.

Living in a Vacuum We live our lives in an ocean of air. Have you ever thought about how things would be different in a vacuum? In a **vacuum**, there is no air to breathe and no air pressure. A vacuum is fatal to people. In outer space there is a near-vacuum. Astronauts must wear specially made space suits for protection. Fig. 18-7.

SECTION 2
TechCHECK

1. What are the effects of microgravity?
2. Why do astronauts and objects appear weightless when the Shuttle reaches orbit?
3. What holds the atmosphere around the Earth?
4. **Apply Your Knowledge.** Research how a change in altitude affects the boiling point of water. Make a graph illustrating your findings.

	GRAVITY FACTOR
Mercury	0.284
Venus	0.907
Moon	0.166
Mars	0.38
Jupiter	2.34
Saturn	0.925
Uranus	0.795
Neptune	1.125
Pluto	0.041

ACTIVITY

Figure out what you would weigh on each of the planets and the moon.

ACTION ACTIVITY

What Does the Space Environment Feel Like?

Be sure to fill out your **TechNotes** and place them in your portfolio.

Real World Connection

Scientists, astronauts, physicians, engineers, and technicians must understand the space environment in order to design safe equipment that is also comfortable. For example, space suits must be made strong enough to protect the astronaut from temperature extremes, small meteors, radiation, and the vacuum of space.

In this activity, you will work as a research technician to investigate the effects of reduced air and air pressure. Fig. A.

Design Brief

Demonstrate the effects of a vacuum on materials using a vacuum chamber and a vacuum pump. Simulate the feeling of wearing a space suit glove in the vacuum of space.

Materials/Equipment

- clamp
- hose clamp, 4"
- rubber gasket sheet
- rubber glove
- Styrofoam plastic foam
- vacuum jar
- vacuum pump and tubing
- silicone caulk
- acrylic plastic, 1/8"
- ABS plastic pipe coupling, 4"
- marshmallows
- beaker
- balloon
- tennis ball
- carbonated soda
- battery-operated buzzer

SAFETY FIRST

Follow all of the general safety rules on pages 42-43. Follow your teacher's instructions on the safe use of power tools. Use power tools only with your teacher's permission and supervision.

Fig. A. For this activity, your teacher will demonstrate the use of the vacuum pump and vacuum jar.

Procedure

In the first part of this activity, you will test different objects or materials in a vacuum. In the second part, you will simulate the feeling of wearing a glove while working in space.

Part 1 · Testing Materials in a Vacuum

You will work in groups of three or four. Your teacher will demonstrate the proper use of the vacuum pump and vacuum jar.

1. Gather the materials you would like to test under reduced pressure. Ask your teacher to approve your test materials.
2. Make a chart of the materials you are testing, your predictions, and the effects you observe. Your chart might look like the one shown in Fig. B.
3. Place each object in the bell jar and draw a vacuum. Watch and listen for the effects of reduced pressure.

Part 2 · Wearing a Space Suit Glove

The entire class or a large group (8-12 students) can work on this part of the activity.

1. Measure and mark a piece of 1/4" acrylic plastic as shown in Fig. C (page 414).
2. Cut the shape with the help of your instructor.
3. Measure and mark the appropriate size for the rubber gasket. Use scissors to cut the gasket to size.
4. Design and cut a plastic foam stand for the vacuum jar. Check the hole in the acrylic plastic for the proper height.
5. Use silicone caulk to attach the ABS plastic coupling to the hole in the acrylic sheet. Let the caulk dry overnight.
6. Stretch the glove over the coupling and attach it with a hose clamp.
7. Assemble all the parts and use a vacuum pump to remove the air around the glove. Put your hand in the glove and try to make a fist. Fig. D (page 414).

Evaluation

1. What happened to the balloon when you placed it in a vacuum? Explain.
2. What do you think would happen to your body in space if you didn't have the protection of a space suit? Explain.
3. Describe how it felt to use the glove in a vacuum.
4. Why do you think tennis balls are sometimes packaged in pressurized cans?

Fig. B

Effects of Vacuum		
Materials	**Prediction**	**Actual**
Marshmallow		
Balloon		
Tennis Ball		
Buzzer		
Soda		

(Continued on next page)

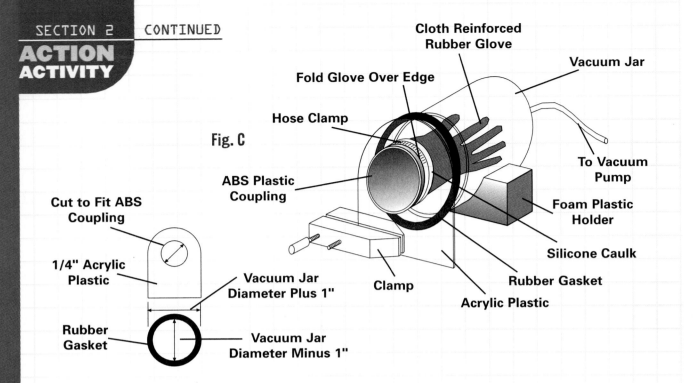

Fig. C

- Cloth Reinforced Rubber Glove
- Vacuum Jar
- Fold Glove Over Edge
- Hose Clamp
- ABS Plastic Coupling
- To Vacuum Pump
- Cut to Fit ABS Coupling
- Foam Plastic Holder
- 1/4" Acrylic Plastic
- Silicone Caulk
- Vacuum Jar Diameter Plus 1"
- Clamp
- Rubber Gasket
- Rubber Gasket
- Vacuum Jar Diameter Minus 1"
- Acrylic Plastic

5. Are sonic booms created in space from rapidly moving space vehicles like the Space Shuttle? Explain.

6. Going Beyond. Make a list of similarities and differences between space suits and deep sea diving suits.

7. Going Beyond. Why do mountain climbers at high altitudes have to boil food longer in order to cook it than they would at sea level? Why does water "boil" at cold temperatures in reduced pressure?

8. Going Beyond. NASA researchers have found that metal parts sometimes weld themselves together when they touch in a vacuum. How could this happen? Describe an experiment that would illustrate "cold" welding.

Fig. D

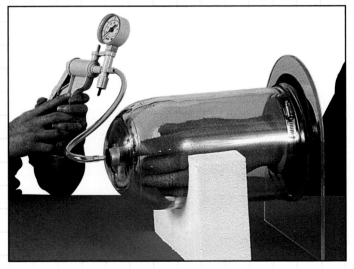

Working in Microgravity

THINGS TO EXPLORE

- Describe the effects of microgravity on the human body.
- Explain how astronauts train for space.
- Create a simulation of neutral buoyancy.

TechnoTerms
grapple point
neutral buoyancy
Remote Manipulator
System (RMS)

Telephone communication and important weather information depend on satellites that work properly. Even space telescopes like the *Hubble* require repair and upgrading. Did you ever think about how equipment might be repaired or replaced in space?

Working in space is a real challenge for astronauts. As you read in Section 2, they must work in a microgravity environment as well as in a near-vacuum. How do you think their bodies are affected?

How Does Space Affect the Body?

Did you know that a microgravity environment will make you taller? You will actually grow an inch or two because gravity isn't compressing (pushing together) the spongy disks between the bones in your backbone. How do you think this affects the design of a spacesuit?

Sometimes your face will get puffy in space because gravity doesn't pull the blood away from your head. Your leg muscles will become smaller because you don't have to work them as much. Doctors have also found you will lose calcium from your bones and they will become weak.

Researchers are working on medicines to help solve some of these problems. But the best thing is exercise— and lots of it! Spacecraft now have equipment such as treadmills and exercise bicycles on board. The astronauts must use the equipment for certain periods daily to keep in shape. Fig. 18-8.

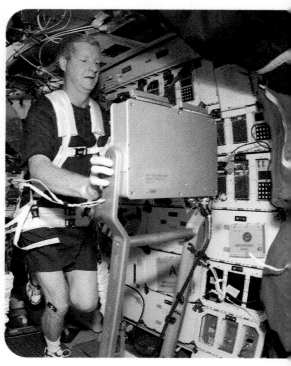

Fig. 18-8. These astronauts are using a treadmill to help counteract the effects of weightlessness. Why do you suppose the astronaut is strapped to the machine?

INFOLINK

See Chapter 17 for more information about Newton's Laws of Motion.

Practicing for Microgravity

Astronauts are expected to perform some specialized jobs while on space walks. Fig. 18-9. They have to think about how microgravity will affect their work. It is sometimes hard to figure out which way is up or down. They and other objects are in free fall and are therefore weightless. They have to be careful how they move. Every time they exert a force their bodies will move in the opposite direction. The wrong move can send them off in a direction in which they might not want to go.

Before they go into orbit, astronauts have to spend many hours practicing how to use special tools and equipment. They train for some special tasks in an underwater facility. Fig. 18-10. Mission specialists also try to simulate the floating effect of neutral buoyancy using *helium* gas. When something floats in air or on water, it is *buoyant*. Positive buoyancy causes something to rise. Negative buoyancy makes it sink. **Neutral buoyancy** means the object doesn't rise *or* sink. It stays where it is.

Fig. 18-9. These astronauts are repairing the *Hubble Space Telescope.* Do some research to learn about repairs made to the *Hubble.*

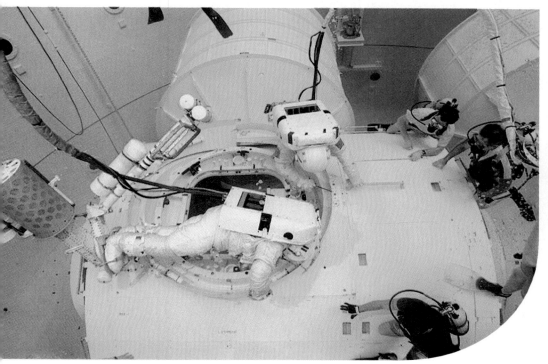

Special Tools for Work in Space

On the real Space Shuttle, astronauts often have to use a mechanical arm called the **Remote Manipulator System (RMS)**. It is used to move payloads (cargoes) in and out of the cargo bay. Astronauts learn how to use the arm in the Manipulator Development Facility (MDF). Fig. 18-11.

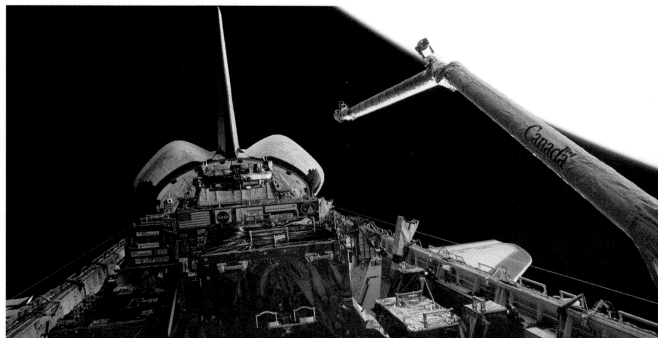

Fig. 18-12. Here the arm is being used for work on the *International Space Station*. Astronaut Tamara E. Jernigan's feet are attached to the *Discovery's* manipulator arm (top) by a mobile foot restraint. She is handling a crane, which was later installed on the space station.

This 50-foot-long robot arm is also used to help launch and retrieve satellites. Astronauts and mission specialists must practice using it to *grapple*, or grab, satellites before they go into orbit. Satellites are built of very lightweight material that could easily be damaged. To make sure no harm is done, strong "handles," called **grapple points**, are designed into the satellite. The RMS is also used for work on the *International Space Station*, which you will learn about in the next section. Fig. 18-12.

SECTION 3
TechCHECK

1. What effects would space have on your body?
2. What is the best way to stay strong in space?
3. How do astronauts train for their jobs?
4. **Apply Your Knowledge.** Choose one of the following:
   - If your school has a robot arm, simulate launching a satellite from a model of the Space Shuttle.
   - Use computer software to simulate movement of a robot arm.
   - Build a working model of a robot arm from scrap materials.

ACTION ACTIVITY

Creating a Simulated Satellite

Be sure to fill out your **TechNotes** and place them in your portfolio.

Real World Connection

On Earth, NASA specialists use *helium* gas to simulate neutral buoyancy, or the "floating" effect of a microgravity environment.

In this activity, you will design a simulated satellite made from a helium-filled balloon. Fig. A. After you have built your satellite, you will see how long it takes to grapple it as it floats in the air.

Design Brief

Design, build, and test a model of a satellite that will float in air. Your simulated satellite should have the following features:

* *Solar array*: A solar array, or panel, is used to change sunlight into electricity to power satellites. In real satellites, the electricity can be used directly or stored in batteries. Your array should be made from lightweight balsa wood and tissue paper. Try to design yours to look like a solar panel on a real satellite.
* *Grapple point*: A grapple point is a type of handle designed to be grabbed by a robot arm. The grapple point prevents damage to the delicate parts of the satellite.

Materials/Equipment

* balloons (12 inch)
* string or thread
* helium
* balsa wood
* tape
* paper
* markers
* straws
* Styrofoam plastic foam
* tissue paper
* markers
* stopwatch or watch

SAFETY FIRST

* **Do not inhale helium!** Your body cannot use helium. Breathing it can reduce or replace the oxygen in your lungs, and you need oxygen to live.
* **Read and follow the warnings on the helium tank. Use helium only with your teacher's permission.**
* **Follow the safety rules listed on pages 42-43 and the specific rules provided by your teacher for tools and machines.**

MICROGRAVITY BALLOON

Helium Balloon
Tape
Balsa Wood
NASA
Thread
Tissue Paper Solar Array
Styrofoam® Grapple Point

Fig. A

(Continued on next page)

ACTION ACTIVITY

Procedure

Part 1 · Building the Satellite

1. Working in small groups of three or four students, make sketches of how your satellite design might look.
2. Make a list of the materials you will need.
3. With your teacher's help, experiment with the weight of the materials and the lifting ability of helium.
4. Modify your design if needed.
5. Build the final satellite.
6. Test your satellite by adjusting its weight until it floats in air without going up or down.

Part 2 · The Grapple Test

If your school has a robot arm, use it to grapple your satellite. Record the time it takes in your TechNotes. If you do not have a robot arm, follow the steps below to use a human arm instead.

1. Choose a member of your team to be the "robot."
2. Clear an area of the classroom large enough for the robot to stand free of obstacles like desks or chairs.
3. Blindfold the robot. Have members of the group form a circle around him or her at arm's length. Fig. B.
4. Ask your human robot to respond to the verbal commands shown in Fig. C the way a real robot would.
5. Place your satellite near the human robot and time how long it takes the robot to grab the satellite's grapple point. Record your time on your TechNotes.
6. Take turns being the human robot and giving voice commands for directions.

Fig. B. Before blindfolding the robot, review the commands shown in Fig. C.

Evaluation

1. List seven different commands needed to move a robot arm.

2. Why is it important for astronauts to practice things on Earth before going into orbit?

3. What do you think would happen to a real satellite in space if it were bumped accidentally?

4. **Going Beyond.** Use a video camera and monitor to simulate remote telemetry control of your real or human robot.

5. **Going Beyond.** Design, build, and launch a weather balloon that will record temperatures at higher altitudes. Keep the balloon tethered with kite string. Use a maximum/minimum digital thermometer. Graph the relation between altitude and temperature.

6. **Going Beyond.** Research the effect of temperature as it relates to the lifting ability of helium. Find information about Boyle's and Charles' laws and explain the laws to your class.

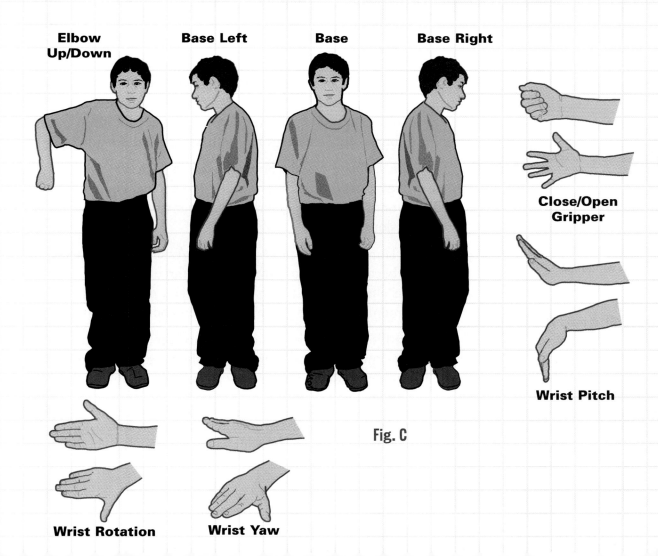

Elbow Up/Down · Base Left · Base · Base Right · Close/Open Gripper · Wrist Pitch · Wrist Rotation · Wrist Yaw

Fig. C

Living in Space for a Long Time

THINGS TO EXPLORE
- Describe the challenges of living in a space colony.
- Name the first space stations.
- Design and build a simulated space station.

Once you're up in space, what would it be like to live there? What's so different about living in space for a long time?

Space Stations

In order to find out more about living in space, both the United States and the former Soviet Union built space stations where people could stay for longer times. Living in space for long periods brought new challenges to both people and equipment.

Salyut 1, launched by the Soviet Union in 1971, was the first space workstation. *Salyut 1* was about 44 feet long. Though it was used only once, it was the beginning for many other successful *Salyut* missions.

Skylab, launched on May 14, 1973, was America's first space station. It became home for three crews of three people each. *Skylab* was almost 120 feet long, about the size of an average three-bedroom house. On Earth, it weighed about 100 tons.

Then in 1986, the Soviet Union launched the **Mir 1** space station. Its computers took over some of the tiresome jobs people once had to do. Cosmonauts on *Mir* proved that people can live in a microgravity environment for over a year!

The *Skylab*, *Salyut*, and *Mir* missions gave us lots of information about ways to make life better for people living and working in space. We want to use this information in building future space stations.

Many countries, working independently and together, already have satellites in orbit above the Earth. In the future, more nations

TechnoFact

SPACE SUIT DESIGN Space suits contain enough water and oxygen to last about 7 hours. The suit has to be able to withstand temperatures from -157°C (-250°F) to +121°C (+250°F). It must also be pressurized, because there is hardly any atmosphere outside the spacecraft or space station. The visor on the helmet is made to protect the wearer's eyes from ultraviolet rays and tiny meteorites.

expect to send people into space. Fig. 18-13. Sixteen nations, including the United States, Russia, and Japan, are now working together to create the *International Space Station*. Building this space station will be an engineering challenge. Many shuttle trips over a period of three to five years will be needed to assemble, outfit, and get the space station ready. The modules (separate compartments) will be taken up separately and attached to the space station. The modules will be like rooms in a house, only smaller. The idea is for you to be as comfortable as possible during your time in space. Power will come from solar cells and batteries.

Fig. 18-13. The next space station will provide a platform for conducting experiments in microgravity, materials testing, and water recycling. Find out which nations are involved in creating the *International Space Station*.

Figure 18-14 (page 424) describes life in space as you might find it today. How would you like to be the first student in space?

Space Colonies

People are also thinking of developing space colonies on the moon or Mars as permanent homes. Living on another world will be different from a short trip in a shuttle or a longer time on a space station. The space colony will be built from materials brought from Earth. It will have an artificial environment that can be controlled for maximum comfort. The colony may have its own farming areas, artificial gravity system, and even shopping centers. Space colonies will have to become almost *self-sufficient.* That means they must produce what they need to survive.

SECTION 4
TechCHECK

1. What were the first space stations called?
2. Why do astronauts have to use liquid salt and pepper?
3. What does it mean to be self-sufficient?
4. **Apply Your Knowledge.** Prepare your favorite meal as if you were planning on eating it in space.

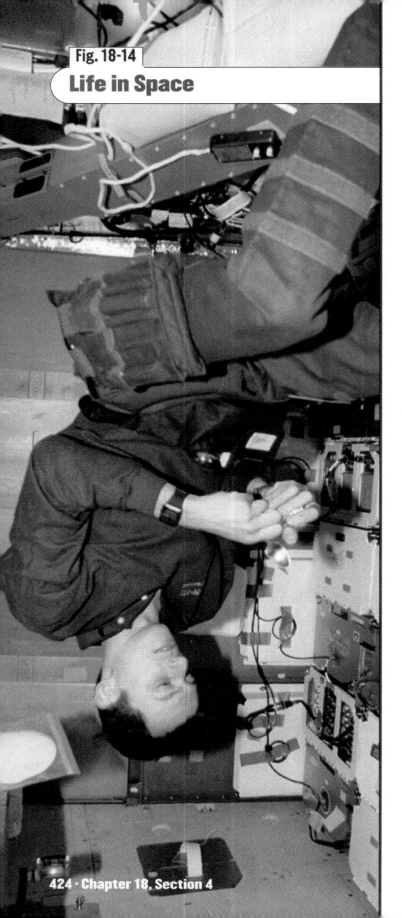

Fig. 18-14
Life in Space

What's for dinner? Breakfasts, lunches, and dinners are packed as meals in containers. Many foods have been vacuum-packed, dehydrated (dried), and deep-frozen. Most drinks, except orange and grapefruit juice, come in powdered form. Luckily some things, such as peanut butter, nuts, and chewing gum, stay in their natural form.

Eating in space. Preparing the food and eating it take special equipment. A punching machine inserts hot or cold water into dehydrated food packs. Velcro tape or magnets hold the food onto your tray. Most food is in a thick gravy or jelly-like substance, so it sticks to your spoon or fork. Salt and pepper must be added in liquid form because crystals would float away.

Sleeping in space. You don't need a bed. You just have to tie your sleeping bag someplace so that you don't float off when you turn over in your sleep. Even your breathing can make you move!

Keeping clean. To take a shower in space, you have to hook your feet to the floor, or you might end up upside down from how you started. Suction devices pick up water so that droplets don't float around. Even to use the toilet you have to strap yourself down with a seat belt!

Building a Space Station

Real World Connection

Space stations provide a place for astronomy, for doing long-term experiments on materials, and for testing the effects of microgravity on people. They also provide a place for careful observation of the Earth. They might even help us locate and fight the harmful effects of air or water pollution.

In this activity, you will work as engineers, drafters, and technicians to design and build a simulated space station. Fig. A.

Design Brief

Design and build a simulated space station that is large enough to hold four student astronauts. At least two exits must be provided. Include a mission control center and a communications system that will connect to the space station. Consider the following:

- Your space station must be strong enough to support the weight of the building materials used and the equipment you put in it.
- The design of your space station must provide for easy entry and exit in an emergency. The exits should not be able to be locked or blocked.
- The design of your space station must provide for electrical safety. Any use of extension cords must be approved by your teacher.
- Your space station will be used to simulate experiments that astronauts would perform in space. Plan what kind of experiments you could do, and make a list of the equipment that you might put into the space station.

Materials/Equipment

- plywood, prefinished hardboard, particleboard
- electrical components: wire, switches, etc. (12-volt DC)
- nails, wood screws, panel adhesive
- band saw, scroll saw, saber saw
- electric soldering gun
- CAD software

> **Be sure to fill out your TechNotes and place them in your portfolio.**

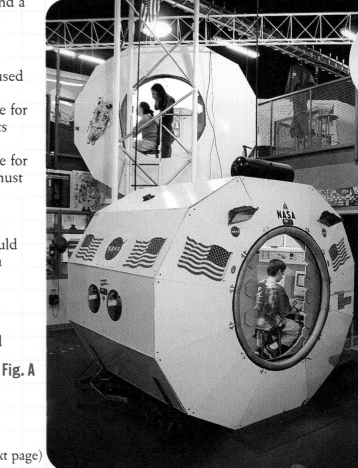

Fig. A

(Continued on next page)

ACTION ACTIVITY

Procedure

This is a long-term activity that will require the help of all students in each class. You may work in groups to make different parts.

1. Brainstorm ideas for the shape of your space station. The space and materials available may be factors to consider.

SAFETY **FIRST**

Follow your teacher's instructions on the safe use of electricity. You must get the approval of your teacher before any electrical connections are made. Follow the safety rules listed on pages 42-43 and the specific rules provided by your teacher for tools and machines.

---

2. Use CAD software to design the shape of your space station. Keep the design specifications in mind.

3. Choose the design that best suits your needs. Plan the structural skeleton that will support the station. Your design might look like the one shown in Fig. B.

4. When the skeleton is complete, create the inside walls of the space station. Leave the outside uncovered for now so you can easily run wires and connect equipment.

5. Design and install the electrical circuits so they will run on 12 volts DC or less. Put a fuse in each circuit to prevent possible fires. Ask your teacher for help.

INFOLINK

See Chapter 5 for more information about CAD.

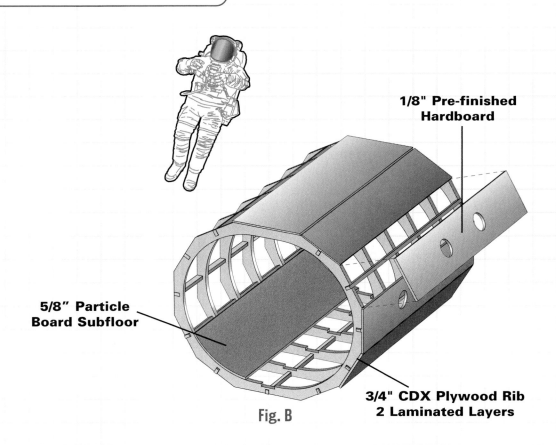

1/8" Pre-finished Hardboard

5/8" Particle Board Subfloor

3/4" CDX Plywood Rib 2 Laminated Layers

Fig. B

6. Complete the outside "skin" of your space station. Keep in mind that you may want to remodel your design or add more equipment in the future. Each outside panel should be removable to give easy access.

7. Use your imagination to think of ways equipment might be used in the simulation of space station experiments. Make a list of the equipment and its use in the space station. Your list might include:

• *Computer.* Can your computer be connected to a modem or a network so you could communicate with other student astronauts?

• *Laser.* Could a laser be used to simulate the alignment of your space station with the Space Shuttle for docking?

• *Video phone.* Can you use video cameras and monitors to let student astronauts see and talk with mission control in another classroom?

• *Cooking equipment.* How would food preparation be different in space? What would it be like to eat in microgravity?

• *Exercise equipment.* Astronauts must exercise in space to keep their muscle tone. How could you simulate exercise activities in your space station?

8. Work in groups to perform space-related experiments in the space station. Switch groups so that everyone has a chance to be at mission control and "in space."

Evaluation

1. List the features that your simulated space station has in common with a real space station.

2. If you could add one more piece of equipment to your space station, what would it be and how would it be used?

3. How could space station research help us on a mission to Mars?

4. Why would it be important to recycle materials on a space station? How would recycling in space be different from recycling on Earth?

5. **Going Beyond.** Research real space stations and those in science fiction. What do the two have in common? How are they different?

6. **Going Beyond.** Contact other schools that have simulated space stations. Communicate with them using a modem or fax machine while you are simulating a space mission.

7. **Going Beyond.** Write to NASA requesting information on space stations. Ask NASA if you could set up a phone call to an astronaut or mission specialist.

8. **Going Beyond.** Download information from SpaceLink (available from NASA) or other space-related bulletin board services. Downlink information related to space from a satellite dish receiving system, such as NASA Select television.

9. **Going Beyond.** Research the design and construction of simulations conducted by NASA to prepare for the building of a space station. Research the former Soviet Union's *Mir* space station.

REVIEW &

CHAPTER SUMMARY

SECTION 1

- Space spinoffs are products, medicines, or other items that can be used on Earth.
- The future of space exploration will be determined by how people want their tax money spent.

SECTION 2

- In a microgravity environment, the effects of gravity are reduced compared to gravity at Earth's surface.
- All orbiting spacecraft are actually falling toward the Earth.
- In a vacuum, there is not enough air to breathe and a lack of air pressure; both are fatal to humans.

SECTION 3

- Daily exercise is needed to reduce the effects of microgravity on the body.
- Mission specialists simulate the floating effect of a microgravity environment using neutral buoyancy.

SECTION 4

- The *Skylab, Salyut,* and *Mir* missions gave us information about living and working in space.
- Many nations are working together to build the *International Space Station*.

REVIEW QUESTIONS

1. What is microgravity?
2. Explain why spacecraft are really falling toward the Earth when they are in orbit.
3. What are space spinoffs? Give examples.
4. Describe how eating in a microgravity environment would be different from eating on Earth.
5. How did the first space stations help with plans for the *International Space Station*?

CRITICAL THINKING

1. Do you think tax money should be spent on space exploration? Explain.
2. What would happen if you spilled water in the vacuum of space?
3. Do you think it is possible to swallow while standing on your head? How does this relate to eating in microgravity?
4. Explain how hot air balloons and dirigibles are made neutrally buoyant.
5. What would happen if you tried to throw a ball while walking in space?
6. Why do you think the *International Space Station* is made up of modules rather than one large piece?

ACTIVITIES 18

CROSS-CURRICULAR EXTENSIONS

1. SCIENCE Make a sketch of what you think life forms from beyond the Earth might look like.

2. TECHNOLOGY Make a model of a space station. Include a living module, lab module, and other important modules.

3. COMMUNICATION Write a science fiction story about a space colony or lunar city of the future.

4. MATHEMATICS Research how air pressure changes with increasing altitude. Make a graph to show your findings.

EXPLORING CAREERS

Currently, space stations are used to conduct scientific experiments and serve as temporary homes for astronauts. Some day space stations may house factories, airports, hotels, schools, and hospitals. Many of the jobs on the space stations will be no different than those on Earth. However, those jobs will offer opportunities unheard of today!

Life Sciences Researcher How would plants and animals fare during a journey in space? To find out the answer to this and many other questions, life sciences researchers conduct many tests. They must be able to understand how one step in a life process affects the other steps. They must also work well in a team environment.

Astronaut On a space mission, each astronaut is responsible for a special area, such as biotechnology research, astronomy, or computer systems. Astronauts are well-qualified in their fields. They undergo hard physical training while continuing their education. They must be team players and be willing to work long hours.

ACTIVITY

Design and draw the home page of a website for a business you might find in space some day.

Exploring Chemical & Bio-Related Technology

Chemical and Bio-Related Technology

THINGS TO EXPLORE

- Explain how chemical and bio-related technologies affect you.
- Describe the changes chemical and bio-related technology advancements might bring.
- Tell how advancements in chemical and bio-related technology raise ethical questions.

TechnoTerms

aquaculture
bionics
bio-related
   technology
ethics
genetic disease
hydroponics

Did you ever stop to think that the apple you ate for lunch or the cereal you had for breakfast may have been changed by technology? Rapid advancements in agriculture, health care, waste management, and other chemical and bio-related technologies affect you.

Chemical technology has to do with the many chemicals used to produce such things as plastics, paint products, and even fertilizers. **Bio-related technology** has to do with living things. Fig. 19-1. Bio-related technologies include

- **Genetic engineering:** The process of changing the gene structure of a plant or animal. It may also involve designing new life forms.
- **Bioprocessing:** The use of microorganisms to change materials from one form to another. One example is to use specially designed microorganisms to clean up oil spills, turning the oil into something harmless.
- **Bionics**, or *biomechanics*: Designing artificial parts, or *prostheses*, for the human body, such as arms, legs, hips, and teeth. Bionics also involves such things as computers and new wheelchair designs that make it possible for people with disabilities or degenerative (continually worsening) diseases to lead more productive lives.

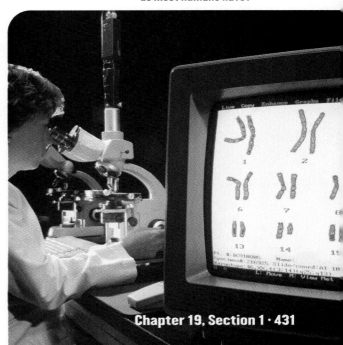

Fig. 19-1. A genetic researcher studies the chromosomes that carry genetic material. Chromosomes come in pairs. **How many pairs do most humans have?**

◀ OPPOSITE Bio-related technology has to do with living things.

- **Biomaterials:** Human-made materials designed to be used inside the body.
- **Hydroponics:** Growing plants without soil. Hydroponics can be used in areas where land is limited, because you can grow more plants in a limited amount of space. In addition, the crops grow faster and produce more. You can even reuse the water and fertilizer.
- **Aquaculture:** Raising fish and food plants in water.
- **Pesticide management:** The use of chemicals to control insects and other pests.
- **Waste management:** The management of garbage and other things we throw away. Some wastes can be hazardous (dangerous) to both humans and the environment.

Uses for Chemical and Bio-Related Technologies

Continued advancements in chemical and bio-related technologies will improve health care, produce new foods, and help us manage environmental concerns. Fig. 19-2.

Vaccines will be produced in the next few years to protect you against many diseases. Many **genetic diseases** (carried from parents to children) are also being studied to find a way to reduce or change their effects.

Finding ways to improve food crops and animals is another important research area. *Growth hormones* make cows grow faster

Fig. 19-2. A scientist collects sperm from this endangered wild salmon. It will be used to fertilize the eggs of salmon on fish "farms." The qualities of the farm salmon will be improved. The genes of the wild salmon will be passed on.

and help them produce more milk than they normally would. Scientists can create a line of leaner, stronger, healthier cattle. You can even have "custom-made" calves. How would you like a turkey with more than two drumsticks? It might soon be possible.

Slowing the aging process, turning seawater into fresh water, and making new fuel sources are all areas of development for these technologies.

Bioethics

Chemical and bio-related technology can improve many things. Many people predict the future of bio-related technology to be bigger than the electronics industry is today. But some people fear what might happen if the technology is used strictly for profit or without thinking about how it could change our world. Fig. 19-3. Many ethical questions are raised. (**Ethics** involves following a high standard of conduct.)

For example, are the changes brought about by bio-related technologies good or bad? Are these technologies being used appropriately to help people and the environment? Changes are coming very rapidly, and many issues need to be evaluated. New standards of *bioethics* must be established that apply to people working in bio-related fields.

Fig. 19-3. Many people are concerned about the effects of bio-related technology. Do you think bio-related technologies should be regulated by the government? Give reasons for your answer.

SECTION 1
TechCHECK

1. Name some chemical and bio-related technologies.
2. List three changes that come from advancements in chemical and bio-related technology.
3. What are ethics?
4. **Apply Your Knowledge.** Research progress in biomechanics. Make a chart of the artificial body parts available for people.

Distillation and Fermentation at Work

THINGS TO EXPLORE

- Define *distillation* and tell how it works.
- Define *fermentation*.
- Describe how distillation and fermentation help people and the environment.
- Distill salt water into fresh water.

TechnoTerms
biofuel
distillation
ethanol
fermentation
methanol
reverse osmosis

Did you know it's just as easy to die from thirst in the middle of the ocean as it is in the middle of the desert? That's because you can't drink salt water.

Distillation

A chemical process called **distillation** can change salt water into fresh water, making it drinkable. Fig. 19-4. Distillation separates substances based on their different boiling points. As salt water is boiled in a container, the water changes to a gas. The gas is collected in a tube where it is cooled and turns into a liquid again. Salt's boiling point is well over 1000°C, so it remains in the original container. However, distillation takes lots of energy to produce fresh water.

A new, less expensive method called **reverse osmosis** is being tested for use by desert nations. In this process a high-pressure pump forces sea water through microscopic openings in special filtering materials. These materials filter out the salt.

Fig. 19-4. Distillation plants, like this one, can turn salt water into fresh water. Name some countries where this process might be needed.

Microorganisms and Fermentation

Fermentation uses microorganisms to turn grains into alcohols. It is a promising source of **biofuels** (fuels made from biomass, which is plant or animal matter). Fig. 19-5. Almost any kind of biomass that contains starch or cellulose, from cornstalks to cow dung, can be changed by fermentation and distillation into **ethanol**, or grain alcohol.

When ethanol is mixed with gasoline, it produces *gasohol*, which can be used as an alternative to gasoline. While ethanol is presently somewhat more expensive to produce than gasoline, it can often be made from resources countries already have, such as sugar cane. In Brazil, for example, over half of the automobile fuel used is gasohol.

Biofuels, like ethanol, produce less air pollution than gasoline does. **Methanol**, a product made from fermented wood-product waste, is not as good an alternative as ethanol because it pollutes more. However, it still does not put as much carbon monoxide into the air as gasoline does. Many car manufacturers are planning to produce cars in the future that burn alternative fuels.

Fig. 19-5. Animal wastes flow into a special lagoon. When wastes can be recycled into fuels, they cause fewer problems for the environment. Is gasohol available in your area?

TechnoFact

FUEL SAVINGS If we replaced all gasoline in the United States with gasohol, we would cut use of gasoline by 10 percent and cut oil imports by 20 percent. Considering that we use about 19 million barrels of oil per day in this country, that's a lot!

SECTION 2
TechCHECK

1. What is distillation?
2. What is fermentation?
3. How do distillation and fermentation help people and the environment?
4. **Apply Your Knowledge.** Research the use of gasohol in cars. What advantages are there to adding ethanol to gasoline for use as a fuel?

ACTION ACTIVITY

Making Salt Water into Fresh Water

Be sure to fill out your TechNotes and place them in your portfolio.

Real World Connection

It seems strange that some countries are running out of drinking water when they are right next to an ocean. But drinking sea water can cause illness and even death. In this activity, you will be working as a chemical engineer to distill salt water into fresh water.

Design Brief

Turn salt water into fresh water using the process of distillation.

Materials/Equipment

- flask, rubber stopper, glass tubing
- food-grade plastic tubing
- ring stand, clamps
- beaker
- condenser
- table salt, water
- electric hot plate or Bunsen burner

SAFETY FIRST

- Follow the safety rules listed on pages 42-43 and the specific rules provided by your teacher for tools and machines.
- All the equipment used in this experiment should be clean and free of any other chemicals.
- Do not start this experiment until your teacher has checked the setup.

Procedure

1. Teams of students should work on the different tasks. Your teacher will make assignments.
2. Mix 50 cc of water with one teaspoon of salt.
3. Set up the ring stand, condenser, and hot plate as illustrated in Fig. A.
4. Connect the plastic tubing to the flask and condenser.
5. When everything is ready, pour the salt water into the flask. Turn on the hot plate or burner. Watch for the first signs of boiling and condensation.
6. Collect the distillate (condensed vapor) in a clean beaker or cup. Allow the distillate to cool.
7. Dip your finger in the distillate and taste the water.

Evaluation

1. Did the distillate taste salty? Explain.
2. Why is it so expensive to make water fresh by distillation?
3. **Going Beyond.** Design a solar-powered desalination plant. How do you think the output of the solar-powered plant would compare with distillation using traditional fuels?
4. **Going Beyond.** Calculate the output of your distillation experiment. How long would it take to make the fresh water you use in one day using the equipment in this distillation experiment?
5. **Going Beyond.** Make a list of five ways your family could save water.
6. **Going Beyond.** Can you filter out the salt in sea water using filter paper? Explain.

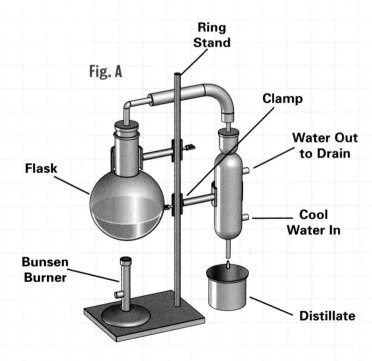

Fig. A

Ring Stand

Clamp

Water Out to Drain

Flask

Cool Water In

Bunsen Burner

Distillate

The Flow of Fluids

THINGS TO EXPLORE

- Explain why scientists are studying fluid flow.
- Describe some ways fluid flow affects you.
- Design, build, and test a fluid-flow device.

TechnoTerms

cholesterol
crystal

In chemistry, the term *fluid* can refer to a gas or a liquid. A fluid is a substance that flows. Unlike a solid, a fluid has no shape of its own. Instead, it conforms to the shape of its container.

Why Study Fluids?

Scientists study the flow of fluids carefully to understand, control, and improve many natural and industrial processes.

INFOLINK

See Chapter 8 for more information about fluids.

The fact that fluid moves is important both in nature and in technology. Your blood is a fluid that brings oxygen and nutrients to your cells. The air we breathe and the water we drink are fluids. What would happen if air couldn't expand? The movement of fluid powers many products of technology, such as automobile brakes, hovercraft, and jackhammers.

Fluid flow affects your everyday life. Studying fluid flow can help us understand such things as

- How chemical processes can build better products
- How **cholesterol** (a fat-like substance) is moved through the blood
- How crystals form to make new medicines. (A **crystal** is a solid whose atoms are arranged in a repeating pattern.)
- How pollutants are carried in water and air
- How car engines run, airplanes fly, home heaters operate, and rocket engines work

TechnoFact

JUST ANOTHER PHASE Did you ever stop to think that three of the four states of matter (gas, liquid, and plasma) are fluids? Even the fourth state (solid) behaves like a fluid under many conditions.

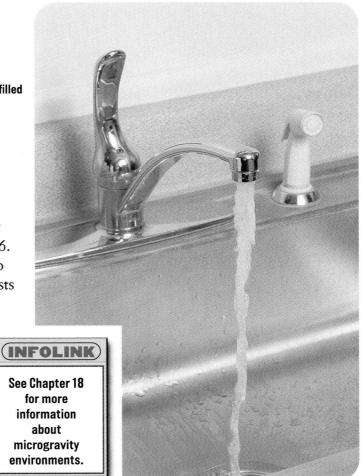

Fig. 19-6. Water rushing from a faucet is filled with movement. Try to describe this flow as accurately as possible in three or four sentences.

Studying fluids is difficult. Look carefully at water running from a faucet and try to study its movement. Fig. 19-6. Can you appreciate how difficult it is to accurately describe? That's why scientists often study fluid flow in a microgravity environment. The lack of gravity changes the way the fluid moves. If the flow of fluids can be made two-dimensional and controlled carefully, it is easier to study. That's what you will do in the activity for this section.

INFOLINK

See Chapter 18 for more information about microgravity environments.

SECTION 3
Tech CHECK

1. What is fluid flow and why do scientists study it?
2. Why is studying fluids difficult?
3. List three ways fluid flow affects your world.
4. **Apply Your Knowledge.** Research fluid flow. Share the results with your classmates.

ACTION ACTIVITY

Studying Fluid Flow

Be sure to fill out your TechNotes and place them in your portfolio.

Real World Connection

Fluid flow is an important field of study for rocket scientists as well as medical researchers. In this activity, you will make a two-dimensional fluid-flow device to investigate how fluids move. You will create a film made of a thin layer of soap on each side of a single layer of water molecules. The film will be so thin that it can be considered to have no thickness.

Design Brief

Design, build, and test a fluid-flow device. Place different shapes in the fluid to study how they disturb the flow.

Materials/Equipment

- 2-liter plastic soda bottle
- utility knife
- fishing line
- rubber stopper with one hole
- medicine dropper
- adjustable hose clamp
- 1/4" washer
- dishwashing soap (such as Dawn liquid soap)
- drip pan
- weight
- plywood (3/4" x 6" x 12")
- scroll saw
- rubber tubing

SAFETY FIRST

- Follow the safety rules listed on pages 42-43 and the specific rules provided by your teacher for tools and machines.
- Use liquid soap to make it easy to push the glass medicine dropper into the hole in the rubber stopper. Be careful not to break the glass.
- Wear safety goggles to avoid contact with soapy water.
- Clean up any spills immediately to prevent someone from slipping.
- Use caution when using a utility knife.

Procedure

1. Work in groups of two or three students. Divide the following tasks among members of your team.
2. Cut off the bottom of an empty 2-liter plastic soda bottle using a utility knife.
3. Remove the bulb from a medicine dropper. With your teacher's help, carefully push the glass part of the dropper through the hole in the rubber stopper.
4. Place a 2"-long piece of rubber tubing on the end of the medicine dropper. Attach the tubing clamp to the rubber tubing.
5. Cut a double strand of fishing line as long as the height of your fluid-flow device. Try to make your device at least as tall as you are.
6. Tie both strands of fishing line to a 1/4" washer.

7. Thread the two free ends through the hole in the bottle and the medicine dropper and rubber stopper as shown in Fig. A.
8. Tie the loose ends to a weight. Tie four short pieces of fishing line to the two main strands as shown in Fig.B on the next page.
9. Cut the plywood to make a support for the bottle. Clamp the support to the top of a door or other steady structure.
10. Mix a weak solution (2%-10%) of dish soap and water. Fill the bottle half full of soap solution.
11. Place the drip pan under the weight to catch the soap solution.

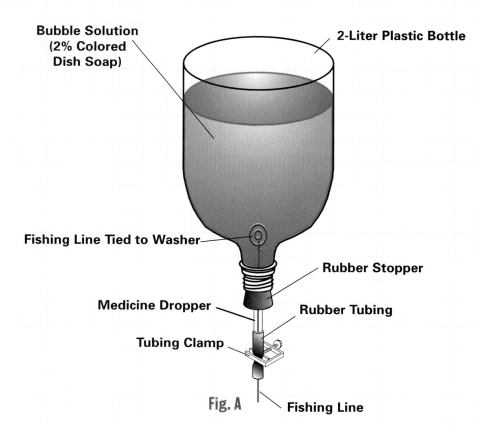

Fig. A

(Continued on next page)

12. Adjust the flow so the soap solution drips slowly down the fishing line. When the solution reaches the weight on the bottom, slowly pull the short strings apart to form a film of fluid. Fig. B.

13. Adjust the size of the soap film and the flow of the solution so it continues to flow.

14. Wet your finger in the soap solution. Poke your finger through the bubble film and watch the flow carefully. Notice the pattern of flow downstream of your finger. Look for a spiral-shaped pattern called a *vortex*.

Evaluation

1. Why is it easier to study a two-dimensional fluid flow than a three-dimensional fluid flow?

2. List some real-world applications for studying fluid flow.

3. What is a vortex?

4. **Going Beyond.** Use a camcorder to tape the vortex patterns produced by different shapes placed in the fluid flow. Experiment with different camera angles, backdrops, and lighting to get the best picture. Play the tape to analyze the patterns.

5. **Going Beyond.** With your teacher's help, take the fluid-flow device to the top of a stairs or ladder. Try to make the fluid flow as large as possible.

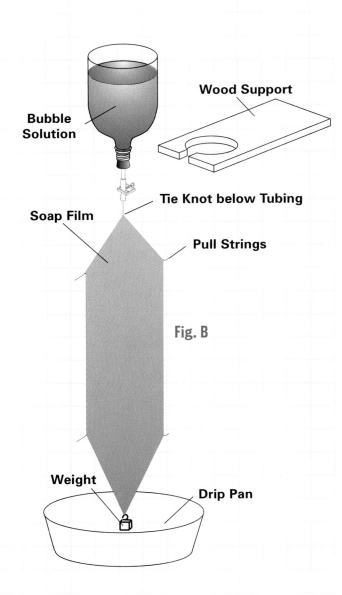

Bubble Solution

Wood Support

Tie Knot below Tubing

Soap Film

Pull Strings

Fig. B

Weight

Drip Pan

Genetic Engineering

THINGS TO EXPLORE

- Define *genetic engineering*.
- Describe ways genetic engineering helps us.
- Explain Mendel's Law of Dominance.
- Use a Punnett square to show the genotypes and phenotypes in a genetic cross.

TechnoTerms

gene
genetic engineering
genotype
Mendel's Law of
    Dominance
phenotype
Punnett square

A **gene** is a basic unit of heredity that carries certain information about the development of a plant or animal. What color is your hair? Do you have blue eyes? Are you color-blind? All these traits (characteristics) are determined by your genes. *Genetics* is the study of how traits are passed on from generation to generation.

Genetic Engineering

Genetic engineering is a process of changing the gene structure of a plant or animal to improve it for human use. Fig. 19-7.

How does genetic engineering work? In most cases, a sample of genetic material is taken from a subject and then grown in cell cultures for use. The cells multiply to produce more of the same kinds of cells. This process is used in the manufacture of antibiotic medicines such as penicillin and streptomycin. The production of insulin, a hormone used to treat diabetes, is also produced using genetic engineering techniques.

Fig. 19-7. This scientist is taking samples of sunflower seeds. In what ways do you think a crop of sunflower seeds could be improved?

The Uses of Genetic Engineering

Genetic engineering can produce plants with new genes that make them grow faster or be more resistant to disease. Fig. 19-8. Researchers are looking for a way to put a gene into plant cells so plants would not need nitrogen fertilizer. Other researchers are trying to isolate (set apart) genes in people that cause hereditary diseases such as muscular dystrophy. They want to be able to change the gene structure early as a baby develops in its mother's body. That might change how the disease affects the person.

Fig. 19-8. Genetic engineering can produce new varieties of fruits and vegetables. Write a paragraph stating your views on the genetic engineering of our food supply. Give reasons for your opinion.

SCIENCE CONNECTION

DNA—Your Own Special Code

Fig. 19-9. Antibodies are produced by cells to fight foreign substances. These were altered to fight cancer and then cloned (duplicated). The cloned antibodies will be injected into a cancer patient.

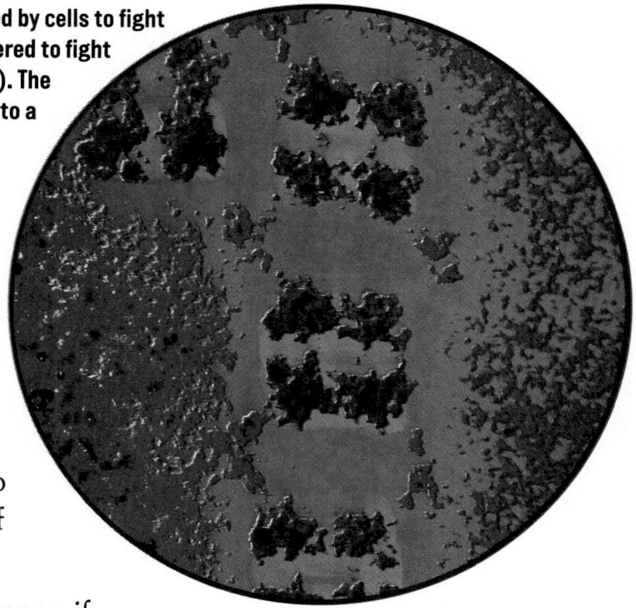

Genetic engineering raises many questions. When it is used in the treatment of diseases, you can see its benefits. Fig. 19-9. But whether technologists and scientists should change or improve humans is another issue. What if we could design the "perfect" person by genetic engineering? Who would decide what traits are the best? What if we could *clone*, or duplicate, a person?

Researchers also don't know what might happen if altered microorganisms are released into the environment. Maybe they will end up multiplying too fast and become a pollution problem. Fig. 19-10. What if genetic engineering produces dangerous bacteria and viruses that could be used in war? This problem is one that many countries have tried to prevent through special agreements.

DNA (deoxyribonucleic acid) is the chemical code of instructions found in the nucleus of every living cell.

DNA controls the way plants and animals look and the way their bodies work. The code transmits characteristics from one generation to the next.

DNA is very complex even though only four base compounds—adenine, guanine, cytosine, and thymine—make up its structure. These bases pair up to form a double strand called a double helix. Adenine always pairs with thymine, and guanine with cytosine. Except for identical twins, no two individuals have the same DNA makeup.

ACTIVITY

Assume that A = adenine, G = guanine, C = cytosine, and T = thymine. Make the code for a double strand of DNA if the base pairs of one strand are A A G C T T G G C C C A T T A.

Fig. 19-10. Cleaning up after an oil spill, like this one in San Francisco, is very difficult. Many animals die in the meantime. But genetically altered bacteria (see inset, above) can eat the oil quickly. What do you think could go wrong with using bacteria in this way?

TechnoFact

IMPROVING CROPS
Corn has been improved to make it higher in nutritional value, more resistant to disease, and able to grow in colder climates. Scientists hope to try the same processes on other key crops such as wheat and rice.

The Law of Dominance

To improve plants for human use, geneticists must find out which plant traits determined by genes are *dominant* (producing traits that show) or *recessive* (producing traits that show only if no dominant gene is present). Each trait requires two genes to express it, one from each parent.

Gregor Mendel (1822-1884), an Austrian monk, was the first person to study how traits are carried from one generation of living things to the next generation. After many experiments with pea plants, Mendel stated the **Law of Dominance**. A dominant gene will always mask, or hide, a recessive gene when they occur together. Dominant and recessive genes make up the **genotype** (what actual genes the organism has) and the **phenotype** (what an organism looks like). Geneticists can use this information to change life forms.

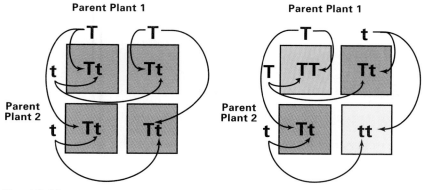

Fig. 19-11. In these Punnett squares, T indicates the gene for tallness and t indicates the gene for shortness. **Which is the dominant gene?**

A grid called a **Punnett square** is often used to show possible gene combinations in offspring (next generation). Figure 19-11 shows the offspring of two plants. The four boxes show the genotypes and the phenotypes of the trait for tallness. The upper-case letter T indicates the dominant gene for tallness and the lower-case t indicates the recessive gene for shortness. Will the plants be tall or short? You can tell just by looking to see if a dominant T appears in the box. If one or both genes are T, then the offspring will be tall. Are there any short plants in this generation? The second Punnett square shows the possible plant sizes if two parent plants with different genes (Tt) are crossed. Are all the plants tall?

TechnoFact

TONGUE TWISTERS
Can you roll your tongue into a distinct U shape? That's a dominant trait in many people. Can you twist your tongue into an S shape? That trait is recessive. Check your classmates for these traits!

SECTION 4
TechCHECK

1. What is genetic engineering?
2. How might genetic engineering help us?
3. What is the Law of Dominance?
4. **Apply Your Knowledge.** Find out more about Gregor Mendel and his plant studies. Make a report to your class.

ACTION ACTIVITY

Recessive or Dominant?

Be sure to fill out your TechNotes and place them in your portfolio.

Real World Connection

Researchers use Punnett squares to show possible gene combinations. In this activity, you will experiment with a Punnett square to show how dominant and recessive genes express themselves.

Design Brief

Use a Punnett square to determine possible gene combinations from crossing parent plants with different traits. A parent plant might be tall and have smooth seeds (TW), tall and have wrinkled seeds (Tw), short and have smooth seeds (tW), or short and have wrinkled seeds (tw). Fig. A.

Materials/Equipment

- 3" x 5" cards (at least 64)
- pencil and paper
- cardboard or tagboard
- colored pencils or markers
- PTC paper (optional)
- computer with graphics software (optional)

SAFETY FIRST
Follow the safety rules listed on pages 42-43 and the specific rules provided by your teacher for tools and machines.

	TW	Tw	tW	tw
TW	TTWW			
Tw				
tW				
tw				

Fig. A

Procedure

1. Work in groups of four students. Each group should make up the following 3" x 5" cards to represent genes:

- 16 cards with a large T to stand for dominant tallness
- 16 cards with a small t for recessive shortness
- 16 cards with a large W for dominant smooth pea seed
- 16 cards with a small w for recessive wrinkled pea seed

2. Put all the cards together in a deck.

3. Make a large 16-box Punnett square to show plant combinations from a cross of two different traits.

4. Each student should draw a card from the deck and place it in the appropriate box on the Punnett square. Refer to the chart in Fig. A.

5. Determine how many plants are

- tall and smooth (TW) ___
- tall and wrinkled (Tw) ___
- short and smooth (tW) ___
- short and wrinkled (tw) ___

You should get a 9:3:3:1 ratio.

Evaluation

1. Did you get a 9:3:3:1 ratio? If not, try again. This is a ratio of dominance for two traits determined by Gregor Mendel.

2. Is a dominant gene necessarily better than a recessive gene? What is the real difference?

3. **Going Beyond.** Research some genetic diseases such as sickle-cell anemia, cystic fibrosis, or Tay-Sachs disease. Find out if they result from dominant genes or recessive genes.

4. **Going Beyond.** Brainstorm ideas about how food crops or animals could be altered to make food production better. Evaluate the possible advantages of crossing a tomato and a watermelon.

REVIEW &

CHAPTER SUMMARY

SECTION 1

• Bio-related technology includes technologies related to living things, such as genetic engineering, bionics, bioprocessing, hydroponics, and aquaculture.

• Chemical and bio-related technologies also deal with pesticides, waste management, biomaterials, hazardous wastes, and other environmental concerns.

• Bio-related technologies raise ethical questions about how we should use them.

SECTION 2

• Distillation can change salt water into fresh water. The process separates substances by their different boiling points.

• Fermentation uses microorganisms to turn grain into alcohols.

• Biofuels, such as ethanol and methanol, are made from fermentation.

SECTION 3

• Scientists study the flow of fluids to understand, control, and improve many natural and industrial processes.

SECTION 4

• Genetic engineering is a process of changing the gene structure of a plant or animal to improve it for human use.

• Genetics is the study of how traits are passed on from generation to generation.

REVIEW QUESTIONS

1. What is bionics?

2. Name two biofuels.

3. Name two types of fluids.

4. What is a gene?

5. Why would people want to alter genes in plants or animals?

CRITICAL THINKING

1. Research the many uses of soybeans as food substitutes for meats and other products. Do a survey to see if people can tell from taste and texture when soybeans are substituted for meat.

2. What effects do you think genetic engineering will have on the future? List three possible changes in your lifetime.

3. If you wanted to develop a plant that would resist a certain disease, would you try to make that trait dominant or recessive? Why?

4. Why were the *Apollo* astronauts kept away from other people for two weeks after their return to Earth? Was this precaution needed?

5. Research the operation of an artificial leg or hand. Draw a diagram of the prosthesis.

ACTIVITIES

CROSS-CURRICULAR EXTENSIONS

1. **SCIENCE** Research the genome project or the discovery of the structure of **DNA**. Write a short report on your findings.

2. **MATHEMATICS** Calculate the number of gallons of fresh water you use in one year based on the following averages per use: bath

or shower—25 gallons; toilet flush—5 gallons; dishwashing—7 gallons; laundry—40 gallons per load.

3. **COMMUNICATION** Ask a doctor or geneticist to talk with your class about genetic engineering and the ethical questions involved.

EXPLORING CAREERS

As we understand more about genetics, scientists can improve the food we eat, the plants we grow, the animals we raise, and the treatment of diseases. The area of genetic research is growing rapidly. Listed below are two careers you will find in this high-tech field.

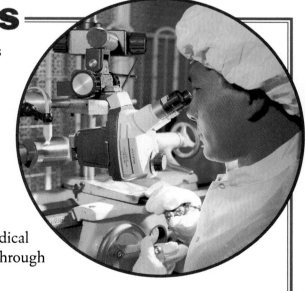

Biotechnology Researcher A biotechnology researcher is a type of high-tech explorer. These researchers combine biology and technology to create medicines, treatments, and medical devices. They help develop the product and see it through testing. Biotechnology researchers work long hours over small details. They often find satisfaction when there is a breakthrough in their work.

Geneticist Geneticists study genes to determine inherited traits, such as color differences, size, and disease resistance in plants. They develop methods for altering or producing new traits using chemicals, heat, or light. Geneticists usually have good mathematics and writing skills. They often prepare technical reports.

ACTIVITY

Research the pros and cons of genetically engineered crops. Present your findings to the class.

CHAPTER 20

Technology and Your Future

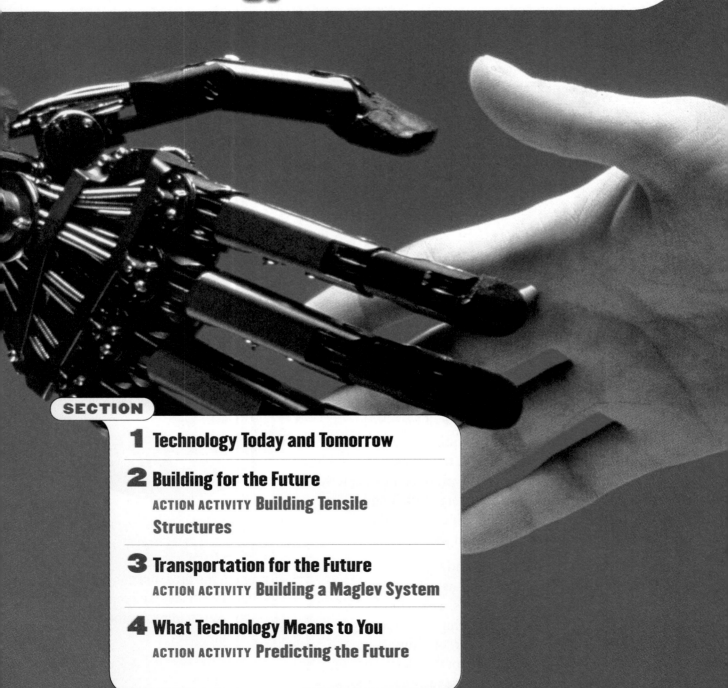

Technology Today and Tomorrow

THINGS TO EXPLORE

- Explain how technology affects you now and how it may affect you in the future.
- Who decides whether or not a technology is beneficial to people and the environment?
- Define what a futurist is and give examples of some futuristic predictions.

TechnoTerms
format
futurist

People decide whether technology is good or bad. The way technology is used depends on people like you! In this course, you have learned how quickly technology is growing and changing. Those changes affect all of us in many different ways. Think how communication and transportation have changed in the last century. You can communicate with people anywhere in the world in a matter of seconds through computers and satellites. You can travel across the Atlantic Ocean in about three hours or travel into space at a speed of 17,500 miles per hour. Fig. 20-1.

Fig. 20-1. One of the most thrilling accomplishments of the twentieth century was our entry into space. This satellite is gathering data on the sun and the solar wind. The Space Shuttle is to the left. Research the next Shuttle mission on the Internet. Report your findings to the class.

OPPOSITE Robots are only one of the changes the future will bring.

Fig. 20-2. This robot fish can swim freely, without wires. It is being used to study how fish can be such efficient swimmers. What benefits do you think this research could bring?

The important thing is that you understand how technology affects you. You can then make intelligent decisions about whether technologies are beneficial or harmful to people and the environment. Fig. 20-2.

Thinking about the Future

Futurists are people who try to predict (make an educated guess about) the future. What do you think our technology will be like in the future? Here are some predictions futurists have made.

- **Computers.** Computers will someday be without keyboards. Instead, you will communicate by voice commands.
- **Telephones.** Instead of using separate telephone numbers for the car, your home, or your parents' workplace like you do now, you will use one telephone number that works for everywhere. Also, a basic wristwatch will have a telephone that shows a video image of the person you are talking to.
- **News media.** Your future newspaper will be personalized so you read only things that interest you. Furthermore, the **format** (way it is presented) might be in video, on the computer, or in print, depending on which format you like best.
- **Medicine.** Research has shown that broken bones heal faster when they are exposed to ultrasound. Small ultrasound devices will be placed in casts to help the healing process. Robots already assist doctors in precise operations. Maybe robots will do the job by themselves in the future. Fig. 20-3.

INFOLINK

See Chapter 4 for more information about computers.

INFOLINK

See Chapter 19 for information about bio-related technology.

TechnoFact

FUTURE ENERGY SOURCE Currently, we use nuclear fission as a source of energy (see Chapter 13). Nuclear *fusion* is another type of nuclear reaction that gives off huge amounts of energy. There is only one problem—fusion reactors consume more energy than they produce. However, if one day it becomes practical, it will provide a limitless source of fuel. It will also produce less radioactive waste.

- **Manufacturing.** Robotic machines will take over more jobs that are too dangerous or too boring for people. Factories will be built in outer space where they can manufacture new alloys and other products without Earth's gravity getting in the way.

You can expect to see many advances in other fields that will affect the way you live. Some new technologies may drastically change your world, much like the invention of the electric light-bulb did in the last century. In the next two sections of this chapter, you will explore ideas in construction and transportation for the future.

Fig. 20-3. An artificial heart (top right), equipment for replacing a knee joint (lower right), and a voice-activated computer (top left) all point the way to the future. How do you think each of these technologies could become even better?

SECTION 1
TechCHECK

1. What do futurists do?
2. According to futurists, how might telephones work in the future?
3. Who decides whether a technology is good or bad?
4. **Apply Your Knowledge.** Make a list of things you would like to see in the future.

INFOLINK

See Chapters 10, 17, and 18 for more information about robots.

Building for the Future

THINGS TO EXPLORE

- Explain how technology has changed the construction industry.
- Describe what buildings and cities of the future might look like.
- Design and construct tensile structures.

Techno Terms

encapsulated
modular unit
smart house

The changes in technology have affected the construction industry. Products constructed with new materials and technology are generally better than those made the old way. We have new lightweight, ultra-strong steel and reinforced plastic and glass materials with which to build things. Fig. 20-4. The processes we use today are also more efficient.

Computers have done much for the construction industry already. Structural designers and engineers can plan, revise, and even test new designs before any actual construction starts on a project. Using computers for these steps saves a company both time and money.

Building Design

What can you expect building in the future to be like? The construction industry must meet the needs of an ever-growing population. You will notice many changes in the design of structures as new materials, building techniques, and construction methods are developed.

Fig. 20-4. New materials are improving construction methods.

Fig. 20-5. Some homes are built in factories, then shipped to the site where they are put together. How can modular homes be improved in the future? Explain.

Better homes must be built at lower cost. Your new home might be mass-produced in **modular** (separate) **units**. Fig. 20-5. Assembly of modular units is simpler and takes less time and money than building a home piece by piece at the site.

More people will live in **smart houses**. In smart houses electronics and computers control energy use. You may have built-in computers that control lights, telephone calls, microwave ovens, televisions, and thermostats. You will probably speak to someone else in the house using special video and voice-activated devices. More people will work at home instead of going to the office, thanks to computers. That will also bring changes in building design.

Cities of the Future

Future cities may look very different from those of today. Maybe you will live in an **encapsulated** (completely enclosed) city. It would be completely air-conditioned and heated with purified air. Some researchers even see the possibility of making domed cities underwater.

Fig. 20-6. When ground space is limited, buildings must climb higher.

INFOLINK

See Chapter 14 for information about hovercrafts.

TechnoFact

SHAKE, RATTLE, AND ROLL How would you do laundry in space? Space washing machines can't use too much water and they can't rattle and shake or they'll ruin experiments. One company has been working on a washing machine that doesn't have an agitator and uses only 20 percent of the water most of them use. A special cleaning fluid replaces the water, and clothes still come out sparkling clean.

Building space in cities will have to be used more efficiently. Fig. 20-6. Buildings will need parking space for cars. You might drive a car that rises vertically (straight up) to a parking spot on the roof! Better yet, other transportation systems that carry more people and are powered by nonpolluting energy sources may be used. Two possibilities are the hovercraft and maglev trains. Maglev trains are discussed in Section 3.

Construction in the future will be an exciting challenge to meet people's needs on Earth and in space. Maybe you will be the engineer who comes up with a plan for these future projects!

SECTION 2
Tech CHECK

1. List three ways technology has changed the construction industry.
2. What is a smart house?
3. What does modular mean?
4. **Apply Your Knowledge.** Design a city playground of the future. Show where it fits in the overall city plan.

ACTION ACTIVITY

Building Tensile Structures

Be sure to fill out your TechNotes and place them in your portfolio.

Real World Connection

New materials and advancements in technology are making it possible to build structures in new ways and in new shapes. In a tensile structure, for example, most of the parts are held in tension with steel cables. The parts that hold up the structure are made of concrete or structural steel that resist compression forces. In this activity, you will design and build a tensile structure model.

Design Brief

Design and build a tensile structure. The roof must be made of materials that are strong in tension. The supporting structure must resist compression. Your design should include an interesting shape and a large unsupported floor area.

Materials/Equipment

- wood
- dowels
- rubberbands or string
- wood glue
- scroll or band saw
- power hand drill or drill press
- hot melt glue gun
- tape measure

SAFETY **FIRST**

- Follow the general safety rules on pages 42-43 for working in the technology lab and the specific rules provided by your teacher for tools and machines.
- Remember to wear eye protection.
- Take your time and work safely with machines and materials.

(Continued on next page)

ACTION ACTIVITY

Procedure

1. In this activity, you will work with a partner to build a tensile structure. Study the designs in Fig. A and brainstorm other possible designs with your partner.

2. Cut the base for your structure. For your model, cut wood dowel rods to the length your design requires.

3. Carefully measure and mark equally spaced holes for the compression parts of your structure. Drill holes in the base of your model to hold up the dowel rods. Cut a small notch in the end of each dowel to hold the string or rubberbands. Glue the dowels into the holes in the base.

4. Assemble the model using string or rubberbands as the tensile part of the roof. You might need to weave the string through the structure to make the design even stronger.

5. When the model is completed, gently press on the roof of your structure to test its strength. Fix any weak spots as needed. Real tensile structures complete the roof with a coating of concrete or a plastic fabric or membrane to keep out the weather. Can you think of a material that would simulate a finished roof on your model?

Evaluation

1. Would you like to live in a tensile structure? Explain.

2. What are the advantages of this type of construction? What are the disadvantages?

3. What materials are good for resisting compression forces?

4. What materials have good tensile strength?

5. **Going Beyond.** Design and build a pneumatic (air-filled) structure with the help of your teacher.

6. **Going Beyond.** Research pictures of tensile structures. Are there any in your area?

7. **Going Beyond.** Use a computer and CAD software to design other tensile structures.

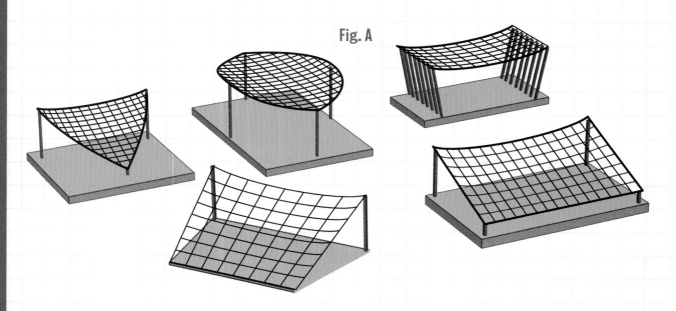

Fig. A

Transportation for the Future

THINGS TO EXPLORE

- Explain how future transportation will be different from today's transportation.
- Define *maglev*.
- Tell how a maglev train is different from other trains.
- Design and build a simulated maglev system.

TechnoTerms
guideway
levitate
maglev

Transportation in the future will be safer, more economical, and more convenient. Fig. 20-7 (page 462). You will also see more vehicles that use alternative energy sources and create less pollution. New frontiers in space and underwater exploration will also bring new challenges to transportation systems as we try to move people and products.

Maglev Trains

One form of mass land transportation being developed is the **maglev** (short for magnetic levitation) train. These trains **levitate**, or "float," above a **guideway** (track) on a magnetic field, instead of rolling on wheels on a steel track. In Germany, you can see one of these trains today.

Maglevs can travel much faster than conventional trains. Speeds of over 300 mph are not uncommon. Less maintenance is required because snow, ice, and heavy rain do not affect the track. Less pollution is created because maglev trains do not burn gasoline or diesel fuel.

INFOLINK

See Chapter 8 and 14 for more on maglev trains.

For the activity in this section, you will build a maglev system. A maglev train is shown on pages 316 and 466.

TechnoFact

WHAT, ME NERVOUS? Are you a nervous flyer? A company in Long Beach, California, is making special safety compartments that fit inside commercial airplanes. In an emergency, passengers inside the individual safety enclosures will be safe from shock, changes in air pressure, and temperatures up to **2,000°F.** A communication system will allow them to contact search-and-rescue teams.

Fig. 20-7. The three modes of transportation are land, air, and water. Examples are shown here. How do you think these vehicles will change in the future?

SECTION 3
TechCHECK

1. How will future transportation vehicles be better than those we have now?
2. What does *maglev* mean?
3. How is a maglev train different from other trains?
4. **Apply Your Knowledge.** Using the Internet and other resources, research the maglev train in Germany. Make a display of your findings.

ACTION ACTIVITY

Building a Maglev System

Real World Connection

As new transportation systems are developed, you will see more vehicles that use alternative energy sources and are less polluting to the environment. In this activity, you will build your own maglev system.

Design Brief

Design, build, and test a simulated maglev train system. Fig. A. Your train must levitate on a magnetic field. The train must move on a track made of nonmagnetic materials such as plastic, wood, or aluminum, using a safe propulsion method.

Materials/Equipment

- permanent magnets
- Styrofoam plastic foam
- acrylic plastic (Plexiglas)
- 2' x 2' x 1/2" plywood sections (1 per group)
- aluminum angle
- hot glue gun
- scroll or band saw
- hot wire cutter
- plastic strip heater

Fig. A

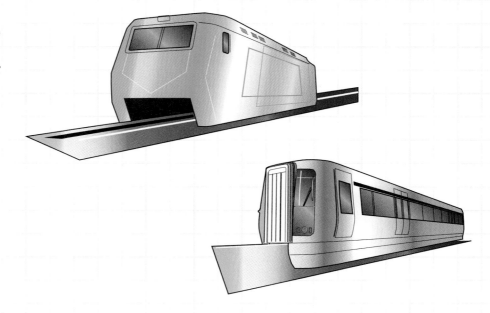

(Continued on next page)

ACTION ACTIVITY

Procedure

1. In this activity, the class will be divided into four groups. Each group will make a section of maglev train track and a simulated train. The sections will fit together to make a complete track. Fig. B. The different designs will be evaluated by the class as a whole, and the best design chosen.

2. Each group should brainstorm ideas for a train and track design. The track base should be made of nonmagnetic materials such as plastic, wood, or aluminum. Some designs you might consider are shown in Fig. C. Remember the steps in problem solving. Experimentation will be necessary to decide which design will work best for all four groups. Keep in mind that the opposite poles of magnets attract each other. Like poles repel each other.

(INFOLINK)

See Chapter 1 for the steps in problem solving.

3. Build a test section of your track 2 feet long. Test a train design on your track to see if it will levitate. Experiment with a propulsion system. The propulsion method you choose must be safe for indoor use.

SAFETY FIRST
- Follow the safety rules listed on pages 42-43 and the specific rules provided by your teacher for tools and machines.
- Keep magnets away from computer disks, videotapes, or any other magnetic information storage. Magnets will erase the information.
- Wear eye protection at all times while working around machines.

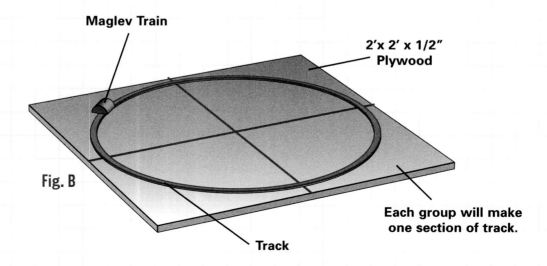

Maglev Train

2′x 2′ x 1/2″ Plywood

Each group will make one section of track.

Fig. B

Track

4. Present your idea for the track design to the class. Evaluate each design as if you were a city council member deciding on spending millions of tax dollars on a mass transit maglev train for your city. Give a maximum ten points for each of the following questions:

- Does the train work on curved sections of track?
- Is the track cost-effective? How many magnets per foot are required?
- Does the design appear to be long-lasting?

- Is the design as effective as possible, or does it need to be improved?
- Does the design team seem knowledgeable? Will they help other groups understand the design?

TechnoFact

WHO STARTED IT?
The idea of maglev technology was discussed by scientists as early as 1909. However, work on developing this technology did not begin until later. In the 1960s, two Brookhaven National Laboratory scientists, James Powell and Gordon Danby, pioneered research in maglev technology. They developed superconducting magnets for maglev trains. This was an important breakthrough because of the supermagnet's power to lift enormous weights.

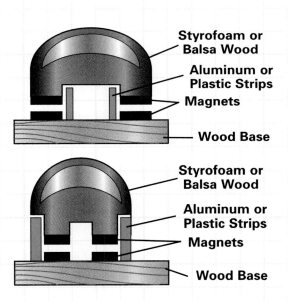

Styrofoam or Balsa Wood
Aluminum or Plastic Strips
Magnets
Wood Base

Styrofoam or Balsa Wood
Aluminum or Plastic Strips
Magnets
Wood Base

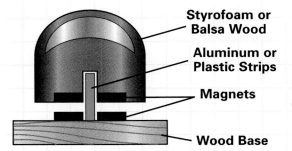

Styrofoam or Balsa Wood
Aluminum or Plastic Strips
Magnets
Wood Base

Fig. C. Here are three possible track designs for a maglev train.

(Continued on next page)

ACTION ACTIVITY

5. Total the points for each team to determine your choice. Compare the points for each group to determine the overall winning design. If your group design wasn't selected, you are still in the running for the best train design. Each group will design and make a train to run on the class track.

6. As a class, make drawings and a list of specifications so each group will be making track sections that will fit together.

7. Divide the tasks needed to complete your section of track and the simulated train. Make a list of all of the tasks. Assign the tasks to members of your group. Your task list might include

- Measure and cut the base for the track.
- Measure and cut the alignment rails according to the track specifications.
- Glue the magnets to the track with the proper polarity.
- Make the simulated train.

8. Construct each track section. Assemble the sections to make a complete track.

9. Give each group three chances to have its train make it around the track. Time each of the tests and take an average. The group with the fastest average time wins!

Evaluation

1. Why do you think the winning train design went the fastest?

2. How could your train design be improved?

3. What do you think it would be like to ride on a maglev train? Explain.

4. Going Beyond. Design a maglev train that uses superconducting magnets. How could the superconductors be kept cold?

5. Going Beyond. Design a way to make your track using electromagnets. Design a method of using electromagnets to propel the train.

6. Going Beyond. Maglev projects are under way in the United States, Germany, and Japan. Fig. D. Find out about these programs and prepare a report.

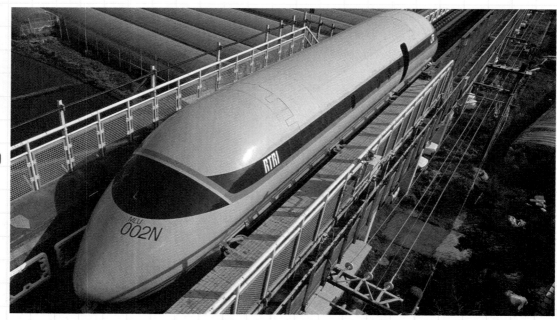

Fig. D

What Technology Means to You

THINGS TO EXPLORE

- Explain how your definition of technology has changed since the start of this course.
- Explain what kind of person can make technological change happen and why people need to keep up with change.
- Creatively express some of your own ideas about the future.

TechnoTerm
risk taker

N ow that you've studied technology, your idea of what it is may have changed. What is your definition of technology now? Did you remember to include problem-solving skills in your definition? You need to be able to think through problems associated with technology as well as those associated with any other subject.

Are There Limits to Technology?

The future of technology is wide open. The only limit is your imagination. As Albert Einstein said: "Imagination is more important than knowledge, for knowledge is limited to all we know and understand, while imagination embraces the entire world and all there ever will be to know and understand." Fig. 20-8.

Fig. 20-8. Albert Einstein was perhaps the greatest genius of the twentieth century. Read a biography of Einstein and share what you learned with the class.

Fig. 20-9. Prosthetic legs allow these men to compete in the Para-Olympics. What characteristics do you think materials used in artificial legs would need?

Many courageous people who are **risk takers**, who are curious about how things work, and who believe in trying have led the way in creating technologies that have changed your life. Technology must continue to fit the needs of the people and the environment. Fig. 20-9. That means you have to plan how new technologies can help make life better for people around the world.

COMMUNICATION CONNECTION

What's New?

Do you like to know what's happening in the world around you? Most people do. You might wonder how much damage was caused by an earthquake many miles away or when the next space launch will take place.

You might also like to find out about your personal interests, such as who won last night's basketball game or when a new music CD becomes available.

For centuries, newspapers were the main source of current information. In the 1950s, television became popular as a means for learning about the news. Because no printing process was needed, TV news reached people more quickly. Today, information is transmitted by satellite, and TV cameras allow us to watch important events while they are taking place.

You are on the road to becoming technologically literate. One thing is certain. Technology will keep changing quickly! It will be your responsibility to keep up with the future changes in technology so that you will always know how your world works.

INFOLINK

See Chapter 1 for a discussion of technology's rate of change.

SECTION 4
TechCHECK

1. What is your definition of technology?
2. Why do we need people with imagination and curiosity?
3. Why do you need to keep up with changes in technology?
4. **Apply Your Knowledge.** Ask your parents what technology they were excited about at your age. Explore how that technology may have changed since then. Share the information with your class.

In recent times, an even faster method for spreading the news has become popular—the Internet. From your living room, you can communicate almost instantly with other people halfway around the world. News of wars and natural disasters can reach you before a TV camera can be brought to the scene. You, too, can deliver news over the Internet.

You can send information about yourself and your community to any of the millions of Internet users with just the click of a mouse. How do you think the news will be delivered in the future? Will newspapers gradually disappear and all news reach us online?

ACTIVITY

With your teacher's help, create your own personalized news page website.

ACTION ACTIVITY

Predicting the Future

Be sure to fill out your TechNotes and place them in your portfolio.

Real World Connection

In this course, you have been learning about how technology is used today and how it might be used tomorrow. It's now time for you to express your ideas of what the future might bring. In this activity, you will work as a futurist to give us a glimpse of what you think the future holds. Fig. A.

Design Brief

Write a story, create a video, make a drawing, or choose some other means of demonstrating what you think the future holds.

Materials/Equipment

- pencil, paper
- computer with word-processing software
- camcorder (optional)

SAFETY FIRST

Follow the safety rules listed on pages 42-43 and the specific rules provided by your teacher for tools and machines.

Fig. A

Procedure

1. Work individually or in small groups to complete this activity. Remember the steps in problem solving? They will help you solve this problem and others you are faced with in the future.

2. Brainstorm ideas of what the future might be like 20, 50, or 100 years from now. Make a list of the things your group thinks of. Consider the following topics:
   - transportation
   - energy
   - production
   - communication
   - computers
   - space colonization
   - bio-related technology

3. Decide how your idea of the future will be presented to the class. Will it be a written story? A video production? A drawing?

4. Gather all the materials and information you think you need. Complete your project.

5. Present your ideas to the class.

Evaluation

1. Which idea of the future did you like the best? Why?

2. In your opinion, which idea seemed the least likely to occur? Explain.

3. How can an understanding of technology help you in the future? Explain.

4. **Going Beyond.** Make a time capsule with the items you think are most representative of the world today. With your teacher's permission, bury the capsule or place it somewhere to be opened in 10 or 15 years.

5. **Going Beyond.** Design a smart house and tell what devices you would have for communication, living comfort, security, and so on.

Fig. A (Cont'd)

20 REVIEW &

CHAPTER SUMMARY

SECTION 1
• Futurists are people who try to predict the future.

SECTION 2
• New materials, building techniques, and construction methods are changing the construction industry.
• Products made from new materials are lighter in weight and stronger.
• The need to use energy efficiently has definitely changed some parts of construction.

SECTION 3
• Transportation in the future will be safer, more economical, and more convenient.
• One form of mass land transportation being developed is the maglev train.

SECTION 4
• Technology must continue to fit the needs of the people and the environment.
• People who have imagination and are risk takers make change happen.

REVIEW QUESTIONS

1. What kinds of predictions does a futurist make? Give examples.

2. Why are modular units easier and cheaper to build?

3. How have communication devices changed in your lifetime?

4. What are the advantages of building maglev trains?

5. Explain in your own words what Einstein meant when he said, "Imagination is more important than knowledge."

CRITICAL THINKING

1. Research one of the topics below and make a model that illustrates your findings:
   • smart houses
   • domed cities
   • new ways to build houses
   • space colonies

2. What do you think will be the most important problem facing technology in the future? Explain.

3. List some disadvantages of maglev trains compared with conventional trains.

4. Which alternative source of power seems to be a good choice for the future? Explain.

5. Do you think bicycles are practical vehicles for transporting people to work in the United States in the future? Explain.

ACTIVITIES CHAPTER 20

CROSS-CURRICULAR EXTENSIONS

1. `COMMUNICATION` Think of a name and logo for your maglev train. Use computer graphics software to make your design. Glue your finished design to your train.

2. `SCIENCE` Make a display showing the magnetic fields created by electromagnets to explain how maglev trains are propelled.

3. `MATHEMATICS` Check with the census bureau to find the fastest growing city or area in the United States. Compute how many people per square mile live in that city or area. Use that information to explain why future cities will be designed with space limitations in mind.

EXPLORING CAREERS

Who knows what the future will hold? We may have great buildings constructed in the ocean and robots that can travel inside our bodies to perform surgery. Such changes will bring many changes in career opportunities.

Internet Storefront Operator

Internet commerce is giving consumers a greater choice of products. They can shop around the clock in the comfort of their homes. Internet "storefront" operators develop and market new services or offer existing products through virtual stores. Technical skills are helpful in maintaining and updating product information. Internet store operators must be willing to take risks in getting their businesses started.

Genetic Research Technician

Research involving DNA has moved to the front in medical technology. Genetic research technicians conduct experiments and collect large amounts of data, usually in a laboratory setting. Successful technicians are intelligent, curious people who work toward a goal.

ACTIVITY

Suppose you are the owner of an online virtual store. Develop a plan for marketing your site. Include information about the products and/or services that you will offer.

Table of Trigonometric Ratios
For use with the rocket activity in Chapter 17.

Angle Measure	sin	cos	tan	Angle Measure	sin	cos	tan
1°	0.0175	0.9998	0.0175	31°	0.5150	0.8572	0.6009
2°	0.0349	0.9994	0.0349	32°	0.5299	0.8480	0.6249
3°	0.0523	0.9986	0.0524	33°	0.5446	0.8387	0.6494
4°	0.0698	0.9976	0.0699	34°	0.5592	0.8290	0.6745
5°	0.0872	0.9962	0.0875	35°	0.5736	0.8192	0.7002
6°	0.1045	0.9945	0.1051	36°	0.5878	0.8090	0.7265
7°	0.1219	0.9925	0.1228	37°	0.6018	0.7986	0.7536
8°	0.1392	0.9903	0.1405	38°	0.6157	0.7880	0.7813
9°	0.1564	0.9877	0.1584	39°	0.6293	0.7771	0.8098
10°	0.1736	0.9848	0.1763	40°	0.6428	0.7660	0.8391
11°	0.1908	0.9816	0.1944	41°	0.6561	0.7547	0.8693
12°	0.2079	0.9681	0.2126	42°	0.6691	0.7431	0.9004
13°	0.2250	0.9744	0.2309	43°	0.6820	0.7314	0.9325
14°	0.2419	0.9703	0.2493	44°	0.6947	0.7193	0.9657
15°	0.2588	0.9659	0.2679	45°	0.7071	0.7071	1.0000
16°	0.2756	0.9613	0.2867	46°	0.7193	0.6947	1.0355
17°	0.2924	0.9563	0.3057	47°	0.7314	0.6820	1.0724
18°	0.3090	0.9511	0.3249	48°	0.7431	0.6691	1.1106
19°	0.3256	0.9455	0.3443	49°	0.7547	0.6561	1.1504
20°	0.3420	0.9397	0.3640	50°	0.7660	0.6428	1.1918
21°	0.3584	0.9336	0.3839	51°	0.7771	0.6293	1.2349
22°	0.3746	0.9272	0.4040	52°	0.7880	0.6157	1.2799
23°	0.3907	0.9205	0.4245	53°	0.7986	0.6018	1.3270
24°	0.4067	0.9135	0.4452	54°	0.8090	0.5878	1.3764
25°	0.4226	0.9063	0.4663	55°	0.8192	0.5736	1.4281
26°	0.4384	0.8968	0.4877	56°	0.8290	0.5592	1.4826
27°	0.4540	0.8910	0.5095	57°	0.8387	0.5446	1.5399
28°	0.4695	0.8829	0.5317	58°	0.8480	0.5299	1.6003
29°	0.4848	0.8746	0.5543	59°	0.8572	0.5150	1.6643
30°	0.5000	0.8660	0.5774	60°	0.8660	0.5000	1.7321

(Continued on next page)

Table of Trigonometric Ratios (Cont'd.)

Angle Measure	sin	cos	tan	Angle Measure	sin	cos	tan
61°	0.8746	0.4848	1.8040	76°	0.9703	0.2419	4.0108
62°	0.8829	0.4695	1.8807	77°	0.9744	0.2250	4.3315
63°	0.8910	0.4540	1.9626	78°	0.9781	0.2079	4.7046
64°	0.8988	0.4384	2.0503	79°	0.9816	0.1908	5.1446
65°	0.9063	0.4226	2.1445	80°	0.9848	0.1736	5.6713
66°	0.9135	0.4067	2.2460	81°	0.9877	0.1564	6.3138
67°	0.9205	0.3907	2.3559	82°	0.9903	0.1392	7.1154
68°	0.9272	0.3746	2.4751	83°	0.9925	0.1219	8.1443
69°	0.9336	0.3584	2.6051	84°	0.9945	0.1045	9.5144
70°	0.9397	0.3420	2.7475	85°	0.9962	0.0872	11.4301
71°	0.9455	0.3256	2.9042	86°	0.9976	0.0698	14.3007
72°	0.9511	0.3090	3.0777	87°	0.9986	0.0523	19.0811
73°	0.9563	0.2924	3.2709	88°	0.9994	0.0349	28.6363
74°	0.9613	0.2756	3.4874	89°	9.9998	0.0175	57.2900
75°	0.9659	0.2588	3.7321				

GLOSSARY

A

acceptance sampling A quality assurance method used in manufacturing when it is not always possible to check each product. A few samples are inspected, and if they meet the standards, the whole batch is approved. If the samples don't pass inspection, the batch is rejected. (Ch. 11)

acrylic A thermoplastic that is clearer than glass. (Ch. 9)

ad lib To make up dialogue; done by performers while a TV or radio program is taking place. (Ch. 16)

administration Department in a company that carries out the decisions made by management. (Ch. 11)

aerodynamic Streamlined. (Ch. 14)

aerodynamics The branch of science having to do with forces created by air. (Ch. 14)

air bag Car safety device that operates automatically in case of a collision. A large bag instantly inflates to protect the driver and passenger. (Ch. 14)

air compressor A machine that squeezes, or compresses, air for use by a pneumatic tool. (Ch. 8)

air cushion vehicle (ACV) Vehicle that rides on a cushion of air created by high-speed fans. (Ch. 14)

airfoil An object, such as an airplane wing, designed to create a certain effect when air moves over it. (Ch. 14)

alloy A mixture of two or more metals or a metal and another material. (Ch. 18)

alternative energy Energy source other than fossil fuels. (Ch. 13)

aluminum A very strong, lightweight metal that conducts heat and electricity easily. (Ch. 9)

amperage The amount of electrons flowing through an electrical circuit. A measure of electrical power. (Ch. 9)

ampere (amp) A unit of electrical power. (Ch. 9)

amplitude modulation (AM) (AM-plih-tood mahd-joo-LAY-shun) Altering the size of an electromagnetic signal to enable it to carry a message. (Ch. 16)

AMTRAK The American Travel Track system that provides all the long-distance rail service in the United States. (Ch. 14)

angle of attack Angle at which an airplane wing cuts through the air; the angle between the chord and the direction of the wind. (Ch. 14)

annual Happening once a year, such as an annual meeting. (Ch. 11)

anthropometric data (an-throh-poh-MEH-trik) Size information collected from many people. (Ch. 9)

apogee (AP-uh-jee) The maximum height of a rocket's flight. (Ch. 17)

application In computer technology, software used for a specific purpose. Also called a *program*. (Ch. 4)

aquaculture Raising fish and food plants in water. (Ch. 19)

assembly line A manufacturing system in which products move past a line of workers, and each worker performs a specific step in making the product. (Ch. 10)

atom The smallest part of an element. (Ch. 2)

audio Things that are heard. (Ch. 16)

automated storage and retrieval system (AS/RS) In manufacturing, a computer-controlled crane that travels between pallet racks and automatically loads and unloads them. (Ch. 10)

automated transit system System of driverless vehicles often used at airports and in remote parking areas. (Ch. 14)

automatic guided vehicle system (AGVS) A series of computer-controlled, driverless carts that follow wire paths built into the floor of manufacturing plants. (Ch. 10)

automation The automatic control of a process by a machine. (Ch. 10)

B

barge A boat with a flat bottom and blunt ends, used to haul cargo on inland waterways. (Ch. 14)

belt drive Mechanical device used when forces have to be transmitted over a longer distance than gears can handle easily. (Ch. 8)

Bernoulli's Principle A basic principle used in designing airplanes which states that as the speed of a fluid increases, its pressure decreases. (Ch. 14)

bill of materials A list of the materials needed for a manufactured product that includes their size and quantity. (Ch. 8)

bimetallic Made of two different metals. (Ch. 8)

binary counting system A counting system in which only two digits, 0 and 1, are used. This system is used in computers and other electronic devices to transmit signals that are either "off" or "on." (Ch. 4)

biodegradable Able to be broken down by natural processes, such as the effects of the weather. (Ch. 3)

bioethics A set of rules or procedures dealing with the ethical implications of bio-related technology. (Ch. 19)

biofuel Fuel made from biomass, which is plant or animal matter. (Ch. 19)

biomass Living or dead plant or animal matter. (Ch. 13)

biomaterials Human-made materials designed to be used inside the body. (Ch. 19)

bionics The making of artificial parts for the human body; also called *biomechanics*. (Ch. 19)

bioprocessing The use of microorganisms to change materials from one form to another. (Ch. 19)

bio-related technology Any technology that has to do with living things. (Ch. 19)

bit The smallest unit of information used by computers. Acronym for *binary digit*. (Ch. 4)

blimp Lighter-than-air vehicle filled with helium gas and used for advertising or photography and sometimes freight. (Ch. 14)

board of directors Officials who set a company's policies and determine the main company goals. (Ch. 11)

boroscope An optical inspection device that goes inside a machine so specialists can look around. (Ch. 11)

brainstorm To offer many ideas for solving a problem. The ideas are not judged as either good or bad until a later time. (Ch. 1)

break-even point The point at which a company has earned enough money selling a product to have paid for the product's manufacture. Money earned beyond that is profit. (Ch. 11)

Bronze Age (3000 B.C. to 1200 B.C.) Historical period during which people learned how to mix copper with tin to make a stronger metal called *bronze*. (Ch. 1)

building codes Laws that set the minimum requirements that a building must meet. (Ch. 12)

burn-in test A quality assurance test in which a manufactured product is allowed to run for several hours; done mostly on electronic products like computers. (Ch. 11)

byte (BITE) Eight bits (binary digits). (Ch. 4)

C

cam A mechanical part that changes rotational (turning) motion into reciprocating (up-and-down) motion. (Ch. 8)

capacitor Electronic device that temporarily holds an electrical charge. (Ch. 8)

capital Money. (Ch. 11)

carpal tunnel syndrome A painful condition caused when nerves in the wrist and arm are damaged by repeating certain movements over and over. (Ch. 9)

catalyst A special chemical added to make a chemical reaction go faster or to change the desired result. (Ch. 8)

cell 1. In a database, a space created by the intersection of a column and a row. (Ch. 4) 2. In animation, one frame, or scene. (Ch. 5)

cellular phone Phone containing a small radio transmitter that sends a radio signal to the base of the telephone. The base is connected by wire to the telephone lines. (Ch. 15)

central processing unit (CPU) A special kind of integrated circuit that forms the "brain" of a computer. (Ch. 4)

chain drive Mechanical device used when forces have to be transmitted over a longer distance than gears can handle easily. (Ch. 8)

character generator Special computer software used to create text for a TV program. (Ch. 16)

chemical technology Technology in which chemicals are used to produce products, such as plastics, paint, and fertilizers. (Ch. 19)

cholesterol A fat-like substance found in the blood of animals. (Ch. 19)

chord An imaginary straight line drawn from the leading edge of an airplane wing to the tail. (Ch. 14)

circuit The complete path along which electrons flow. (Ch. 8)

claymation (clay-MAY-shun) A specialized form of animation in which clay figures seem to come alive. Artists adjust the scenes and characters by small increments, and a computer is used to store and quickly play back the individual pictures. (Ch. 5)

client The person a job is done for. (Ch. 12)

clip art Pre-drawn images used to create design layouts. (Ch. 5)

clone An organism that is the genetic duplicate of another. (Ch. 19)

closed system A system that is self-contained and that receives no inputs from outside. The Earth, for example, is a closed system and gets no air or water from outside sources. (Ch. 7)

commission A certain percentage of the purchase price that a salesperson makes on the product sold. (Ch. 11)

communication The process of exchanging information either by sending it or by receiving it. (Ch. 15)

communication satellite A satellite that serves as a relay station for television and radio signals. (Ch. 15)

community A group of people, usually living in the same area, who have common interests. Also refers to the area itself. (Ch. 12)

company An organized group of people doing business. (Ch. 11)

component 1. One part of a larger device or system. 2. An electronic part. (Ch. 8)

composite Material created by combining two or more different materials in order to obtain new and often better properties. (Chs. 9 & 14)

compression A force that squeezes an object. (Ch. 12)

compression strength The ability to resist being squeezed or smashed. (Ch. 9)

computer-aided design (CAD) Design done on a computer using special software. (Ch. 5)

computer-aided manufacturing (CAM) Manufacturing done using CAD drawings that are sent directly from the computer to automated machines that make the parts. (Ch. 5)

conductor A material that lets electricity pass through it. (Ch. 8)

consensus Agreement. (Ch. 2)

conserved Saved. (Chs. 7 and 13)

constructed Built. (Ch. 12)

construction The act of building structures, such as houses, bridges, and airports. (Ch. 12)

consumers People who buy products or services. (Ch. 3)

control In experiments, the part for which all conditions are kept the same. (Ch. 3)

conventional Common; traditional, as in conventional energy sources. (Ch. 13)

conveyor A device with wheels, rollers, or a belt that moves items from one place to another. (Ch. 10)

coordinates Numbers giving the longitude and latitude of a particular location. (Ch. 15)

corporation A company organized and owned by stockholders. (Ch. 11)

corridor A numbered route used by airplanes in a three-dimensional traffic network. (Ch. 14)

cost-effective Term used to describe something that is economical to do or use. (Ch. 10)

crane In manufacturing, a hoist that moves in a limited area. (Ch. 10)

crop In photography, a term used to describe the removal of unwanted background. (Ch. 5)

crude oil Oil as it comes from the ground before processing. (Ch. 8)

crystal A solid whose atoms are arranged in a repeating pattern. (Ch. 19)

cue card In television broadcasting, a large sheet containing dialogue written with large letters that the talent can read at a distance in case they forget their lines. (Ch. 16)

custom manufacturing Making one item at a time. (Ch. 10)

D

data Information. (Ch. 4)

database A computer application that lets the user find and organize information quickly and easily. (Ch. 4)

dead air Unplanned, momentary silence during a radio program. (Ch. 16)

dead load In construction, a static force that includes the entire weight of the structure itself—the beams, floors, walls, insulation materials, columns, and ceilings of a building, or the deck of a bridge, and so on. (Ch. 12)

decompose To break down into simpler parts; to decay. (Ch. 7)

defect Flaw. (Ch. 11)

design brief A detailed definition of a

problem when presented to a designer for solving. (Ch. 1)

design patent A patent that protects the invention of a new design for an item that will be manufactured. (Ch. 6)

desktop publishing (**DTP**) Organizing text and graphics into publishable form using special computer software. (Ch. 4)

destructive testing Testing a material until it breaks or is destroyed. (Ch. 9)

detail drawing In construction, a drawing that gives more information about a particular part of a structure. (Ch. 12)

development plan A plan for a community that includes the type of construction, where it will be located, and the laws that control what kinds of construction can take place. (Ch. 12)

dialogue In a script, written conversation between two or more people. (Ch. 16)

digit (DIHJ-it) A number. (Ch. 5)

digital image processor Computer software that makes it possible to combine elements of two or more photos or graphics. (Ch. 15)

digitizing (DIHJ-ih-tyz-ing) To change into *digits,* or numbers, used in computer code. (Ch. 5)

dimensions Sizes. (Ch. 11)

diode An electronic device through which electrons flow in only one direction. (Ch. 8)

direct sales Selling a product directly to the customer. (Ch. 11)

dirigible (dihr-IHJ-ih-bul) Lighter-than-air vehicle filled with hydrogen and used to carry passengers and freight. (Ch. 14)

distillation A chemical process used to change salt water into fresh water. (Ch. 19)

distribution In manufacturing, getting goods to the buyers. (Ch. 11)

dividend A payment given to stockholders of a company. (Ch. 11)

DNA Deoxyribonucleic acid; the chemical code of instructions found in the nucleus of every living cell. (Ch. 19)

dominant In genetics, a term used to describe genes that produce traits that show. (Ch. 19)

downlink A transmission from a satellite to an Earth station. (Ch. 15)

download To receive a document electronically over phone lines using a computer modem or direct Internet connection. (Ch. 5)

drag The resistance of an object to the flow of a fluid, such as air or water. (Ch. 14)

drawing to scale Drawing an object larger or smaller than it really is while keeping its parts in the correct proportions. (Ch. 9)

dry cell A battery that uses a paste of chemicals. (Ch. 8)

dynamic load In construction, a force acting on a structure that changes rapidly, like vibrations from people walking along a floor. (Ch. 12)

E

ecology (ee-KAHL-uh-jee) The study of how things interact with the environment. (Ch. 3)

edit To check and correct. (Ch. 15)

electric generator A machine that converts mechanical energy to electrical energy. (Ch. 13)

electricity The flow of electrons (small, negatively charged parts of an atom) through a material. (Ch. 8)

electrolyte A chemical used to produce electrical voltage. (Ch. 8)

electronic communication system A communication system, such as radio or TV, that uses electronic or electromagnetic signals to carry messages through cables or through the air. (Ch. 15)

electronic noise Interference, such as static, caused during transmission of an electronic signal. (Ch. 15)

electronics The part of technology concerned with the movement of electrons through conductors, insulators, semiconductors, and superconductors. (Ch. 8)

elevation drawing In construction, a view of a structure from the front, the sides, or the back. (Ch. 12)

e-mail Messages sent electronically from one computer to another using a modem and telephone lines. (Ch. 15)

encapsulated Completely enclosed. (Ch. 20)

end effector A device at the working end of a robot arm that helps the robot perform some task, such as gripping. (Ch. 10)

energy The ability to do work. (Ch. 13)

energy-intensive Using large amounts of energy. (Ch. 13)

entrepreneur Someone who starts a company. (Ch. 11)

ergonomics The study of how the human body relates to things around it, such as furniture and clothing. (Ch. 9)

escape velocity The speed at which a vehicle must travel in order to escape Earth's gravity; 25,000 miles per hour. (Ch. 17)

estimate To figure closely but not exactly. (Ch. 9)

ethanol Grain alcohol that is used as a fuel. (Ch. 19)

ethics Standards of conduct. (Ch. 19)

evaluate Judge. (Ch. 3)

expansion joint A line (as in a concrete sidewalk) designed to allow for expansion when sun heats the concrete. (Ch. 12)

exponential rate of change (ex-poh-NEHN-shul) A rate of change in which a quantity is multiplied by itself. (Ch. 1)

extravehicular mobility units (EMU) Space suits. (Ch. 9)

F

fad Something that is temporarily popular. (Ch. 6)

fastening Holding materials together, such as with nuts and bolts, welding, or adhesives. (Ch. 7)

fatigue strength The ability to resist breakage after being bent back and forth. (Ch. 9)

fax (facsimile) machine Device that quickly sends text or graphics electronically over telephone lines. (Ch. 15)

feedback 1. Information about the output of a system. 2. A loud hum produced by sound feeding back into a microphone. (Ch. 8)

fermentation Chemical process in which microorganisms turn grains into alcohols. (Ch. 19)

ferrous Term used to describe a metal containing iron. (Ch. 9)

fiberglass A material made of very thin glass strands glued together with a liquid plastic. (Ch. 9)

fiber-optic cable A cable made out of many thin strands of glass fibers that can carry light for long distances without loss of power. (Ch. 15)

fiber-optic system System of cables that uses light to carry information. (Ch. 15)

field In a database, one part of the data, such as a person's first name. (Ch. 4)

fifth wheel A large, disk-shaped hitch that hooks a semitrailer to the tractor. (Ch. 14)

file In a database, a group of records. (Ch. 4)

finishing Adding a coating, such as paint, varnish, or plastic, to an item. (Ch. 7)

floor plan An accurate, detailed, scale drawing of one floor of a structure. (Ch. 12)

flowchart A type of graph containing symbols that is used to illustrate the steps in a process. (Ch. 3)

fluid A substance that flows. (Ch. 8)

flywheel A metal wheel used in engines that is heavy enough to keep spinning once it is set into motion. (Ch. 8)

font Style of type characters (letters and numbers). (Ch. 4)

footprint In communications, the area over which satellite signals can be picked up. (Ch. 15)

force The effort applied to push or pull something. (Ch. 8)

form A mold used to shape a material, such as concrete, while it hardens. (Ch. 7)

format Way in which something is set up or presented. (Ch. 20)

forming In manufacturing, changing the shape of a material without adding anything or taking anything away. (Ch. 7)

fossil fuel Fuel produced deep in the Earth over time from decaying animals and plants; coal, oil, and natural gas. (Ch. 7)

frame Still picture in live TV or on videotape. (Ch. 5)

framegrabs In claymation, the individual pictures that were taken. When quickly played back, these give the impression of movement. (Ch. 5)

free fall Falling through space. Astronauts are in free fall because they can't rest against the floor of a spacecraft; the floor is falling at the same speed as they are. (Ch. 18)

frequency (FREE-kwin-see) 1. The number of cycles per second at which a radio or TV broadcast transmission travels. (Ch. 16) 2. Vibration speed. (Ch. 2)

frequency modulation (FM) (FREE-kwin-see mahd-joo-LAY-shun) Altering the frequency of an electromagnetic signal to enable it to carry a message. (Ch. 16)

fulcrum The pivot point of a lever. (Ch. 8)

futurist Person who tries to predict the future. (Ch. 20)

G

gasohol A fuel made from a mixture of ethanol and gasoline. (Ch. 19)

gauge Measuring tool used for quality inspections. (Ch. 11)

gear Mechanical device used to transmit forces from one part to another. Gears can be used to change the speed or direction of spinning parts. (Ch. 8)

gene A basic unit of heredity that carries certain information about the development of a plant or animal. (Ch. 19)

genetic disease Disease passed from parents to children through their genes. (Ch. 19)

genetic engineering The process of changing the gene structure of a plant or animal to improve it for human use; also, the designing of new life forms. (Ch. 19)

genetics The study of how traits are passed on from one generation of living organisms to the next. (Ch. 19)

genotype In biotechnology, the classification of an organism based on the combination of its genes. (Ch. 19)

geostationary Term used to describe an orbiting object that stays in the same place relative to the Earth at all times. (Ch. 15)

geosynchronous Term used to describe the orbit of an object that is synchronized with the turning of the Earth. (Ch. 15)

geothermal energy Heat from beneath the Earth's crust. (Ch. 13)

global positioning system (GPS) A system of satellites and electronic devices that can relay an object's location anywhere on Earth in terms of latitude, longitude, and altitude. (Ch. 15)

graphics Illustrations or design elements. (Ch. 5)

grapple point "Handle" built into a space satellite that allows it to be grabbed by a robot arm. (Ch. 18)

gravity The force of attraction between objects. (Ch. 17)

greenhouse effect A natural buildup of heat on the Earth's surface caused by atmospheric gases, mainly carbon dioxide. (Ch. 9)

growth hormones Naturally occurring chemicals in a living organism that cause it to grow. (Ch. 19)

guideway Track. (Ch. 20)

H

hardness The ability to resist dents. (Ch. 9)

hardware The parts of a computer system, exclusive of the software. (Ch. 4)

hardwoods Woods that come from trees that have broad leaves, such as walnut and maple. (Ch. 9)

heavier-than-air vehicle Vehicle that must supply power to fly. (Ch. 14)

hertz One cycle per second. Unit used to measure radio or TV signal transmission. (Ch. 16)

hoist Mechanical device used to lift heavy loads. (Ch. 10)

hovercraft Vehicle that rides on a cushion of air created by high-speed fans. (Ch. 14)

Hubble Space Telescope A large telescope in orbit around the Earth. (Ch. 17)

human resources A department in a company responsible for hiring, training, and rewarding workers. Also called *personnel*. (Ch. 11)

hydration (hi-DRAY-shun) A chemical reaction that makes concrete harden. (Ch. 7)

hydraulic pump (hi-DRAW-lik) A pump used to send a fluid through a hydraulic system. (Ch. 8)

hydraulic system (hi-DRAW-lik) A fluid system that operates using a liquid, usually oil. Fluid systems apply pressure on the fluid to do work. (Ch. 8)

hydroelectric power Power created when water stored behind a dam passes through a turbine. (Ch. 13)

hydrofoil A passenger ship that moves above the surface of the water. (Ch. 14)

hydroponics (hi-droh-PAH-niks) Growing plants without soil. (Ch. 19)

hypothesis (hi-PAH-thuh-sis) An educated guess about the solution to a problem. (Ch. 1)

I

idea bank All the ideas presented for solving a problem. (Ch. 1)

impact Effect. (Ch. 3)

in-betweening An animation process in which the basic cells are drawn on the computer by an artist, and then software is used to automatically create the cells needed in between. Also called *tweening*. (Ch. 5)

in-camera editing Producing a video program by planning each shot in sequence and taping shots in order. (Ch. 16)

independent product testing Testing done by government agencies or companies not involved in manufacturing the tested products. (Ch. 3)

Industrial Revolution (1750 to 1900) Historical period during which many inventions brought changes that affected all of society. Factories were set up that could produce goods cheaper and faster. (Chs. 1 & 10)

industrial robot Robot used in industry. (Ch. 10)

innovate To create a new idea, device, or way of doing something. (Ch. 6)

input Something that is put into a system. (Ch. 8)

insulated Covered by a material to prevent the transfer of heat or cold. (Ch. 13)

insulator A material that does not allow electrons to flow easily from one atom to another. (Ch. 8)

integrated Put together, as in integrated subject areas. (Ch. 2)

integrated circuit A small "chip" of silicon that contains many electronic circuits. Also called a *microchip*. (Ch. 3)

interference In communications, anything that gets in the way of a message being understood. (Ch. 15)

International Space Station A space station to be built by sixteen cooperating nations, including the United States, Russia, and Japan. (Ch. 18)

International System of Units (**SI**) Parts of the metric system used internationally. (Ch. 9)

Internet The world's biggest computer network. (Ch. 15)

inventory Items in storage. (Ch. 11)

Iron Age (1200 B.C. to A.D. 500) Historical period during which people made tools from iron. (Ch. 1)

isometric drawing Pictorial drawing that shows an object from an edge and tilted slightly toward the viewer. Lines that show the width and depth of the object are drawn at 30° angles from the horizontal. (Ch. 15)

J

joystick A device that changes hand movements into actions on a computer screen. (Ch. 4)

just-in-time manufacturing (JIT) Cost-saving manufacturing method in which materials and parts arrive at the factory just in time for production. When the product is finished, it is not stored but is immediately shipped to the customer. Also called *synchronized production*. (Ch. 11)

K

kinetic energy Energy of motion. (Ch. 13)

knowledge base All the facts known about something. (Ch. 1)

L

landfill Area set aside for burying discarded items; garbage dump. (Ch. 3)

laser A high-intensity beam of light. Acronym for light amplification by stimulated emission of radiation. (Ch. 15)

Law of Dominance Mendel's law stating that a dominant gene will always mask, or hide, a recessive gene when they occur together. (Ch. 19)

leading edge The point on an airplane wing that is farthest forward; the spot where wind hits the wing first. (Ch. 14)

lever A bar-like device used to help move heavy loads. (Ch. 8)

levitate Float. (Ch. 20)

lift A force created as air flows over an airplane wing. It allows the plane to overcome gravity. (Ch. 14)

light-emitting diode (LED) An electronic device that gives off light when voltage is applied to it. (Ch. 15)

lighter-than-air vehicle Vehicle that floats in air. (Ch. 14)

linear editing Fast-forwarding or rewinding a videotape to find a certain segment and copying it onto another tape. (Ch. 16)

line of golden proportion. An invisible line created in a drawing by dividing the height of the drawing paper into three equal sections. The line one-third down from the top is the line of golden proportion. (Ch. 15)

linkage A kind of lever used in a machine. Also called a *crank*. (Ch. 8)

live load In construction, a static force that a structure supports as it is used under normal weather conditions. Live loads include people, furniture, equipment, or stored materials. (Ch. 12)

loads In construction, forces working on structures. (Ch. 12)

local area network (LAN) Many computers linked together in a nearby area. (Ch. 15)

logo Company symbol. (Chs. 5 & 11)

M

machining Changing the shape of materials by cutting away pieces, or "chips." (Ch. 7)

maglev Train that floats on a magnetic field above a track instead of rolling on wheels; short for *magnetic levitation*. (Ch. 20)

management People who run a company. (Ch. 11)

manually By hand, as in the manual operation of equipment. (Ch. 10)

manufactured Made by humans, usually in factories. (Ch. 7)

manufacturing Making products. (Ch. 10)

marketing plan Strategy for advertising and selling a product. (Ch. 11)

Mars Pathfinder An unstaffed spacecraft that landed on Mars. (Ch. 17)

mass production Manufacturing method that uses assembly lines. (Ch. 10)

materials handling Moving and storing materials. (Ch. 10)

materials testing Testing of materials in order to determine their characteristics before they are used to make products. (Ch. 3)

Mendel's Law of Dominance A principle of genetics that says a dominant gene will always mask, or hide, a recessive gene when they occur together. (Ch. 19)

methanol A type of alcohol that can be made from fermented wood-product waste. (Ch. 19)

metric system A base-10 system of measurement. (Ch. 9)

microchip An integrated circuit (IC); a piece of silicon with hundreds of tiny electronic parts linked together to form circuits. (Ch. 4)

microgravity Reduced gravity in comparison to that on Earth. (Ch. 18)

microwave A short electromagnetic wave. Usually refers to waves between 1 millimeter and 1 meter long. (Ch. 15)

MIDI (musical instrument digital interface) A device that lets music be input into a computer using an electronic musical keyboard. (Ch. 4)

Mir 1 A space station launched by the Soviet Union in 1986. (Ch. 18)

mode Way or method. (Ch. 14)

model A representation or pattern (usually a miniature) of something to be made. Also, a description used to help people understand something that cannot be seen. (Ch. 5)

modem A device that lets a computer communicate with another computer over a telephone line. (Chs. 4 & 15)

modular unit A section of something, such as a house, that can be put together with other units to create the complete structure. (Ch. 20)

mother board The main circuit board that contains the CPU and connectors to other parts of a computer. (Ch. 4)

N

NASA National Aeronautics and Space Administration. (Ch. 17)

negative A reverse image produced on photographic film. (Ch. 5)

network Several computers connected together. (Ch. 15)

neutral buoyancy The ability of an object to float in one place without rising or sinking. (Ch. 18)

Newton's Laws of Motion Three scientific laws that describe physical motion. (Ch. 17)

Nichrome wire. Wire made from an alloy of nickel, chromium, and iron. Nichrome is often used in the heating elements of electrical appliances. (Ch. 8)

non-linear editing Jumping by means of a computer to different parts of a videotape without having to fast-forward or rewind. (Ch. 16)

nonrenewable Term used to describe resources, such as oil and gas, that cannot be replaced. (Ch. 7)

nuclear energy The energy found in atoms. (Ch. 13)

nuclear fission A kind of nuclear reaction in which atoms of materials such as uranium are split, releasing huge amounts of heat energy. (Chs. 13 & 20)

nuclear fusion A kind of nuclear reaction in which hydrogen atoms are fused, or joined, releasing large amounts of heat energy. (Ch. 20)

O

oblique drawing (oh-BLEEK) A pictorial drawing that shows one surface of an object straight on. Two other surfaces are shown at an angle. (Ch. 15)

ohm A unit of measurement of electrical resistance. (Ch. 9)

Ohm's Law Amps = volts ÷ ohms. (Ch. 9)

orthographic drawing Drawing in which several views (the top, front, and side) of an object are shown as if the viewer were looking straight at each one. (Ch. 15)

orthographic projection Drawing method in which several views of the object (usually top, front, and side) are shown at right angles to one another. (Ch. 12)

OSHA Occupational Safety and Health Administration, a part of the U.S. Department of Labor. OSHA inspects workplaces to make sure they meet safety standards. (Ch. 7)

output What comes out of a system after processing has taken place. (Ch. 8)

P

palmtop A pocket-sized computer. (Ch. 3)

parallel circuit An electrical circuit arranged so that other components continue to work even if one component fails. (Ch. 8)

parameters (puh-RAM-ih-ters) Specific details of a design problem. (Ch. 1)

partnership A business owned by two or more people. (Ch. 11)

passive car safety system Safety device that doesn't require the passengers in a car to do anything. (Ch. 14)

patent A special government license that protects an invention from being copied without permission for 20 years. (Ch. 6)

peripherals (purr-IF-ur-uhls) Useful devices, such as printers, that can be connected to a computer system. (Ch. 4)

persistence of vision An effect created when still pictures are changed so quickly that the viewer's brain thinks the viewer is watching moving pictures. Persistence of vision is what makes animation possible. (Ch. 5)

perspective drawing A realistic pictorial drawing in which parts of objects that are receding into the distance look smaller. (Ch. 15)

pesticide management The use of chemicals to control insects and other pests. (Ch. 19)

petrochemical A product produced from oil, or petroleum. (Ch. 8)

petroleum Oil. (Ch. 8)

phenotype (FEE-noh-tipe) In biotechnology, the classification of an organism based on its appearance. (Ch. 19)

photovoltaic cell (foh-toh-vohl-TAY-ik) A device that makes electricity directly from sunlight; also called a *solar cell*. (Ch. 13)

pick-and-place maneuver Motion in which a robot picks up a part and moves it somewhere else. (Ch. 10)

pictorial drawing Drawing that shows an object in three dimensions. (Ch. 12)

piggybacked Term used to describe semitrailers carried on railroad flatcars. (Ch. 14)

piston A disk or cylinder that is moved back and forth inside a hollow cylinder by the pressure of a fluid. (Ch. 8)

pixel Tiny rectangle of light used to create the image on a TV or computer screen; acronym for *picture element*. (Ch. 5)

plant patent A type of patent that protects the invention or discovery of new varieties of plants. (Ch. 6)

pneumatic system (new-MAT-ik) A fluid system that operates using a gas, usually compressed air. Fluid systems apply pressure on the fluid to do work. (Ch. 8)

polycarbonate A clear, strong thermoplastic material that can be bent or shaped with heat. (Ch. 9)

potential energy (poh-TEN-shul) Energy at rest waiting to do work. (Ch. 13)

power The amount of work done in a certain time period; the *rate* of doing work. (Ch. 13)

precipitation Rain, snow, or sleet. (Ch. 13)

precision Accuracy; the degree to which something, such as a measurement, conforms to a standard. (Ch. 9)

Pre-Industrial Age (A.D. 500 to 1750) Time period during which products were still made by craftspeople using their own tools in their home workshops. During the second part of this period, technology and science began to bring changes. (Ch. 1)

preliminary First, as in preliminary sketches. (Ch. 12)

pre-production The work that goes into planning and rehearsing a radio or TV program. (Ch. 16)

print Image produced by shining light through a photographic film negative onto light-sensitive paper. (Ch. 5)

printer A machine that outputs text (words) and graphics (pictures) on paper. (Ch. 4)

problem-solving strategy A step-by-step method for solving problems. (Ch. 1)

problem statement The definition of a problem. (Ch. 1)

process 1. A method by which raw materials are turned into manufactured products. (Ch. 7) 2. What is done with the inputs to a system. (Ch. 8)

profit Money left over after all the bills a company owes are paid. (Ch. 7)

programmed Computer controlled. (Ch. 10)

proprietorship A business owned by just one person. (Ch. 11)

prosthetic devices Artificial body parts. (Ch. 9)

prototype A full-scale model of a product to be manufactured. The prototype is usually a working model. (Ch. 6)

public service announcement (PSA) Special announcement on TV or radio that informs the public about something they should be aware of. (Ch. 16)

Punnett square In biotechnology, a grid used to show possible gene combinations in offspring. (Ch. 19)

Q

quality assurance Method of assuring that a product is manufactured according to specific plans and standards. Also called *quality control (QC)*. (Ch. 11)

quality control Method of assuring that a product is manufactured according to specific plans and standards. Also called *quality assurance*. (Chs. 7 & 11)

R

rack and pinion gear A gear used in a car's steering mechanism. (Ch. 8)

RAM Temporary computer memory used for applications and documents. Acronym for *random access memory*. (Ch. 4)

recessive In genetics, a term used to describe genes that produce traits that "show" only if no dominant gene is present. (Ch. 19)

record In a database, all of the fields put together to create one entry. (Ch. 4)

recycling Reusing an item to prevent unnecessary waste. (Ch. 3)

refining Process by which a natural product, such as oil, is changed to a more usable form. (Ch. 8)

reinforced concrete Concrete that has been strengthened with steel or another metal. (Ch. 12)

Remote Manipulator System (RMS) A mechanical arm on the Space Shuttle that is used to move payloads (cargoes) in and out of the cargo bay. (Ch. 18)

renewable Term used to describe something that can be replaced, such as a renewable resource. (Chs. 7 & 13)

research and development (R&D) A department in a company that improves existing products or designs new products. (Ch. 11)

resistance In mechanics, the load moved by a lever. (Ch. 8)

resistor Electronic device that resists the flow of electricity. (Ch. 8)

resource Anything that is used in the production of a manufactured product. (Ch. 7)

retail sales Selling of products in department or discount stores to consumers. (Ch. 11)

reverse osmosis (ahs-MOH-sis) Process in which sea water is pumped through microscopic openings in special filtering materials to remove the salt and make the water drinkable. (Ch. 19)

revolutions per minute (rpm) A measure of the speed at which a gear, or other rotating device, turns. (Ch. 8)

risk taker Person who is willing to encounter a certain amount of danger; courageous person. (Ch. 20)

robot A highly advanced, computer-controlled machine. The word comes from a Czech word that means "to work." (Ch. 10)

robotic vision The ability of a robot to "see" using television cameras for eyes. (Ch. 10)

ROM Permanent memory built into a computer's electronic circuits and used for operating instructions. Acronym for *read only memory*. (Ch. 4)

route Path. (Ch. 14)

S

sales forecast Prediction of how many products a company will sell. (Ch. 11)

Salyut 1 The first space workstation to orbit the Earth; launched by the Soviet Union in 1971. (Ch. 18)

scanner A machine that copies text and graphics from paper into a computer. (Ch. 4)

schematic diagram (skee-MAT-ik) Drawing of an electronic circuit that uses a set of symbols to represent real parts. Also called a *schematic*. (Ch. 8)

scientific method A process of problem solving in which (1) the problem is recognized; (2) an hypothesis is formed about the solution; (3) a prediction about the outcome of the solution is made; (4) experiments are made to test the solution; (5) the hypothesis, prediction, and results of experiments are organized into a general rule. (Ch. 1)

search engine Internet service that helps users find information by taking a key word or phrase and finding matching files. (Ch. 15)

section view In construction, a drawing of the inside of a structure as if a part had been removed. (Ch. 12)

semiconductor A material that conducts electricity only under certain conditions. (Ch. 8)

sensor An electronic input device that senses such things as heat and light. (Ch. 10)

sensor probe A device that can measure such things as temperature and radiation. (Ch. 17)

serendipity (sair-uhn-DIP-ih-tee) Having a lucky accident. (Ch. 6)

series circuit An electrical circuit in which components are connected in line, one after the other; when one component fails, the circuit will not work. (Ch. 8)

series-parallel circuit An electrical circuit that is a combination of series and parallel circuits. (Ch. 8)

shear A push-pull force that can cause an object to tear apart. (Ch. 12)

shelter A place out of the weather. (Ch. 12)

short circuit A situation in which electrons bypass the proper path. (Ch. 8)

silence sensors Electronic devices that detect silence in a radio program and automatically trigger a different audio source, such as a recording of music. (Ch. 16)

simulate Imitate. (Ch. 5)

site In construction, the land on which a structure is built. (Ch. 12)

site plan In construction, a drawing that shows the property boundaries and the exact location of a structure. (Ch. 12)

Skylab The United States' first space station. (Ch. 17)

smart house House in which electronics and computers control energy and appliance use. (Ch. 20)

software Coded instructions that tell a computer what to do. (Ch. 4)

softwoods Woods that come from trees that have needles, such as pine and fir. (Ch. 9)

solar array A panel of solar cells used to change sunlight into electricity. (Ch. 18)

solar cell A device that makes electricity directly from sunlight; also called a *photovoltaic cell*. (Ch. 13)

solar energy Energy from the sun. (Ch. 12)

space spinoff A new technology or product developed during a space project that can be used for additional purposes. (Ch. 18)

spreadsheet A computer application that helps keeps track of calculations, such as those for budgets. (Ch. 4)

spring Coiled metal used to store mechanical energy. (Ch. 8)

sprocket Toothed wheel used in a chain drive. (Ch. 8)

Sputnik The first human-made satellite to orbit the Earth; launched by the Soviet Union in 1957. (Ch. 17)

standard 1. A value or condition that a person or product must meet. (Ch. 3) 2. Term used to describe an exact unit of measurement used by everyone. (Ch. 9)

standard size Size of an item that is commonly available. (Ch. 8)

static load In construction, a force acting on a structure that is unchanging or slowly changing. (Ch. 12)

statistical process control Quality-assurance method used to make sure that a manufacturing process is being done right. If the process is correct, then the product itself doesn't need to be inspected. (Ch. 11)

stockholder A person who has bought shares in a company. (Ch. 11)

Stone Age (2,000,000 B.C. to 3000 B.C.) Prehistorical period during which people used tools made mostly of stone, animal bones, and wood. (Ch. 1)

storyboard A simple drawing showing the main "scenes" in a video production. (Ch. 16)

structural drawing In construction, a drawing that shows structural parts. (Ch. 12)

subsystem A system that is part of a larger system. (Ch. 8)

superconductor A material that can conduct electricity perfectly, with no resistance. (Ch. 8)

synchronized production Cost-saving manufacturing method in which materials and parts arrive at the factory just in time for production. When the product is finished, it is not stored but is immediately shipped to the customer. Also called *just-in-time manufacturing (JIT)*. (Ch. 11)

synthetic Term used to describe a material made by humans that cannot be found in nature. (Ch. 9)

system A combination of parts that work together as a whole. (Ch. 8)

systematically Term used to describe the use of a system to solve a problem in an organized way. (Ch. 8)

T

talent Performers at the radio microphone or in front of the TV camera. (Ch. 16)

teach pendant A remote keypad used to program a robot. (Ch. 10)

technical illustrator Artist who makes technical drawings that help people understand the sizes and shapes of objects. (Ch. 15)

technical writer Writer trained to write technical manuals and instructions. (Ch. 15)

technologically literate Term used to describe someone who understands the impacts of technology. (Ch. 1)

technology The use of knowledge, tools, and resources to help people solve problems. (Ch. 1)

teleprompter A computer used to display TV dialogue in front of a camera. The talent can read the dialogue in case they forget their lines, but it doesn't show on camera. (Ch. 16)

tensile strength The ability of a material to resist stretching or being pulled apart. (Ch. 9)

tension A pulling force that stretches materials; the opposite of compression. (Ch. 12)

theory An idea about something. (Ch. 2)

thermal Having to do with heat. (Ch. 8)

thermal expansion Expansion caused by heat. (Ch. 12)

thermocouple A device similar to an electric thermometer. (Ch. 8)

thermoplastic Plastic that can be melted and remelted many times using heat. (Ch. 9)

thermosetting plastic Plastic that changes chemically when it sets. It cannot be remelted and reshaped. (Ch. 9)

thesaurus (thuh-SAW-ruhs) A reference book or computer program that provides synonyms (words that have the same meaning) for the word looked up. (Ch. 4)

thrust The push that comes from a rocket's engines. (Ch. 17)

timeline A graph showing developments over time or through history. (Ch. 2)

trailing edge The back edge of an airplane wing. (Ch. 14)

transistor An electronic switch or amplifier made with a semiconducting material. (Ch. 8)

transportation The movement of people or goods from one place to another. (Ch. 14)

trend A current preference people have. (Ch. 6)

troubleshooting Trying to find the problem in a system. (Ch. 8)

turbine A type of generator that is turned by the force of a gas or liquid striking its fanlike blades. (Ch. 13)

tweening An animation process in which the basic cells are drawn on the computer by an artist, and then software is used to automatically create the cells needed in between. Also called *in-betweening*. (Ch. 5)

U

uniform resource locator (URL) Address of a site on the World Wide Web portion of the Internet. (Ch. 15)

universal gripper An end effector that allows a robot to grasp any object. (Ch. 10)

uplink A communication transmission from a station on Earth to a satellite in space. (Ch. 15)

utility patent A type of patent that protects inventions considered to be "new and useful". (Ch. 6)

V

vacuum The absence of air or air pressure. (Ch. 18)

variable In experiments, the part that is being tested. (Ch. 3)

video Things that are seen, as on a television program or videotape. (Ch. 16)

video conferencing A communication system in which video cameras are hooked to computers in a network. This allows people from all over the world to talk to each other at the same time. (Ch. 5)

visualizing Picturing an idea in the mind. (Ch. 6)

volt A unit of electrical pressure. (Ch. 9)

voltage Electrical pressure. (Ch. 9)

Voyager A U. S. space mission begun in 1977 that launched two spacecraft. *Voyager 1* is the most distant human-made object in space. (Ch. 17)

W

waste management The management of things we throw away. (Ch. 19)

waypoint Each location along a path that helps a global positioning receiver keep track of an object's exact movement. (Ch. 15)

website A location on the World Wide Web portion of the Internet. (Ch. 15)

wet cell A battery that uses liquid chemicals. (Ch. 8)

wholesale sales Sales of products by people or companies who buy large quantities from manufacturers and sell them to other businesses in large quantities. (Ch. 11)

wind tunnel An enclosure used for testing in which strong winds can be created. Structures are placed inside the wind tunnel to see how they are affected by wind currents. (Ch. 9)

word processing Creating documents containing words, or text, using a computer. (Ch. 4)

work envelope The maximum range of a robot's movements. (Ch. 10)

World Wide Web (www) A part of the Internet developed to present Internet information in a format that is easy to use. (Ch. 15)

Z

zero gravity Reduced gravity; also called *zero-g*. (Ch. 18)

zoning laws Laws that regulate the type and size of buildings that can be constructed in a neighborhood. (Ch. 12)

INDEX

DNA (deoxyribonucleic acid), 444–445, 480
Dominance, Law of, 446–447
dominant genes, 446, 448–449, 480
downlink, 354, 480
download, 115, 480
drag, 329, 480
drawing
    with computers, 110–112
    to scale, 194, 480
dry cell, 183, 480
dual-in-line package (DIP), 72
Dudley, Charles, 75
dynamic loads, 276, 480

E

Earth, gravity of, 388, 408, 455
earthquakes, 61, 62
    simulating, 62
    and structural design, 274–277
earth-sheltered construction, 281
ecology, 78, 480
economics
    and entrepreneurship, 239
    and exponential growth, 29, 30
    and figuring costs and profits, 258–259
    and human resources department, 242
    and marketing, 242, 257
    and mass production, 218
    and R&D department, 242
    and starting companies, 243–245,
        248–250
    and use of robots, 228
Edison, Thomas A., 130
edit, 344–345, 480
educational robots, 230
Einstein, Albert, 135, 467–469
electrical circuits, 174–175
electrical energy, 287
electrical system, 165
electric generator, 291, 480
electricity, 173, 302, 480

conventional energy sources, 291–292
    measuring, 194–195
electrolytes, 183, 480
electromagnet, 232
electromagnet spectrum, 362–363
electromechanical machines, 167
electronic communication, 351–355, 480
    fax (facsimile) machines, 353
    fiber optics, 351
    modems, 352
    satellite, 353–354
    telephones, 352
electronic noise, 338, 480
Electronic Numerical Integrator And
    Computer (ENIAC), 87
electronics, 173, 481
    components in, 176
elevation drawings, 271, 481
e-mail (electronic mail), 340, 481
emerging technology. *See* technology,
    emerging
employability skills. *See* job skills
encapsulated city, 457–458, 481
end effector 232, 481
energy, 148
    alternative, 301–307
    changes in form of, 287
    chemical, 287
    conservation of, 151–152, 268, 281
    conserving, 297–300
    defined, 287, 481
    dependence on, for technology, 290
    efficiency, 297–298
    electrical, 287
    geothermal, 304
    kinetic, 289
    nuclear, 291
    and pollution, 288
    potential, 289
    shortages of, 281
    solar, 210, 281, 290, 301–302

hydraulic systems, 179, 483
hydroelectric power, conventional energy
    sources of, 292, 483
hydrofoils, 322, 483
hydroponics, 432, 483
hypermedia, 341
hypersonic plane, 311
hypothesis, 26, 483

I

idea bank, 38, 483
ideas
    borrowing of, 39
    exploring, 34
    protecting new, 139–143
    sources of, 134–136
image builder, 108
impacts, 67, 483
in-betweening, 120, 483
in-camera editing, 375, 483
independent product testing, 75, 483
industrial areas, 283
Industrial Revolution, 27, 221, 483
industrial robots, 228, 229, 483
information, 148
    evaluating, 343, 349
    exploring, 34
    integration of, 60
    modeling, 30
    writing and drawing, 349–350
innovation, 483
    impact of serendipity on, 127
    process of, 127–128
    prototypes in, 208–209
    reasons for needing, 128–129
    research and development in, 34, 135, 242
    sources of ideas for, 134–136
    tools in, 137–138
    visualizing in, 130–131
inputs, 163, 483
insulation, 298, 484

insulators, 173, 484
insulin, 443
integrated circuits, 86, 484
    analyzing, 72–73
integration of subject areas, 60, 484
interchangeable parts, 166
interference, 338, 484
International Space Station, 418, 423, 484
International System of Units, 193, 484
Internet, 469
    defined, 340–341, 484
    and safety, 342
    searching, 342–343
Internet Explorer, 341
inventors, 130, 131, 132–133
    women as, 132–133
inventory, 247, 484
iron, 26
Iron Age, 26, 484
isometric drawings, 345, 484

J

Jenner, Edward, 134
jet aircraft, 330
Jet Propulsion Laboratory, 405
jig, 249
jobs. *See also* careers
    impact of robots on, 229
job shadowing, 263
job skills
    applying for a job, 248-250
    brainstorming as, 38
    creative thinking as, 39
    integration of, 60–61
    problem solving as, 32–35
    training for, 229, 242
    technology research as, 36–37
    working in groups and teams, 38-41
joystick, 90, 484
just-in-time manufacturing, 246–247, 484

water cycle, 290
water transportation, 321–326
Waypoint, 355, 492
websites, 341, 492
weightlessness, 409
wet cell, 183, 491
"what-if" questions, 39
Whitney, Eli, 218
wholesale sales, 256, 258, 492
wind bracing, 276
wind energy, 302, 311, 312
wind power, 321
winds, 289
wind tunnel, 209, 491
  testing airfoil design in, 331–333
women, as inventors, 132–133
woods, 203
word processing, 91–95, 491
  software for, 91–92, 93
work envelope, 228, 491

working drawings, 269–271
work in process (WIP), 251
World Wide Web (WWW or Web), 341,
  492
Wright Brothers, 27, 328
Writing
  for radio, 366–367
  technical, 344–345
  for TV, 366–367

X

X-33, 136

Y

Yeager, Chuck, 311

Z

zero gravity, 408, 492
zoning laws, 270, 492
Zworykin, Vladimir, 362

CREDITS

Cover and Interior Design:
  Bill Smith Studios

Cover Photos: Image Bank;
  NASA; PictureQuest; Stone,
  CAD/CAM Techniques

Ball Aerospace & Technologies
Corporation/Scott Kahler, 23
Keith Berry, 154
Alelia Bundles/Madam
    Walker Family
    Collection/Alexandria,
    Virginia, 132
Ancient Art & Architecture
R. Sheridan, 26
AP Wide World Photos, 133
Charles Bennett, 433
Archive Photos, 7, 130, 384, 384,
    386, 398
    American Stock, 352
    Reuters/NASA, 133, 384
    Popperfoto, 467
Arco Solar Incorporated, 305
Arnold &Brown, 42, 43, 62, 107,
    114, 117, 118, 139, 140, 163,
    167, 168, 169, 176, 177, 178,
    183, 185, 191, 192, 199, 219,
    220, 240, 257, 276, 288,
    293, 298, 299, 300, 301, 356,
    364, 373, 378, 436, 439
Art Macdillos/Gary Skillestad,
    269, 270, 271, 362
Artville, Burke, Triolo, 449
Ball Aerospace Corporation, 6
Blue Sky Studios, 8, 39, 106, 119,
    120, 123
Corbis Images
    Natalie Fobes, 432
    Owen Franken, 455
    Colin Garratt, Milepost 92/180
    INP/Corbis-Bettman, 132
    Steve Raymer, 247
    Ted Spiegel, 435

Tim Wright, 320
    Michael S. Yamashita, 466
Howard Davis, 270
Everett Collection, 27, 312
Eyler Auto Center Inc., 170, 295
Curt Fischer, 42
FPG International
    Harry Bartlett, 451
    Tom Campbell, 125
    Jim Cummins, 164
    DL-FC1992, 22
    FPG International 2000, 218
    Derald French, 10
    Dennis Halliman, 66, 310
    Mason Morfit, 145, 430, 456
    Eric Pearle, 238
    Bryan Peterson, 285
    C. Stephen Simpson, 65, 190
    Stephen Simpson, 146
    Jeffrey Sylvester, 84, 164
    Telegraph Colour Library, 7,
      87, 161, 360, 403, 444
    P. Thompson, 134
    Tom Tracy, 264
    VCG, 473
David R. Frazier Photolibrary
    Inc., 12, 15, 24, 25, 32, 51,
    56, 57, 58, 60, 63, 73, 86,
    89, 99, 111, 112, 115, 127, 128,
    138, 148, 162, 165, 171, 179,
    184, 205, 209, 215, 216, 221,
    222, 246, 254, 258, 265,
    266, 267, 286, 291, 303,
    304, 309, 315, 322, 325,
    327, 335, 340, 345, 348,
    351, 357, 365, 371, 375,
    377, 381, 391, 399, 414,
    420, 425, 457, 458, 468,
    470
Ford Motor Company Inc., 313
Iowa Department of Natural
    Resources/ Ken Formanek,
    151

Steve Karp, 26, 29, 30, 33, 34, 40,
    41, 43, 48, 53, 70, 74, 87, 88,
    90, 94, 100, 134, 137, 143,
    147, 159, 164, 175, 178, 182,
    186, 193, 196, 197, 200, 201,
    207, 225, 230, 239, 241, 243,
    244, 245, 249, 255, 259, 261,
    273, 275, 276, 278, 294,
    296, 306, 307, 326, 328,
    329, 331, 332, 333, 338, 346,
    347, 353, 354, 362, 363,
    367, 370, 378, 385, 394,
    397, 409, 413, 421, 426, 437,
    447, 448, 460, 463, 464, 465
The Makita Corporation, 42
Kevin May, 42, 203
NASA, 14, 49, 67, 129, 171, 229,
    289, 382, 383, 384, 385,
    386, 387, 388, 389, 397,
    398, 399, 400, 404, 407,
    408, 409, 410, 415, 416, 417,
    418, 423, 424, 429, 453
NASA's *Spinoff* Magazine, 405
Photo Edit
    Apollo News Services, 12, 316
    Bill Aron, 115, 352
    Robert Brenner, 153
    Cindy Charles, 108
    Mary Kate Denney, 50
    Myrleen Ferguson, 204
    Tony Freeman, 79, 369, 374,
      412, 462
    Robert Ginn, 468
    Jeff Greenberg, 7, 31, 371
    Will Hart, 46, 48
    Richard Hutchings, 198
    Bonnie Kamin, 379
    Dennis McDonald, 302
    John Neubauer, 317
    Michael Newman, 42, 43, 52,
      80, 101, 116, 195, 221, 260,
      281, 287, 344, 368, 395,
      444

Jonathan Nourok, 38, 248
Mark Richards, 152, 323, 372
David Young-Wolff, 6, 31, 32, 91, 95, 102, 110, 131, 150, 164, 166, 290, 321, 337, 343, 444, 462
Photo Researchers
  Michael Abbey, 446
  Dr. Jeremy Burgess/Science Photo Library, 406
  CNRI/Science Photo Library, 445
  De Malglaive/Explorer, 312
  Department of Energy, 173
  David Ducros/Science Photo Library, 13
  European Space Agency/Science Photo Library, 396
  David R. Frazier Photolibrary Inc., 13
  Bruce Frisch, 202, 203
  Bruce Frisch/Science Photo Library, 233
  Lowell Georgia, 15, 443
  Klaus Gulbrandsen/Science Photo Library, 203
  James Hanley, 292
  Adam Hart-Davis/Science Photo Library, 173
  Phil Jude/Science Photo Library, 194
  Calvin Larsen, 265, 288
  Pat and Tom Leeson, 217
  R.Maisonneuve/Publiphoto, 173, 253
  Jerry Mason/Science Photo Library, 232
  Maximilian Stock Ltd./Science Photo Library, 203, 221, 228
  Tom McHugh, 446
  Will and Deni Mcintyre, 9, 155, 251

Hank Morgan, 455
Hank Morgan/Science Source, 8, 10, 140, 234, 455
G. Muller, Sruers GMBH/Science Photo Library, 206
Joseph Nettis, 11, 280
Novosti/Science Photo Library, 384
Sam Ogden/Science Photo Library, 454
David Parker, 600 Group Fanuc/Science Photo Library, 237, 309
David Parker/IMI/University of Birmingham High TC Consortium/Science Photo Library, 174
Carleton Ray, 322
Rosenfeld Images Ltd./Science Photo Library, 252, 156
Joseph Sohm/Chromosohm, 151
Volker Steeger/Science Photo Library, 227, 149
Sheila Terry, Lecroy/Science Photo Library, 164
David Weintraub, 274
Queensborough Public Library/Long Island Division, 132
The Stock Market
  Phillip Bailey, 471
  Roger Ball, 28, 55
  Paul Barton, 51
  Peter Beck, 208
  Claude Charlier, 9
  George DiSario, 282
  John Feingersh, 83
  Firefly Productions, 434
  Brownie Harris, 361
  John Henley, 61
  Michael Heron, 256
  Ted Horowitz, 11
  Ray Juno, 214

Ronnie Kaufman, 141
Matthias Kulka, 189
Marrkkuu Lahdesmaki, 51
Lester Lefkowitz, 24, 35, 47, 72, 78, 431, 470, 471
Lightscapes, 3, 45, 213
John Madere, 164
R. Markova, 75
Mendola Ltd., 336
Mug Shots, 452
Jose L. Pelaez, 36
David Pollack, 314
Baron Sakiya, 330
Jaime Salles, 126
Pete Saloutos, 359
Sanford/Agliolo, 173
Chuck Savage, 132, 133
Seman Design Group, 122
ML Sinibaldi, 32
Ariel Skelley, 263, 444
Tom Stewart, 69
William Taufic, 85, 135
William Westhiemer, 105
Brad Thode, 73, 77, 156, 181, 211, 224, 235, 319, 392, 393, 401, 414, 419, 441, 442
Courtesy of Tuan Vo-Dinh, 133
Western Reserve Historical Society, 133

Acknowledgements: Blue Sky Studios, Ball Aerospace & Technologies Corporation, Alelia Bundles, Rob Hillyer and Eyler Auto Center Inc., Rushville, Illinois, Iowa Department of Natural Resources, NASA and NASA's Spinoff Magazine, Tuan Vo-Dinh.

Models and fictional names have been used to portray characters in stories and examples in this text.